Rare Plant Conservation:
Geographical Data Organization

Edited by Larry E. Morse
& Mary Sue Henifin

The New York Botanical Garden
Bronx, New York
1981

Preface

The confusion, controversy, emotion, and histrionics that characterized public concern over the "environmental crisis" during the past decade are gradually giving way to the sober realization that Olympian fulmination over erroneous cultural shibboleths does not set things right. Subjugating the still-dominant spirit of Prometheus to the inexorable responsibilities of Atlas in the interest of extending the era of human presence in this world will require sound, careful analysis of the multifarious interactions people have with the world ecosystem within which we all function, and, by quantifying those interactions, erecting models which will serve as bases upon which predictions of environmental perturbation by proposed environmental change can be made. Thus, the empirical, idiosyncratic, almost seat-of-the-pants judgment inherent in so many of today's excessively voluminous, boringly pedestrian, distressingly sophomoric forecasts of environmental impact, which characteristically fail to offer hoped-for means by which trade-offs can be weighed, will be replaced with sound cause-and-effect data drawn from painstaking environmental characterization and hard mathematical models simulating ecosystem dynamics. Accordingly, predictive ecology augurs well as a focus by which the rank hedonism in Prometheus will be tamed by the sober responsibilities upon Atlas.

Requisite to achieving this new and necessary state are measurements of the components of the ecosystem presently at hazard. As no one can say how many plant species should people this earth to assure the genetic evolution of the plant kingdom in response to natural and unnatural environmental change, and thus assure that people in distant millenia will not have suffered by our hand any loss of means for their biological existence, it behooves extant mankind to take stock of the effects wrought in the plant catalog by human endeavor. With land-clearing in the Amazon region alone proceeding at the rate of some 100 acres per hour, it is not surprising to hear that one plant species per day, somewhere in the world, passes over the brink to extinction. Although the historic pace of plant evolution remains heavily speculative, it is obvious that the world plant register is diminishing before our eyes. As there is no known redemption from species annihilation, and it is likely that, as mankind necessarily strives to balance the environmental budget, genetic diversity will be an essential resource for the production of food and other plant products, it follows that conservation of extant botanical diversity should be urgently promoted.

This volume is based on the presentations and discussions at a conference organized to further that end, summarizing what is known, outlining remediation under way, projecting what is needed—the premise throughout being that the human pressures snuffing out more plant species than are evolving in nature constitutes a trend toward botanical impoverishment whose critical limits for ecosystem functioning, and therefore human well-being, are not yet clear. If we cannot determine where the critical threshold lies, or whether future botanical diversity has lost (or even gained) by some fraction in recent decades, then some research needs to be done, and the sooner the better. We have here, thanks to the steadfast efforts of Larry Morse, Jane Lawyer, Mary Sue Henifin, and their collaborators, a valuable summation and a catalyst for further work.

Howard S. Irwin
April 1980

i

Published by The New York Botanical Garden, Bronx, New York 10458, U.S.A.

Issued 7 May 1981

Library of Congress Cataloging in Publication Data

Main entry under title:

Rare Plant Conservation:
Geographical Data Organization

    Based on the presentation and discussions at a
conference sponsored by the U. S. National Park
Service and held in Nov. 1977 at the New York
Botanical Garden.
    Bibliography: p.
    1. Plant conservation—Congresses. 2. Rare
plants—Geographical distribution—Congresses.
3. Plant conservation—Data processing—Congresses.
4. Rare plants—Geographical distributions—Data
processing—Congresses. 5. Data base management—
Congresses. I. Morse, Larry E. II. Henifin, Mary
Sue, 1953- III. United States. National Park
Service. IV. New York (City). Botanical Garden.
V. Title.
QK86.A1G46   333.95'3   80-19361
ISBN 0-89327-223-X

Table of Contents

Section 4. Representative Projects

Appendices

Introduction

Larry E. Morse[1,2] and Jane I. Lawyer[1,3]

In the past decade, the science of rare plant conservation has emerged at the leading edge of biological conservation, now providing novel insights into habitat maintenance, population biology, coadaptation and coevolution, and ecosystem conservation. Although still lacking the broad popular following of the conservation programs for many higher vertebrates, the idea that there are endangered species of plants is now familiar to the general public. Questions of rare plant conservation are now routinely considered in resource management and land-use planning.

Effective plant population conservation requires *geographical data* specifying the places where rare plants grow, for few species can be maintained as self-sustaining breeding populations in places different from present or historically occupied habitat. The low geographical mobility of plant populations allows geographical data to remain valid over a reasonable timescale of management activities and planning processes, permitting compilation and dissemination of information about the special places where these special plants do grow.

Another consideration in conserving populations of rare plants is that in most cases threatened or endangered plant species occupy incredibly small ranges, often less than a square mile altogether and in many cases less than an acre. Site-specific knowledge of these habitat areas and the biological reasons the plants are restricted to them allows management and conservation efforts to be focused directly on these sites and not spread less productively over other areas. In many cases, minor and inexpensive modifications to a prospective human activity can allow the activity to proceed much as planned and nevertheless conserve the affected plant population. The more precise and accurate the site-specific population data, the more effective can be the consideration of this information in conservation and planning work.

The highest priority in conservation is generally accorded to habitats of species endangered or threatened with extinction throughout all or virtually all of their presently extant geographical range. Of greater concern in this group are species representing monotypic genera or subgenera, since they provide particularly important components of the overall genetic diversity of the world's biota. The next level of concern is accorded to conservation of subspecies and geographically distinct varieties of plant species, since these reflect major aspects of the genetic variation within a species but contribute less overall to genetic diversity. Lower concern is given to disjunct and peripheral populations of species otherwise in little need of conserva-

---

[1]   Cooperative Parks Study Unit, The New York Botanical Garden, Bronx, N.Y. 10458.
[2]   Present address: The Nature Conservancy, 1800 N. Kent St., Arlington, Va. 22209.
[3]   Present address: 45 Sturgis Road, Bronxville, N.Y. 10708.

Pages 1-6 *in:* Larry E. Morse and Mary Sue Henifin, eds., *Rare Plant Conservation: Geograph-ical Data Organization*, The New York Botanical Garden, Bronx, N.Y.

tion protection; such peripherals and disjuncts may provide the opportunity for evolutionary innovations, but have not yet diverged as significantly as the already recognizable taxa. Hence, ordering plant conservation priorities requires awareness not only of the local status of a taxon's populations, but also of the significance of those populations viewed against the overall *extant* range of the taxon. The emphasis on extant range is essential, for extirpated populations can no longer contribute to the future biota.

There are many sources of the geographical information pertinent to rare plant conservation. Current population status reports done on a site-by-site basis are of course most important, but as yet are available for few species. More general conservation status reports are now being developed for many of the rare plants in the United States, and are also being used in some other countries and by international conservation organizations. Recent management reports, impact statements, or similar documents sometimes provide further information on current status, but in many cases the only information readily available is in museum collections and the scientific literature. Considerable effort is required both to locate and consolidate such geographical information, and to bring older data up to date.

Conservation agencies, scientific researchers, land-use planners, and regulatory agencies are among the major users of rare plant information. Their needs have been stimulated by legislation, public concern, and specific conservation interests. In the United States, passage of the Endangered Species Act of 1973 and publication in 1975 of the Smithsonian Institution's report on threatened and endangered plants brought heightened concern for conservation of the nation's heritage of plant species diversity. The many state plant conservation laws promote further protection of many species, usually including peripheral or disjunct populations of species which are rare or threatened in a state even though common and of less concern elsewhere. Effective implementation of all these state and federal laws requires current information on the status and geographical distribution of thousands of plant species.

* * *

Throughout the United States, hundreds of researchers are developing state, regional, and national data bases of rare plant information. These people are now gathering considerable quantities of biogeographical information and other data relating to the designation, location, and conservation of rare plant species in the United States. Many federal and state agencies, conservation groups, universities, professional societies, consulting firms, and volunteer organizations have become involved in these efforts, leading to the creation of a diversity of information priorities and data-collection procedures in these various projects. The lack of coordination and effective communication in the plant conservation field has resulted in much duplication of effort; some species and some geographical areas have been studied and reviewed repeatedly, and others not yet at all. Clearly the time has come to consider from many viewpoints the topic of *Rare Plant Conservation: Geographical Data Organization* , with particular emphasis on prospects for project coordination.

To address these points of geographical data organization and project coordination, a symposium sponsored by the U.S. National Park Service was convened in

November, 1977, at the New York Botanical Garden, as part of a cooperative study on synthesis of plant distributions information. The symposium topic was chosen for two major reasons. First, rare plant conservation is of high management concern in the National Park Service and many other agencies, yet opportunities to discuss current techniques and ideas had been limited. Second, the urgency and priority for plant conservation programs has led to the recent development of many information-synthesis techniques of interest more generally in the handling of biogeographical data. The symposium thus considered a sampling of techniques for geographical data handling in natural history, with emphasis on the recent innovations developed specifically to enhance awareness of rare plant locations and conservation concerns. Limitation of the scope of the symposium in this way assured lively discussion by a group already generally familiar with the background and purpose of each other's work.

The various chapters of this book are all based on the presentations or discussions at the New York meeting. Little attention is given here to the question of why or whether to conserve rare or depleted plant species; these topics have been treated at length elsewhere, and familiarity with them was assumed of all the participants in the meeting. Instead, the participants were invited to address the details of their work, particularly regarding their development of priorities and implementation of techniques. Two broad topics of particular interest were treated intensely in invitational workshops—handling of specimen label data and prospects for project coordination.

The keynote address for the symposium, "Rare Plant Conservation through Elements-of-Diversity Information," was delivered by Dr. Robert E. Jenkins, Vice President for Science Programs of The Nature Conservancy. Discussing the uses his organization has for site-specific information, Dr. Jenkins emphasized that the selection of natural areas for conservation is greatly facilitated by awareness of occurrences and population status for rare species. The use of site-specific information on element occurrences is the basis of The Nature Conservancy's well-known State Natural Heritage Programs.

Sessions the next two days considered four topics: Information Sources, Information Priorities, Major Rare Plant Data Syntheses, and Data Resources of State Rare Plant Projects.

Several speakers considered the range of useful information sources, including data from specimen labels, geographical and ecological knowledge useful for locating rare plant populations, and the scientific literature. Stephen R. Edwards described how the wealth of information in scientific collections supports botanical research, noting how the Association of Systematics Collections seeks to stimulate interest in these materials. Theodore M. Barkley and Michael Lamson both discussed the use of specimen labels in distribution mapping; misidentifications and outdated classifications as well as incomplete or inaccurate locality information were regarded as major difficulties in the direct use of label data.

A number of workers favored the use of site-specific field surveys as a primary information source, especially in poorly botanized areas. Some of these studies are now being done by academic groups or land-management agencies, but much more information is known only to individuals who are generally not in contact with rare plant projects. Duane Atwood, of the U.S. Fish and Wildlife Service, described how his use of environmental information on the water, temperature, elevation, and exposure relating to geologic and other factors provided clues for finding plant localities. Jean Siddall from the Oregon Rare and Endangered Plant Task Force described her efforts at obtaining current field information by organizing many volunteers in Oregon.

Theodore J. Crovello discussed the published and unpublished literature relevant to rare plant conservation, defining "literature" broadly to include data banks, environmental impact statements, and semi-formal publications such as park checklists, as well as the traditionally published floras and taxonomic revisions or monographs. Crovello cautioned against overlooking useful botanical information that is often obscured in non-traditional materials.

Among major users of plant distribution information in conservation is the National Natural Landmarks Program, now part of the Heritage Conservation and Recreation Service. As discussed by Gary Waggoner, theme studies are used to assess characteristic features of regional significance in evaluating the national significance of particular areas.

The information needed for plant population conservation in U.S. national parks and forests was discussed by several speakers. Occurrence of a rare or endangered plant in such an area does not necessarily guarantee its protection, since specific information on the location and status of the populations of a species is needed for proper consideration in planning and management. Drawing primarily on her experiences in the floristically rich but heavily visited Great Smoky Mountains National Park, Susan P. Bratton spoke of such management concerns as visitor impacts and threats from spreading exotics. U.S. Forest Service representatives Andrew Robinson, L. E. Horton, and Ed Schlatterer considered the importance of information on rare plants in management of multiple-use and wilderness forest lands.

Bruce MacBryde of the U.S. Fish and Wildlife Service discussed the information needed to implement protection measures for plants within the jurisdiction of the federal government under the Endangered Species Act of 1973. In considering the legal requirements of the Act, MacBryde outlined the data needed for listing of endangered and threatened plants and for the assignment of critical habitat to protect species and their habitats from adverse effects of federal activities.

James L. Reveal complemented MacBryde's discussion by reviewing the information needed to assess rarity and threat to determine to what degree a plant is endangered. He pointed out that rarity is a relative term and can be assigned only if the flora of a region is well-known. Threat, he continued, must be studied through site-specific observations to determine the present and expected pressures on individual populations.

Since the needs for, and challenge of, data organization for rare plant conservation vary, systems developed to handle such information differ. The organizational strategies discussed at the symposium ranged from simple manual procedures to complex computer systems. Of particular interest were the detailed data-organization strategies developed by Paul D. Whitson and J. R. Massey for species biology information, and by Albert E. Radford for characterizing and classifying habitats.

A. V. Hall, of the University of Cape Town, South Africa, described his use of computer-indexed manual files in the Southern African Rare Plant Project. In order to handle data on the 1500 plant species considered vulnerable, endangered, or extinct in Southern Africa, the project uses a computer system to search and sort an abbreviated version of the more complete information in its manual files.

Dr. Crovello also explained the use of a fully computerized information system to support the handling of rare plant information by the Indiana State Biological

Survey for a variety of uses including scientific, educational, and planning activities. The Plant Information Network (PIN), based at Colorado State University, also emphasizes computer searching, providing considerable detail in the computer files themselves, as dicussed by Philip L. Dittberner. PIN is currently used primarily to provide information for environmental inventories, impact assessment, and planning and design of reclamation projects in Colorado, Wyoming, and Montana.

Helmut Moyseenko of The Nature Conservancy reviewed a series of suggestions regarding the efficient organization of information. He cautioned that one planning an information system should consider carefully the issues of anticipated data requirements, data sources, user categories, and frequency of use, to produce an information strategy appropriate for practical and efficient handling of a group's everyday work.

The topic of information transfer was discussed in several papers. John T. Kartesz stressed the importance of improving accessibility to rare plant information at the state level; maintaining awareness of ongoing listings and rare-plant projects at the state level is difficult because of the diversity of agencies, organizations, and individuals involved in such work. Joe Pardue of the Oak Ridge National Laboratory stressed the need for information transfer from researchers to planners, for example in using rare plant information in assessing impacts of energy development projects.

Synthesis and transfer of rare plant information from the local and state levels to the regional and national level characterizes a number of projects supported by the U.S. Fish and Wildlife Service. William D. Countryman, for example, described how members of the New England Botanical Club are preparing state surveys which will then be synthesized into a New England regional report.

Thomas Duncan spoke about the California Native Plant Society's role in providing information through its Rare Plant Study Project to federal and state groups involved in planning and land management. In 1968 the Society first prepared lists of rare plants that were later complemented with information on distribution and survival potential.

During the symposium, an ad-hoc group met evenings to develop an initial draft of a comprehensive outline of the information topics required for official listing, critical habitat assessment, and management and recovery of threatened or endangered vascular plant species. These guidelines were subsequently reviewed, revised, and field-tested, and are presented as an appendix to this volume.

* * *

This book is not a symposium proceedings in the strict sense. Written transcripts of the spoken presentations would be difficult to read, and the reader could not benefit from the rich discussion and exchange of knowledge that took place at the meeting and subsequently among many of the participants. Nor could a written transcript include the many fine illustrations used as integral parts of most presentations.

Recognizing these facts yet desiring a published work reflecting the discussions at our meeting, we invited each speaker to submit a written paper developing the ideas presented at the meeting. Most readily agreed, and provided papers treating their research or management activities in a more thorough way than possible in a brief symposium presentation.

5

Some of the papers include work which took place following our meeting, and some in their written form are by co-authors who were unable to participate in the symposium itself. A few of the present chapters, such as the one on status report guidelines, consider subjects only briefly addressed at the symposium, but given full treatment here because of their importance to the topic of this book. On the other hand, it proved difficult to develop separate chapters from several of the presentations, and their content is incorporated in the "Overview" chapter which follows.

We thank the following 48 people for serving voluntarily as reviewers for one or more of the various papers submitted: Ken Baker, John Ballman, Bill Burley, Jonathan Coddington, Carolyn Corn, Robert A. DeFilipps, Thomas Duncan, Richard W. Dyer, Cynthia A. English, Barbara Ertter, Eileen C. Fairbrother, Katharine G. Field, Martha N. Fisher, Don Gardner, Andrew M. Greller, Jim Grimes, Geraldine A. Guppy, A. V. Hall, Alice Q. Howard, Howard S. Irwin, Robert E. Jenkins, Walter Judd, John T. Kartesz, Jane I. Lawyer, Bruce MacBryde, Gerard McKiernan, Wayne L. Milstead, Helmut P. Moyseenko, Gilbert Muth, Peter J. O'Connor, Joe W. Pardue, James T. Peters, Ghillean T. Prance, Donald W. Reeser, James L. Reveal, Martha S. Salk, Ben A. Sanders, Julie Schaller, Charles J. Sheviak, E. LaVerne Smith, Irene M. Storks, Roy L. Taylor, Alice F. Tryon, Gary S. Waggoner, Russell R. Walton, Margaret Williams, Thomas A. Zanoni, and Chris Zimmer. Only through the assistance of these reviewers were we assured each and every one of these contributions stands as an important statement of the principles, priorities, procedures, and pitfalls of geographical data organization for rare plant conservation.

Finally, for their sponsorship of our Symposium and the scientific editing and reviewing of the written contributions, we thank the National Park Service, U. S. Department of the Interior.

Section 1.

OVERVIEW

Geographical Data Organization in Botany and Plant Conservation:

A Survey of Alternative Strategies[1]

Larry E. Morse,[2,3] Mary Sue Henifin,[2,4] John C. Ballman,[2,5] and Jane I. Lawyer[2,6]

Major strategies for organizing information on the geographical distributions of plants are reviewed, with particular attention to the uses and limitations of point-locality data, distributions mapped by counties, and grid systems such as 7-1/2-minute latitude/longitude quadrangles. It is stressed that data collection and synthesis must be selective in phytogeography, with intensive mapping needed only for areas of site-specific scientific studies and for such high-interest situations as monitoring of aggressive exotics and conservation of rare or endangered species.

Geographical data about plant species distributions are used in three major ways: to report the places particular species occur, to list the species which occur at a particular place, and to determine the significance of the occurrence of a particular species at a particular place. For most of the past century, data concerning distributions and localities of plant species have been primarily of interest to research botanists, and have been documented and disseminated through such well-developed media as specimen labels, taxonomic monographs, and floristic manuals and atlases.

In recent decades, the need for geographically organized data about plant species has broadened considerably, primarily because of increased interest in plant conservation and habitat restoration and the expanding challenges of nature interpretation and environmental education for an increasingly mobile public. In these same years, the development of electronic computers and other modern information-processing tools has provided new options for organizing and documenting the

---

[1] Research supported in part by the U.S. National Park Service through its cooperative research agreement with the New York Botanical Garden.

[2] Cooperative Parks Study Unit, The New York Botanical Garden, Bronx, N.Y. 10458.

[3] Present address: The Nature Conservancy, 1800 N. Kent St., Arlington, Va. 22209.

[4] Present address: Div. Environmental Sciences, Columbia University School of Public Health, New York, N.Y. 10032.

[5] Present address: 4727 Felton St., San Diego, Calif. 92116.

[6] Present address: 45 S urgis Road, Bronxville, N.Y. 10708.

Pages 9-29 in: Larry E. Morse and Mary Sue Henifin, eds., *Rare Plant Conservation: Geographical Data Organization*, The New York Botanical Garden, Bronx, N.Y.

data of natural history, and brought into question the continued usefulness of traditional methodologies in the natural sciences. Also contributing to the increased interest in plant distributions is the increasing integration of phytogeographical aspects of research botany, agriculture, silviculture, horticulture, landscape architecture, and urban ecology.

This review of the state of the art in geographical data organization in botany and plant conservation is based primarily upon a study conducted by the Cooperative Parks Study Unit of the New York Botanical Garden, done at the request of the U.S. National Park Service as part of Project 8 of the U.S. Man and the Biosphere Program, and draws strongly upon the presentations at the Garden's 1977 symposium. In this review, we consider traditional and novel needs for geographically organized data in botany and plant conservation, sources of such data, currently used methodologies for its organization and presentation, and practical considerations pertinent to choices for particular situations.

History of Needs for Geographically Organized Botanical Data

Questions about where various kinds of plants grow can be found in many aspects of human knowledge and social and cultural experience. People of all cultures have an intense curiosity about the world around them as well as real needs to understand natural diversity in locating sources of food, shelter, and medicines. The human inclination to organize and classify information has led repeatedly to the development of local folk taxonomies that not only classify and name the local flora and fauna but furthermore provide a framework for information on places the various kinds of local plants and animals can be found. Raven, Berlin, and Breedlove (1971) and Berlin (1973) provide further discussion of these beginnings.

Wider travel and trade led to increasingly broadened and refined classifications, culminating in the global perspective developed in the eighteenth century by Linnaeus and others who could draw upon the observations and collections of those returning from the great voyages of exploration of those times. However, these early syntheses provided little detail on plant or animal distributions, for too little was known yet of specific occurrences. Linnaeus' North American records, for example, were generally presented with such vaguely defined areas as "Virginia," "Carolina," or "Canada," these having only approximate identity with the early North American colonies.

As the economic and aesthetic values of many of the newly discovered plants became more widely appreciated by Europeans, the need to locate sources of them intensified and led to better recording of locations and directions as well as more detailed locality data in technical publications. Often this information was presented as a broad statement of suspected general range (such as "Rocky Mountains" or "Great Lakes Region"), documented by a short list of specific localities where plants of the species had been seen or collected.

Charles Darwin's presentation in 1859 of his theory of evolution by natural selection provided a new basis for studying and comparing patterns of distribution among various species and various areas, and suggested the need for more precise recording of habitat and environmental information in documenting species' distributions, so the factors limiting a distribution could be more clearly understood. "Why are there so many species?" is still a fundamental question of ecology, invoking the related question of why various particular species are limited to certain geographical areas. Evolutionary studies have also shown that species are not everywhere uniform, providing a framework for studies of infraspecific variation and genetic polymorphism, both fascinating topics in which new studies continue to suggest even more new questions (*cf.* Endler, 1977). Quantitative studies of geographical variation are now common, leading to the realization that effective conservation of genetic diversity involves far more than simply protecting one population of each species (*cf.* Eckholm, 1978; Hawkes, 1978; Committee on Germ Plasm Resources, 1978; National Plant Genetics Resource Board, 1979; Soulé and Wilcox, 1980).

Current Practices

The data requirements for scientific studies of plant systematics and geography are generally standardized. For most purposes, a summary statement of general distribution, accompanied by a map sketch of the geographical range, serves to provide the source data for comparative studies. Selected lists of localities of representative collections, typically emphasizing disjunct and peripheral populations, provide the naturalist and experimental biologist with prospective sources of future study material. For widespread species, it is accepted practice to list a small sampling of specific localities in research publications, without attempting to cite all specimens seen, much less reporting all localities. More detailed mapping, however, is required in some phytogeographical studies, such as correlating distributions with substrates, periglacial landforms, altitudinal zones, or various factors of climate, and precise mapping is also needed to identify instances of local sympatry or parapatry.

A different perspective is developed on plant distribution information in floristic studies, particularly at the state and county levels, where the plants of a relatively small area are catalogued comprehensively. Quite popular in the nineteenth century, these studies have by now developed a wealth of detailed knowledge about the local diversity of America's natural landscape, and provided local people with a more manageable listing of local plants found in regional manuals and floras. However, in recent years, there has been relatively little professional encouragement for intensive floristic studies of local areas, and at present much of this local research is conducted by amateurs and students. Nevertheless, local floristic publications provide a valuable and essential complement to regional and national syntheses, since relatively little local information can be incorporated directly into more broadly based but less detailed works. State and provincial floras and related literature are particularly important in bridging the gap between local study and national synthesis; Lawyer *et al.* (in press) provide a bibliographic guide for such works of state or regional significance in North America north of Mexico as well as other areas under U.S. jurisdiction.

In the present century, and particularly the past decade, conservationists and environmental educators have become both major users and significant producers of data on plant species' distributions. Managers of national parks and other natural areas are mandated to protect and conserve the natural features of their areas, but cannot do so comprehensively unless they know which of the species present have special management requirements (Bratton, this volume; Bratton and White, in press; Morse and Lawyer, 1979). Naturalists and interpreters must also know the species of their areas in order to develop educational activities and respond to visitor inquiries. Thousands of plant species are now subject to conservation or protection by the U.S. federal government or various state governments (Kartesz and Kartesz, 1977), and such activities as land-use planning and environmental impact assessment require detailed knowledge of the status of threatened or endangered species, along with other species having disjunct or peripheral populations deemed deserving of conservation protection. Access to such information is essential to informed land-use planning (Pardue, this volume).

The locality data in the traditional botanical literature is usually inadequate for these newer conservation requirements, primarily because localities are rarely identified in sufficient detail and population status is rarely reported. New field studies and precise mapping of older reports must be undertaken to document significant localities with the precision necessary for management, planning, and legal protection. Precise mapping and site visitation are essential to plant protection in areas subject to development or land-use changes, as well as in parks or other natural areas where effects of visitation, invasions by exotics, or indirect impacts of developments may unknowingly threaten rare plant populations.

The needs for setting priorities in plant conservation work and further botanical research require that a continental context be available for assessing the importance of each plant population of interest or concern (Reveal, this volume; Reveal and Broome, 1979). Knowledge of both the historical range and the current conservation status are required for such purposes, since the degree of depletion of historical range is often a major indicator of threat or endangerment to a species. Consideration of prehistoric distribution (*e.g.,* Davis, 1976) can also be useful in understanding present ranges and trends.

In detailed plant conservation work, it is essential to compile a comprehensive report on the conservation status of the various known populations of a species; without such a summary, action cannot be coordinated, and urgent problems can be overlooked (Bradshaw and Doody, 1978). Such status reports should not only consider present and historical distribution, but also the degree of reproduction and persistence of the plants at each locality, the existence of identifiable threats to them, and the management and land-use policies appropriate for the species' conservation. Recently, comprehensive guidelines for the compilation of such status reports have been developed for population studies (Whitson and Massey, this volume; Henifin *et al.,* this volume), for species-level status reports (Henifin *et al.,* this volume), and for ecosystem studies (Radford, this volume; Hall *et al.,* in press).

Sources of Phytogeographical Data

Human knowledge of plant species distributions generally occurs in three forms: unorganized information known by specific individuals; organized, species-specific information accessible to the research community; and other information indirectly pertinent to a species.

Information known only to individuals is useless to others until communicated personally or incorporated into the general base of scientific knowledge, such as through deposition of collections in public museums, writing or reviewing of research papers or reports, collaborating in research efforts that produce reports or publications, or responding to specific requests for information. Unless vouchered by specimens, photographs, or written notes, such information cannot be objectively evaluated. Nevertheless, in many cases an individual's recollection may provide the only suggestion that some locality should be searched for an individual species.

The direct sources of phytogeographical information are generally well-known to botanists and plant conservationists, and need be only briefly summarized here. *Label data* accompanying museum specimens are often regarded as the most authoritative kind of site-locality documentation, since the identification of the specimen can be verified at will, and the specimen is available for further study in case the taxonomic classification of its genus is revised. However, as Barkley (this volume) and many others have cautioned, specimen data should not be used indiscriminately. *Field notes* and collecting reports are important supplements to label data, particularly for early collections bearing little information on their labels. *Local floras* and checklists, and other sources of precise locality data in the literature, are also important (Crovello, this volume), although the identifications of the plants concerned may be subject to some question unless voucher specimens are available for study.

Ecological studies involving population monitoring or indicator-species surveys may also provide specific information on species distributions, as can some forestry reports on species of commercial timber value. *Conservation status reports* are also becoming important tools for documenting primary data on site occurrences; the population status reports discussed by Henifin *et al.* (this volume) are particularly valuable because of the detail they can provide. Useful primary information may also appear in other research reports such as theses or class projects, but are seldom available readily. *Environmental impact statements,* park management reports, and similar documents also may include species-occurrence data, but are seldom widely disseminated to scientific libraries and hence are rarely consulted in the preparation of biogeographical syntheses. Informal or incidental references to sites or localities in other publications may also provide some useful primary information, but are seldom available readily.

The value of the various indirect sources of plant distribution information depends on the degree of correlation of a species' occurrence with various habitat characteristics that can be independently studied; most such inferences should be viewed as hypotheses suggesting a specific area be searched for a particular species. The general ecological literature, for example, can provide such indications as occurrences of vegetation types or plant communities in which the species of interest might be expected.

Recent developments in *remote sensing,* particularly the use of satellites, can now provide precise mapping of dominant vegetation in many areas still relatively unknown botanically (Schanda, 1976). Landsat images are also very useful in locating areas of appropriate habitat for possibly extant occurrences of species in developed areas. The geological and physiographic literature can also prove valuable in predicting species occurrences, particularly for species characteristically limited to a particular habitat type and substrate, and inventory methods combining ecological and geophysical approaches have proven particularly powerful (Gimbarzevsky, 1976, 1978; Radford, this volume; Radford, Massey, and Whitson, 1978). When a high fidelity of association between two species can be demonstrated, then occurrences of one can be used as a predictor of the other. Biogeographical theories can also be employed on occasion to identify localities meriting study, for example through studies of centers of endemism, concentrations of relicts, or ecologically isolated habitats.

Major Strategies of Phytogeographical Data Synthesis

The variety of methods currently used to organize and synthesize phytogeographical information reflects the diversity of sources and users for such knowledge. Methods such as specimen-label indexing and production of local floras and checklists are closely related to traditional needs of museum curation and local botanical research, while the production of generalized range summaries and maps is intended more to convey selected information concisely.

Computer-based techniques for specimen label indexing, proposed long ago by Grassl (1936a), have been subsequently reviewed by Cohen and Cressy (1969), Cutbill (1971), Hall (1974), Morris (1974), Brummitt (1975), Chenhall (1975), and Brenan, Ross, and Williams (1975); no agreement on the many diverse aspects of this topic is apparent, and techniques range from simplified indexing of selected data (e.g., Morse *et al.*, this volume) to complete label-data transcription (*e.g.*, Crovello, 1972; Morris and Glen, 1978), and even combinations of herbarium and bibliographic research (Shetler, 1973). The related question of plotting distribution maps from label data has been surveyed by Adams (1974); Wilcott and Gates (1978) review the various kinds of computer systems available for handling geographical data.

In an excellent tutorial, Clayton (1971) has reviewed the major strategies for organizing geographical data. The use of the hierarchy of political geography— nation, state, and county, for example—bridges the descriptive and synthetic realms, since a hierarchical approach allows detailed locality data to be summarized with increasing degrees of abstraction. Another widely used hierarchical approach is the group of grid systems, such as latitude/longitude quadrangles or the Universal Transverse Mercator (UTM) grid system; these also allow detailed data to be summarized with various degrees of abstraction. On the other hand, lists of site localities that are not keyed to a hierarchical system are of little value in summary and synthesis, or in comparative studies.

In the United States, the strong tradition of local botanical research has led to the development of large amounts of data organized by political geography but very few specimen labels or other reports giving localities by latitude and longitude or by UTM notation. Grid systems are often used in local, detailed mapping for such purposes as population monitoring, but these local maps are rarely referenced directly to one of the global coordinate systems. Mack and Pyke (1979), for example, discuss seedling mapping to one-millimeter accuracy, using local (rather than global) benchmarks.

Approximate conversions between political-areas and latitude/longitude data are readily done through manual or computer-based methods, but can rarely translate such information with sufficient precision for land-management decisions or conservation purposes. Experience in this area is growing, particularly in the methodologies of the state-level natural heritage surveys and data systems for land-use planning.

The variety of ways in which one locality can be described is a major cause of the difficulty in synthesizing distribution data from a variety of sources with a high degree of resolution. Site-specific mapping is the only long-term solution to this problem (in cases where the effort can be justified), using local topographic maps or other high-resolution methods. When detailed mapping of old reports is required, local residents familiar with the geographical and political history of the area should be consulted if possible.

Phytogeographical data presentation generally takes one of three major forms: *point-locality mapping, synthetic approaches* such as grids or the hierarchy of political geography, and mapping by *intrinsic features* of the landscape, such as habitat types. Each of these approaches has its advantages and its disadvantages, and none of them can be recommended exclusively for general use.

Point-locality mapping is commonly done in taxonomic monographs and some floristic presentations, generally using map scales so small that a "point" as mapped actually is an area several miles wide. While giving an excellent idea of the known distribution of a taxon in a fairly precise way, such maps alone can rarely serve to guide one to the exact sites reported. Instead, these maps serve primarily as a visual summary of the detailed localities cited elsewhere in a publication or documented through specimen collections consulted.

Detailed mapping precise enough to lead one directly to a population is rarely seen in published works, local vegetation studies and park guides excepted. (Indeed, in some cases, maps this detailed may direct unscrupulous collectors to vulnerable populations, particularly for commercially exploited plant species such as orchids, cacti, or insectivorous plants.) Detailed data in any case quickly become obsolete, particularly when done plant-by-plant rather than population-by-population, because the vegetation changes, individual plants grow and die, and landforms themselves develop, erode, or are disturbed (*cf.* White, 1979). Under certain conditions, local colonization and extinction rates may be quite high (MacArthur and Wilson, 1967; Simberloff, 1974), further complicating detailed biogeographical data recording. Natural changes in ranges as responses to climate changes may also appear in long-term studies.

In any case, the intensity of effort required to map thousands or millions of individual plants per tract precludes applying precise detailed plant mapping to any broad-scale or long-term problem. What is more often required is a *synthetic approach* in which distributions are summarized in ways in which the data-collection and documentation effort is reasonable compared with the amount and importance of intended use of the information.

Synthetic approaches to the compilation and organization of plant distribution information differ in details but share the property that a single report serves to document the occurrence of a species in an area larger than occupied by the individual plant upon which the report is based. In brief, synthetic approaches map the distributions of species (or other taxa) rather than merely mapping the distributions of the specimens or other reports available.

For example, many state floras map distributions by counties, marking the pertinent county on a state map whenever one or more specimens of a species are known from that county. The marking of a county in such a work is not intended to imply the species is found throughout the county, or is everywhere equally abundant in that county, but only that it is reported to occur there, or at least to have occurred somewhere in that county in historical times. Other sources must be consulted to determine just where in the county a species occurs, how common it is, whether it still grows there, its preferred habitat, and so forth—the county dot map merely synthesizes county distribution data in a form convenient for conveying an initial impression of the geographical distribution of a species within a state. County maps can also be used effectively in summarizing distributions in regional or national works, since detail to this level can be readily shown on a large-format printed page (*e.g.,* Great Plains Flora Association, 1977; Little, 1971).

Since these areas of political geography are hierarchical, distributions within a county can be recorded by smaller units such as towns or townships, as is often done in New England and the U. S. Midwest. Alternatively, distributions within counties can be detailed by a non-political system, either by intrinsic features of the landscape or by extrinsic methods, such as arbitrary lines. Treatment of portions of larger counties as "subcounties" is appropriate in some applications of data handling when the variation in size of the counties creates difficulties in information retrieval or analysis. The use of arbitrarily defined subcounties has been explored in detail by the Oak Ridge National Laboratory (Goff, Olson, and Hicks, 1975; Olson, Goff, and Stephenson, 1975; Olson and Goff, 1977).

Clustering of counties or other small governmental subunits into more homogeneous regions can also aid in information presentation—for example, distributions of species in the Carolinas are more meaningful biologically when county records are collected into Coastal Plain, Piedmont, and Mountain areas, than when they are merely listed by state. Crovello and Keller (1974) discuss further uses of county data in quantitative floristics.

The *grid* system is another widely employed synthetic method for presenting phytogeographical information, particularly for the oceans and for lands lacking a stable, widely recognized hierarchy of political subdivisions. Grids can be informally defined as regular patterns of straight or gently curving lines; those used in mapping are generally based on the lines of latitude and longitude, but occasionally defined by some other means, such as the Universal Transverse Mercator (U.T.M.) grid used in *Atlas Flora Europaeae* (Suominen, 1974) or the range-township-section system employed for legal land descriptions in many states of the U.S.

A well-defined grid can be used hierarchically, since its smallest cells can be clustered into larger cells of similar shape, producing a larger or smaller number of respectively smaller or larger cells as needed by the requirements of a particular situation. Phipps (1975) discusses criteria for selection of such cell sizes. As in the case of political geography, the tradeoffs of greater or lesser resolution must be considered in selecting a grid appropriate for the situation at hand. A common strategy in using grid systems is to record or index the data to the highest degree of resolution practicable, such as the nearest minute or the nearest five seconds of latitude and longitude, then synthesize such reports to a lesser degree of resolution (greater cell size) as needed in searches, reports, or publications. The high-resolution data are thus available when needed, but do not detract from the overall patterns present when the data are collected together into cells of coarser resolution.

Other methods of primary data synthesis can be developed, but more often are used to supplement a political-geography or grid method. Combinations of methods are also possible; an example of such a strategy is the indexing system used by various state natural heritage surveys (e.g., Fisher and Buttrick, in press; Oklahoma Natural Heritage Program, 1978; Moyseenko, Woodall, and Woodall, 1978; see also Hoose, 1980). In these systems, data are plotted to high resolution on 7-1/2 minute topographic maps, then initially summarized by these map quads, producing a grid synthesis with the resolution of 64 cells per square degree. However, since use of the data by counties, watersheds, physiographic provinces, planning districts, and other non-grid areas is also desired, each site mapped is also specifically indexed by these topics so the data file can be searched by them as well. Such combination systems are readily implemented with computer-based data-processing systems, but require careful work in data recording and indexing to locate each point on the map precisely enough to determine its grid cell, county, watershed, and other pertinent properties. Direct multi-factor searching and synthesis from detailed data bases is also possible, but feasible only in sizeable projects; this topic is discussed briefly by Wilcott and Gates (1978) and at length by Nagy and Wagle (1979) and in a workbook prepared by Comarc Design Systems (1978).

Mapping by such intrinsic features of the land as natural regions and physiographic areas is yet another way of presenting plant distributions data. Informal statements of this kind often appear in the floristic literature—"common in the mountain counties," "found only in the Shenandoah Valley," etc.—but are rarely used in mapping except to supplement other techniques. One example of a detailed data system developed from natural areas information is that of the ecological data base of the James Bay Hydroelectric Development in Quebec (Legendre and Gagnon, 1977). In this project, areas are defined by local terrestrial and aquatic vegetation types, and the unit cells are the intersections of these, each cell being a contiguous area of relatively homogeneous vegetation within a single terrestrial vegetation type and a single aquatic ecosystem type.

At a more general level, the vegetation maps of Küchler (1964) and Bailey (1976) are widely used in the United States, for example in mapping areas within counties or within national parks or other management areas. However, the boundaries of such nationally defined areas are inherently vague, so they cannot be used precisely in local studies, even if agreement could be reached on a single national system. Further discussion of this topic is provided in a recent symposium volume (Lund *et al.*, 1978).

Comparison of the Major Strategies

It is obvious that each of these data-synthesis strategies has its strengths and weaknesses, requires somewhat different source information, and cannot completely serve the purposes better served by the others. Yet each method is a means of presenting and summarizing information from the same ultimate source, the actual distribution of plant populations in nature, so there should be considerable similarity in the information presented by these various strategies, suggesting some degree of interconversion should be possible. A data-synthesis strategy may accordingly be selected for the primary purpose of a project, and interconversion methods relied upon to adapt it to incidental needs of other users. Adding further complexity to the process of choosing among these strategies is the choice of degrees of resolution available for each.

Well-defined, point-locality reports are theoretically convertible into any kind of synthesis, since high-resolution point-locality data should serve almost as well as the plant populations themselves as an information base for any kind of mapping or other biogeographical synthesis. However, there are three major limitations on the exclusive use of point-locality data, either in a direct presentation or as a basis for compilation of information to be presented by some other strategy.

The first limitation to the use of point-locality data is that of the area of the "point" being represented; almost any "point locality" in natural history research implies an area surrounding the identified point, perhaps on the scale of yards or meters, perhaps as large as a square mile or more, in which another occurrence of the same taxon will not be knowingly recorded separately. Localities quite near, but not clearly within, the area of interest should be considered when searching or synthesizing point-locality data. Search and retrieval strategies as well as interconversion procedures should therefore use a margin of error appropriate to the precision and accuracy of the source data and the requirements of the search or synthesis being conducted.

The vagueness of many site reports complicates the precision problem, since the degree of uncertainty in mapping or plotting such sites can be high. This limitation on the precision of locating a point is the second major difficulty in the use of point-locality procedures. Few field reports or specimen labels are so precise that the pertinent locality can be found accurately on a topographic map quadrangle, and errors in initial plotting can place a locality in the wrong township, wrong watershed, wrong vegetation type, or wrong grid quadrangle. A second limitation on point-specific plotting can be the limited accuracy of the available maps.

However, the most important limitation on the indiscriminate use of point-locality mapping procedures in phytogeography is the fact that *there are far more plants in the wild than anyone wants mapped*. Selectivity is essential to point-locality mapping, and even for the rarest species a point of diminishing returns is reached for repeated reports from the same small population. Criteria for such selectivity cannot be defined by any general rules, since the degree of precision, the density of plotting, and the frequency of revision are considerations all dependent on the intended use of the information and the amount of time and effort available for obtaining it.

In general, mapping of a greater number of points gives a better representation of the geographical range of a species, and increases the confidence in the assumption that the species does not occur in the places where observations are not plotted. However, increasing the amount of detail recorded for any species will also increase the size and the cost of the data base recording it. For each species and each geographical area, the benefits derived from increasing the intensity of data recording should be compared with the increased costs of obtaining, editing, recording, maintaining, and presenting the data; Crovello (1967, 1977), Morse (1974), and Shetler (1974) suggest further factors to be considered in such decisions. One useful strategy is to provide intense handling of point-locality data in only those parts of a species' distribution range where site-locality mapping of the species is of high interest, for example in the conservation of disjunct or significant peripheral populations or in promoting awareness of invasive exotics near natural areas.

The difficulties of generalizing point-locality data to a political-areas, grid, or other synthesis are minor compared with the problems of interconversion between these synthetic methods themselves. It should be obvious that county or grid-cell data provide no precise information about specific site localities, and hence conversion is impossible from any of these synthetic systems to areas significantly smaller than the units of data collection. However, if the original specimen label or other source of information can be traced through an indexing or documentation system, then interconversion to that degree of resolution should still be possible, although perhaps expensive. Microfilming of specimen labels during data collection is one economical means of such documentation (Adams and Weber, 1976).

For many purposes, approximate conversions may suffice, for examples in initial exploration of scientific theories or in satisfying general curiosity about geographical ranges, but such approximations are useful in land management and impact assessment only in suggesting areas needing site exploration. Whenever possible, data-collection projects should be designed so sufficient resolution is maintained in the initially collected data to meet the requirements of anticipated use.

Further Considerations

The value of the information in a geographical data base depends on many factors, including the quality of the source information, the efficiency with which new reports are added, the degree to which the data base provides pertinent information beyond mere historical presence, and the extent to which confidential data can be protected.

Quality of source data is the primary limitation on geographical data synthesis in natural history. The identification of specimens and field sightings is a particularly significant problem, especially in areas for which there is no recent taxonomic manual, and in taxonomic groups for which there is no widely accepted general classification. If voucher specimens are collected, their identification can be reconsidered if needed, and pertinent reports corrected; many innovative identification processes are now available (Morse, 1975; Pankhurst, 1978) and may prove useful in reassessing the identifications of historically important vouchers. On the other hand, the identifications of unvouchered field reports can rarely be reassessed later, and must often simply be accepted as unverifiable data. King, Watson, and Gould (1967) have developed a thought-provoking list of seven degrees of reliability for unvouchered sightings.

Another class of ambiguous data concerns vaguely described localities—to what extent does a specimen labelled "Mountains of North Carolina" imply the existence of an additional locality for a species otherwise known from only one mountain ridge in a particular North Carolina county? Sometimes careful historical research can be applied to such reports, but most often they must be left ambiguous. Accordingly, data systems must be designed to accept such vague reports when necessary, yet distinguish them from more precise records; point localities should never be assigned to ambiguous reports without also indicating the possibility of error.

Confusion among the uninitiated can also be caused by past changes in the names and boundaries of political units—many counties of the U.S., for example, were formed by subdivision of earlier, larger counties, with the name of the former large county being kept for one of the smaller units. Errors in interpreting old reports can often be avoided by consulting someone familiar with the state's history; for example, Voss (1967) provides many instructive examples of misinterpreted localities among old reports for Michigan plants.

A comprehensive information-synthesis system should also provide procedures for the handling of corrections to erroneous reports. For example, if a specimen is reidentified subsequent to recording of its label data, the appropriate record should be revised so that subsequent reports will reflect the corrected information. Revision of locality information should also be properly reflected in subsequent uses of the data base. If a natural history data base is anticipated to serve a lifespan greater than that needed for prompt production of a particular work, then such provisions deserve particularly serious consideration.

In some cases, data on the mode of occurrence and the conservation status of a plant taxon and its habitat may also be important. Generally, occurrences in cultivation should be distinguished from occurrences in the wild, and for wild occurrences some provision should be made to distinguish presumed native populations from those appearing naturalized or merely persisting after cultivation; Webb (1963) and Shetler *et al.* (1973) summarize customary practice for such cases. For a few species such as timber trees, it may be appropriate to distinguish deliberate introductions from natural stands, since different gene pools may be represented.

A current concern in the handling of site-specific data is the issue of publicity sensitivity, particularly for localities of rare or vulnerable populations. On the one hand, widespread availability of detailed locality information on species of such groups as orchids and cacti can lead to their exploitation and extirpation (Benson, 1977). Also, in a few cases, suggestions that the occurrence of a rare plant (or animal) might preclude a planned development or land-use change have reportedly led to deliberate destruction of the interfering population or irreversible modification of its habitat (Lipske, 1978). On the other hand, however, there are numerous instances of the innocent local extirpation of rare or disjunct plant populations because the local landowner or land manager was unaware of their existence, much less their importance.

Most people working day-to-day with rare plant data feel that generally the best overall policy regarding publicity-sensitive information is to avoid its widespread dissemination yet provide access to it on a need-to-know basis to landowners and managers, to regulatory agencies, and to conservation interests and watchdog groups as well. Such an attitude assures the opportunity for decision-makers to give careful consideration to the continued survival of plant species. Only when indicated by contrary circumstances should strict confidentiality be invoked; in such cases it may often prove easier to handle the data only in a generalized and nonsensitive form, rather than try to record the specifics and maintain restrictions on their access and use.*

Choosing Among the Major Options

Once the purpose of a project has been well defined and the sources of readily available information are known, a data-synthesis strategy can be developed to address the project goals within the limitations of available and anticipated resources. Short-term projects intended to produce widely useful results quickly must be closely constrained to information already at hand, while longer-term programs can depend more heavily upon information not yet available but expected to be collected during

---

* Two recent U.S. laws also apply to some aspects of the information in plant distributions data banks, particularly those operated by the U.S. government. The Privacy Act of 1974 pertains to records kept by the government about individuals, and thus may require the withholding of information about the plant species occurring on an individual's property, particularly when the data were obtained from a source other than the landowner. The Freedom of Information Act, on the other hand, provides general public access to many government documents. Under most circumstances, Privacy Act conflicts excepted, it is difficult if not impossible to deny a request made under the Freedom of Information Act, so the best defense against undesired disclosure of publicity-sensitive information from government data bases is simply avoiding unnecessary deposition of such detailed information in government files in the first place. Many states also have laws similar to these national ones, and compliance with them may be further considerations in the design and implementation of data syntheses pertaining to particular states.

In New York State, locality information on habitats of certain rare or endangered plants is exempted from the provisions of the state's freedom of information law (Chap. 345 of the Laws of 1979).

development of the data base. Mixed strategies are also possible, in which the readily available knowledge is summarized promptly, then further information sought and added as it becomes available.

The following discussion of the point-locality, political-areas (nation-state-county-subcounty), and grid methods is intended to compare their strengths and weaknesses with respect to short-term and long-term synthesis of geographical distribution data for botanical and plant distribution research in the United States and Canada. Somewhat different considerations pertinent to other nations would need to be evaluated before these observations could be applied to such areas.

Point-locality information is essential to population-specific management and conservation programs, such as endangered species recovery efforts and environmental impact assessment. Knowledge of specific localities is also necessary in scientific or educational activities when visitation to natural populations is important. However, for most purposes general statements of distributions and habitat preferences suffice, since populations of common plants can be located in the field from time to time as needed, without reference to detailed prior reports.

Point-locality mapping can be justified economically only when the cost of recording and maintaining information on particular sites is less than the cost of the fieldwork necessary for locating such sites anew as needed, or less than the value at risk should an unrecorded population be unknowingly destroyed. Such detailed mapping can be clearly justified for threatened or endangered species, for genetically important peripheral and disjunct populations, and for stands of economic importance, but not for the vast majority of plant populations at large. Occurrences of aggressive exotics may also merit site mapping, particularly when present in or near established natural areas. Type localities and other historically important study sites may also merit point-locality mapping under some circumstances, as may other sites of scientific, aesthetic, or historical importance, such as occurrences of unusual species assemblages or of champion-sized trees.

In many of the cases where point-locality mapping can be justified, preparation of a local population status report would also be appropriate, minimally indicating the size, condition, and vigor of the local population of the species, but where pertinent also including information on site history, management policy, and recommendations for further monitoring or other research. Consideration of the presence and abundance of seed in the soil can also be important in understanding the dynamics of rare plant populations (Baskin and Baskin, 1978). Over time, the synthesis of such population status reports into population status reviews should be anticipated.

Future point-locality or detailed-grid mapping can be greatly enhanced by encouraging researchers to include relocatable site identifications in their specimen labels, field notes, and research reports, consistent with any necessary requirements of confidentiality. In general, sufficient data should be solicited so an individual familiar with the general area could locate the collecting or study site promptly with minimal ambiguity, either in the field or on a topographic map.

Prompt development of point-locality data bases should thus be promoted for threatened and endangered plant species, for significant disjuncts and peripherals, for populations of generally rare species, for threatening populations of aggressive exotics, and for other populations of identifiable high interest. As shown by the experiences of such groups as the California Native Plant Society (Powell, Duncan, and Howard, this volume); The Nature Conservancy (Jenkins, this volume); and South Africa's Endangered Species Project (Hall, this volume), such detailed mapping is effectively done on local topographic map quadrangles; any generalized point-locality mapping project is likely to need a full set of the topographic maps for the pertinent area. These maps should not be cluttered with numerous reports of widespread or less interesting species; more generalized mapping of these by a grid or county system seems more appropriate. One possibility for selective dissemination of the detailed maps is annual production of sets of microfiche cards.

In summary, point-locality mapping is challenging and expensive, but also the most productive method of data organization for information on local populations of species of high interest and concern. Use of point-locality mapping should always be justified on a case-by-case basis, with careful attention to the need to keep less interesting or less valuable data out of the point-locality system.

The *political geography approach* (sometimes called the politico-geographic or geopolitical hierarchy) is currently the best known and most widely used system of geographical data organization in natural history in the United States and southern Canada. A vast quantity of information is already recorded in this way, both in the botanical literature and in museum collections, and most research libraries also have floristic works geographically organized. Utilization of this vast base of existing geopolitically organized data is the only hope for prompt production of any phytogeographical data synthesis in the United States and southern Canada without significant quantities of expensive primary field research.

The continued recording and filing of data by the state and county system is also compatible with continuation of the traditionally state-centered approach to biological surveys and conservation inventories in the United States, since the information pertinent to each state can be isolated for review and revision by researchers specializing in the local flora. Continued state-oriented data collection and evaluation is also central to the draft program strategy of the recently formed U.S. Heritage Conservation and Recreation Service (*cf.* Smith and Pritchard, 1978; HCRS, 1979). However, supplementation of state-and-county data with latitude/longitude coordinates should be encouraged to promote the future prospect of more detailed mapping and information synthesis.

Direct use of the county as a unit in data indexing also provides an efficient means of concentrating such information to a degree useful in local planning and impact assessment, particularly in conveying an easily understood indication of the significance of local and disjunct taxa (*e.g.,* Shreve, Calef, and Nagy, 1978). The typical Midwestern county—about 25 by 25 miles—is also an appropriate size for distribution mapping on a state or national scale, providing the degree of detail necessary for showing distribution patterns.

In projects that do not also utilize a grid-cell capability, the largest counties, such as many in the Far West, may merit subdivision for data-gathering purposes, using a locally appropriate system such as town boundaries or vegetation-type regions. Also, biologically disjunct islands should generally be treated independently as county equivalents, as, for example, Isle Royale in Lake Superior, politically part of Keweenaw County, Michigan. Further subdivision of the immense administrative units of the Far North of Canada and Alaska would also be necessary for meaningful data presentation; use of the latitude/longitude grid would be appropriate for these sparsely settled regions.

Determination of land ownership is also important to conservation work, and can rarely be done by indirect means because ownership frequently changes, particularly in the private sector. For many conservation and management purposes it is sufficient to distinguish classes of ownership only, such as federal, state, and private, and supplement these by recording the degree of conservation protection given particular tracts (Moyseenko, this volume). Subdividing county reports by land-ownership classes provides a particularly useful perspective on the conservation prospect for extant populations of high-interest species. National parks, national forests, and other major areas of conservation-oriented management may also for some purposes be treated as county-equivalent units separate from the remainders of their counties.

Thus, the state-and-county hierarchy appears to be the most appropriate technique for short-term geographical data organization in American plant systematics, and furthermore has long-term value in promoting information transfer from field-workers and taxonomic specialists to planners, land managers, and the general public. However, some kind of subdivision of large counties is needed in some areas, particularly Alaska, parts of the far western United States, and most of Canada, to supplement county reports with sufficient resolution for meaningful national or continental mapping and data synthesis.

Other useful supplements to the presentation of data by counties include tagging of land ownership and management status for significant populations of conservation concern, mapping species localities by vegetation types or landform areas within counties, and treating national parks or other major natural areas as distinct county subunits separate from the remainder of their counties. Use of the state/county/subcounty hierarchy is compatible with current U.S. floristic and conservation research, and would lead to coordination of the data synthesis needed to support continued refinement of local floristic works.

The *latitude/longitude grid hierarchy* has the advantage of producing nearly uniform coverage, but presently has the major limitation that very little information is now available in a form that can support prompt development of grid-based distributions records providing meaningful resolution within counties. (Much coarser grids, on the other hand, can be easily produced to a fair approximation by combining county reports as needed.) Two further advantages of grid systems over the political-areas hierarchy are that grids are independent of political change, and that distribution patterns plotted by grids can be more easily compared since the unit cells represent areas of similar shape and size. The distances between grid-plotted localities can also be directly calculated without measurement on a map (Phipps, 1975a). Perring (1971) provides further discussion of the usefulness of grid systems, drawing upon his experiences with the *Atlas of the British Flora* and the Biological Records Centre.

In the United States, the most obvious candidate and only real contender for a grid hierarchy for floristic synthesis is the system of latitude and longitude coordinates. This is available to any desired degree of resolution and is already familiar to virtually all field researchers. Furthermore, the availability of the U.S. Geological Survey's topographic map quadrangles provides a basis for detailed mapping of populations of interest within these 7-1/2 minute-square quadrangles, each containing about 100 km^2 or 40 square miles. It is the availability of these map quadrangles that makes the 7-1/2 minute quadrangle appropriate as a basis for first-level floristic data synthesis; other kinds of basic map quadrangles have been used in other countries, for example 10 km^2 maps as the basis for the grid size in Great Britain (Perring and Walters, 1962) or Poland (Zając, 1978) or the 6' x 10' grid used in West Germany (Seybold, 1972). Use of the 7-1/2-minute quadrangle in the United States as the unit cell of a latitude/longitude grid provides an immediate basis for detailed mapping projects and also allows synthesis at higher levels of abstraction as needed for mapping, searching, and comparative studies. Synthesis of phytogeographical data by one-degree quadrangles is effective for comparative studies at the continental scale, as pointed out by Grassl (1936b) and more recently by Prance (1977), amongst others. Simple labeling systems for 15-minute quadrangles have been developed by Edwards and Leistner (1971) and for 7-1/2-minute quadrangles by Weber and Gregory (1975).

The major disadvantage to use of a grid system is that extremely few specimens and field reports are presently identified with grid coordinates or map quadrangle names, so considerable retrospective study would be necessary to locate collecting sites and supply their coordinates. Indeed, for most species it would be easier to map occurrences anew in the field by making site visits to each grid square, than to try to determine the species present in these areas by retrospective searching of herbaria and the literature, a lesson learned early in the preparation of the grid-based *Atlas of the British Flora* (*cf.* Perring, 1963, 1971) as well as in production of the more detailed grid-based *Flora of Warwickshire* (Cadbury, Hawkes, and Readett, 1971; *cf.* Hawkes, Kershaw, and Readett, 1968). Only for such high-priority situations as rare or endangered species can retrospective grid-cell assignment be justified, and for cases of this kind a slightly higher degree of effort can produce within-quad detail as well, having far greater usefulness in conservation and planning. Comprehensive implementation of a grid system in the United States thus appears feasible and justifiable only if done in coordination with both a major quad-by-quad field effort of the kind done in Great Britain and a program of detailed mapping of historical reports of high-interest species. In any case, a grid-based phytogeographical synthesis in the United States would require considerable time and effort for accumulation of meaningful quantities of data.

Geographical-Report Registers

Invariably the publication of a state floristic atlas or other set of state plant distribution maps leads to a series of briefer publications reporting additional county records. A typical report of this kind—the classic by Moran (1962) is atypical only in its presentation—gives the taxon name, the county or other area to which its known range is extended, the year of the collection or sighting, the evidence substantiating the report, and the name of the person reporting the record. A site locality is also often mentioned in such reports, and the phytogeographical significance of a disjunction may be discussed. The frequency with which such information is offered for publication has led at least two journals (*Castanea* and *the Canadian Field-Naturalist*) to propose recently that such information be standardized in tabular form (Clovis, 1977; Smith, 1977). The data-indexing procedures developed by the New England Botanical Club also reflect similar considerations (Morse *et al.,* this volume).

Any series of phytogeographical records of this county-record kind must be based on a recognized taxonomic classification and a familiar geographical framework, such as a county-and-state hierarchy or a particular grid system; reported additions to floristic knowledge would be meaningless otherwise. Implementation of any such system, whether as a series in a journal or as a computerized data base, should specify the initial taxonomic and geographic framework to be followed, and also provide mechanisms for reviewing these as needed in the future. Individual reports should be reviewed before final acceptance both by a taxonomic specialist familiar with the group and by a floristic expert familiar with the plants and the political history of the pertinent state or local area. Such reviewing networks have proven essential to the success of *Flora Europaea* (Webb, 1978) and the *Atlas Flora Europaeae* (Suominen, 1972). Tagging of records as provisional could be considered when delays are encountered in reviewing or when no appropriate expert is available. Editorial procedures should also be provided in such a system for responding to requests that particular reports be studied further.

Further Prospects

These three main strategies of phytogeographical data synthesis—points, counties, and quads—are actually complementary rather than competitive, and thus development of each to the extent appropriate should be encouraged. Certain species should be promptly mapped in detail and their populations periodically monitored; these will generally be species of high scientific interest, conservation significance, economic value, public curiosity, or management concern. Other species, including most common plants, should be mapped initially to the county level, documenting each report as to its source and nature; these county maps should be expanded as soon as feasible to include separate consideration of portions of counties significantly different in some sense, such as offshore islands or major natural areas, treating these as county-equivalent units.

Meanwhile, gathering of field data can be encouraged for a grid-based synthesis employing the widely available U.S.G.S. topographic map quadrangles, with the long-term goal of eventual synthesis of within-county distributions by these quadrangles on a national scale. Early attention should, of course, be given to detailed mapping of high-interest species within these quadrangles; for common species, unvouchered sightings from qualified observers may suffice. Considerable involvement of students and amateur naturalists may be needed to obtain comprehensive reports for the geographical distributions of widespread species. To provide further support for this prospect of quadrangle-based grid mapping, field botanists should be strongly encouraged to include latitude and longitude as precisely as meaningful on their specimen labels.

Implementation of a substantial number of the ideas and strategies discussed here can lead to many improvements in the quality of baseline floristic data and in information transfer to land managers and resource-development planners. For example, one major advantage of a phytogeographical data synthesis on a national or continental scale would be its value in reviewing patterns of past fieldwork and hence in prioritizing future field studies and focusing local floristic work.

In conservation research, the separation of extant from historically known populations is essential, for the significance of a particular population needs to be judged against both the historically reported range and the known extant distribution of the species. The documentation of the conservation status of populations and species is another exciting prospect, bringing together information from many traditionally independent fields of study, such as population biology, impact assessment, and land-use management. Recognition of significant stands is essential to the biological conservation of a species.

Improved documentation of the geographical distributions of common and widespread species will lead to greater appreciation of their habitat requirements, particularly the factors limiting their distributions. Such information should not only prove of interest to phytogeographers and ecologists, but also help in refining procedures for assessing impacts on their environments, and provide better awareness of species assemblages or unusual species concentrations indicating areas appropriate for conservation protection.

Literature Cited

Adams, R. P. 1974. Computer graphic plotting and mapping of data in systematics. *Taxon* 23: 53-70.

Adams, R. P., and W. A. Weber. 1976. An efficient method for the capture and transmission of specimen label information. *Taxon* 25: 479-482.

Bailey, R. G. 1976. *Ecoregions of the United States*. U.S. Forest Service, Odgen, Utah.

Barkley, T. M. 1981. Use and abuse of specimen labels in distribution mapping. *(This volume, pages 79-82.)*

Baskin, J. M., and C. C. Baskin. 1978. The seed bank in a population of an endemic plant species and its ecological significance. *Biol. Conserv.* 14: 125-130.

Benson, L. 1977. Preservation of cacti and management of the ecosystem. Pages 283-300 *in* Prance, G. T., and T. S. Elias, eds. *Extinction is Forever*. The New York Botanical Garden, Bronx, N.Y.

Berlin, B. 1973. Folk systematics in relation to biological classification and nomenclature. *Ann. Rev. Ecol. Syst.* 4: 259-271.

Bradshaw, M. E., and J. P. Doody. 1978. Plant population studies and their relevance to nature conservation. *Biol. Conserv.* 14: 223-242.

Bratton, S. P. 1981. Information storage and population monitoring within Great Smoky Mountains National Park. *(This volume, pages 63-68.)*

Bratton, S.P., and P. S. White. *In press*. Rare plant management—after conservation, what? *Rhodora*.

Brenan, J. P. M., R. Ross, and J. T. Williams. 1975. *Computers in Botanical Collections*. Plenum, London and New York. *x* + 216 pp.

Brummitt, R. K. 1975. Some thoughts on the computerization of label data in major herbaria, with particular reference to type specimens. *Boissiera* 24: 403-410.

Cadbury, D. A., J. G. Hawkes, and R. C. Readett. 1971. *A Computer-mapped Flora: A Study of the County of Warwickshire*. Academic Press, London and New York.

Chenhall, R. G. 1975. *Museum Cataloging in the Computer Age*. American Assoc. State and Local History, Nashville, Tenn. *viii* + 261 pp.

Clayton, K. 1971. Geographical reference systems. *Geogr. J. (London)* 137: 1-13.

Clovis, J. F. 1977. Note from the editor. *Castanea* 42: 326.

Cohen, D. M., and R. F. Cressy, eds. 1969. Natural history collections: past, present, future. *Proc. Biol. Soc. Wash.* 82: 559-762.

Comarc Design Systems. 1978. *Workbook for a Course on the Successful Application of a Geo-based Information System*. Comarc Design Systems, San Francisco. 116 pp. + appendices.

Committee on Germplasm Resources. 1978. *Conservation of Germplasm Resources: An Imperative*. Comm. Germplasm Resources, National Research Council, National Academy of Sciences, Washington, D. C. 118 pp.

Crovello, T. J. 1967. Problems in the use of electronic data processing in biological collections. *Taxon* 16: 481-494.

——————. 1972. Computerization of the Edward L. Greene Herbarium (ND-G) at Notre Dame. *Brittonia* 24: 131-141.

——————. 1977. Computers as an aid to solving endangered species problems. Pages 337-346 *in* Prance, G. T., and T. S. Elias, eds. *Extinction is Forever*. The New York Botanical Garden, Bronx, N.Y.

——————. 1981. The literature as a rare plant information resource. *(This volume, pages 83-93 .)*

Crovello, T. J., and C. Keller. 1974. Use of computerized floristic data of Indiana for plant geography. *Proc. Indiana Acad. Sci.* 83: 399-406.

Cutbill, J. L., ed. 1971. *Data Processing in Biology and Geology*. Academic Press, London and New York. *xv* + 346 pp.

Darwin, C. 1859. *On the Origin of Species by Means of Natural Selection.* John Murray, London. 502 pp.

Davis, M. B. 1976. Pleistocene biogeography of temperate deciduous forests. *Geosci. Man* 13: 13-26.

Eckholm, E. 1978. *Disappearing Species: The Social Challenge.* Worldwatch Paper 22. Worldwatch Inst., Washington, D.C. 38 pp.

Edwards, D., and O. A. Leistner. 1971. A degree reference system for citing biological records in southern Africa. *Mitt. Bot. Staatssamml. München* 10: 501-509.

Endler, J. A. 1977. *Geographic Variation, Speciation, and Clines.* Princeton Univ. Press, Princeton, N.J. 246 pp.

Fisher, M. N., and S. C. Buttrick. *In press.* The Massachusetts Natural Heritage Program. *Rhodora.*

Gimbarzevsky, P. 1976. Integrated survey of biophysical resources in national parks. Pages 257-269 *in* Proc. 1st Meeting Canad. Comm. Ecol. (Biophys.) Land Classif., May 25-28, Petawawa, Ont., Canada.

_____. 1978. Land classification as a base for integrated inventories of renewable resources. Pages 169-177 *in* Lund, H. G., *et al.,* eds. *Integrated Inventory of Renewable Natural Resources: Proceedings of the Workshop.* U.S.D.A. For. Serv. Gen. Tech. Rep. RM-55.

Goff, F. G., R. J. Olson, and S. D. Hicks. 1975. Delineation of the Eastern Region and definition of Eastern county-sub-county units (CSCU's). *Oak Ridge Natl. Lab. Geoecol. Proj. Memo* GPM-FGG 75-6.

Grassl, C. O. 1936*a.* Visualizing our herbaria by the application of mechanical methods of tabulation and indexing. *Mus. J.* 36: 373-384.

_____. 1936*b.* An international system of botanical districts. *Bull. Torrey Bot. Club* 63: 519-524.

Great Plains Flora Association. 1977. *Atlas of the Flora of the Great Plains.* Iowa State Univ. Press, Ames, Iowa. 600 pp.

Hall, A. V. 1974. Museum specimen record data storage and retrieval. *Taxon* 23: 23-28.

Hall, A. V., M. de Winter, B. de Winter, and S. A. M. van Oosterhout. *In press. Threatened and Critically Rare Plants of Southern Africa.* S. Afr. Natl. Sci. Prog. Rep., C.S.I.R., Pretoria, South Africa.

Hawkes, J. G. 1978. The taxonomist's role in the conservation of genetic diversity. Pages 125-142 *in* Street, H. S., ed. *Essays in Plant Taxonomy.* Academic Press, London.

Hawkes, J. G., B. L. Kershaw, and R. C. Readett. 1968. Computer mapping of species distributions in a county flora. *Watsonia* 6: 350-364.

Henifin, M. S., L. E. Morse, S. Griffith, and J. E. Hohn. 1981. Planning field work on rare or endangered plant populations. *(This volume, pages* 309-312.)

Henifin, M. S., L. E. Morse, J. L. Reveal, B. MacBryde, and J. I. Lawyer. 1981. Guidelines for the preparation of status reports on rare or endangered plant species. *(This volume, pages* 261-282.)

Heritage Conservation and Recreation Service. 1979. *The National Heritage Policy Act.* [Briefing Book]. U.S. Dep. Interior Heritage Conservation and Recreation Service, Washington, D.C.

Hoose, P. M. 1980. *Building an Ark: Tools for the Preservation of Natural Diversity.* Island Press, Covelo, Calif.

Kartesz, J. T., and R. Kartesz. 1977. *The Biota of North America. Part 1. Vascular Plants. Vol. 1. Rare Plants.* Biota N. Amer. Comm., McKeesport, Pa. 361 pp.

King, W. B., G. E. Watson, and P. J. Gould. 1967. An application of automatic data processing to the study of seabirds, I. Numerical coding. *Proc. U.S. Natl. Mus.* 123(3609): 1-29.

Küchler, A. W. 1964. *Potential Natural Vegetation of the Conterminous United States.* Spec. Publ. 36, Amer. Geogr. Soc., New York.

Lawyer, J. I., A. M. Miller, L. E. Morse, and J. T. Kartesz. *in press.* A guide to selected current literature on vascular plant floristics for the contiguous United States, Alaska, Canada, Greenland, and the U.S. Caribbean and Pacific Islands. *Mem. Torrey Bot. Club.*

Legendre, P., and M. Gagnon. 1977. The ecological data bank of the James Bay hydroelectric development. Pages 305-309 *in* Dreyrus, B., ed. *The Proceedings of the Fifth Biennial International CODATA Conference.* Pergamon Press, Oxford and New York.

Lipskc, M. 1978. The withering wreath. *Defenders* 53: 298-304.

Little, E. L., Jr. 1971. *Atlas of United States Trees. Vol. 1. Conifers and Important Hardwoods.* U.S. For. Serv. Misc. Publ. no. 1146.

Lund, H. G., V. J. LaBau, P. F. Ffolliott, and D. W. Robinson, eds. 1978. *Integrated Inventory of Renewable Natural Resources: Proceedings of the Workshop.* U.S. For. Serv. Gen. Techn. Rep. RM-55. *vi* + 482 pp.

MacArthur, R. H., and E. O. Wilson, 1967. *The Theory of Island Biogeography.* Princeton Univ. Press, Princeton, N.J. *xi* + 203 pp.

Mack, R. N., and D. A. Pyke. 1979. Mapping individual plants with a field-portable digitizer. *Ecology* 60: 459-461.

27

Moran, R. 1962. *Cneoridium dumosum* (Nuttall) Hooker f. collected March 26, 1960, at an elevation of about 1450 meters on Cerro Quemazón, 15 miles south of Bahía de Los Angeles, Baja California, México, apparently for a southeastward range extension of some 140 miles. *Madroño* 16: 272. [The text of the article reads simply, "I got it there then"; unusually detailed acknowledgments follow.]

Morris, J. W. 1974. Progress in the computerization of herbarium procedures. *Bothalia* 11: 349-353.

Morris, J. W., and H. F. Glen. 1978. PRÉCIS, the National Herbarium of South Africa (PRE) computerized information system. *Taxon* 27: 449-462.

Morse, L. E. 1974. Computer-assisted storage and retrieval of the data of taxonomy and systematics. *Taxon* 23: 29-43.

_____. 1975. Recent advances in the theory and practice of biological specimen identification. Pages 11-52 *in* Pankhurst, R. J., ed. *Biological Identification with Computers.* Academic Press, London and New York.

_____. 1977. Some information-management problems raised by the international nature of systematic biology data. Pages 323-327 *in* Dreyfus, B., ed. *The Proceedings of the Fifth Biennial International CODATA Conference.* Pergamon Press, Oxford and New York.

Morse, L. E., M. Lamson, A. F. Tryon, and R. R. Walton. 1981. Specimen locality indexing by the New England Botanical Club. *(This volume, pages 185-191.)*

Morse, L. E., and J. I. Lawyer. 1979. Prospective contributions of a computer-based continental flora to scientific information needs of the National Park Service. Pages 95-96 *in* Linn, R. M., ed. *Proceedings of the First Conference on Scientific Research in the National Parks.* U.S. National Park Service, Washington, D.C.

Moyseenko, H. P. 1981. Limiting factors and pitfalls of environmental data management. *(This volume, pages 237-253.)*

Moyseenko, H. P., J. L. Woodall, and S. R. Woodall. 1978. The balanced ecogeographical information system: a vehicle for data collection, systematization, and dissemination. Pages 569-592 *in* Marmelstein, A., ed. *Classification, Inventory, and Analysis of Fish and Wildlife Habitat.* Rep. FWS/OBS 78/76, U.S. Fish and Wildlife Service, Washington, D.C.

Nagy, G., and S. Wagle. 1979. Geographic data processing. *Comput. Surv.* 11: 139-181.

National Plant Genetic Resources Board. 1979. *Plant Genetic Resources: Conservation and Use.* U.S.D.A. National Plant Genetic Resources Board, Washington, D.C. 20 pp.

Oklahoma Natural Heritage Program. 1978. *The Guide.* Oklahoma Natural Heritage Prog., Norman, Okla.

Olson, R. J., and F. G. Goff. 1977. Development and applications of a regional environmental data base for the southeastern United States. Pages 289-293 *in* Dreyfus, B., ed. *The Proceedings of the Fifth Biennial International CODATA Conference.* Pergamon Press, Oxford and New York.

Olson, R. J., F. G. Goff, and R. L. Stephenson. 1975. Geoecology county/subcounty unit data base. *Oak Ridge Natl. Lab. Geoecol. Proj. Memo* GPM-RJO 75-4.

Pankhurst, R. J. 1978. *Biological Identification.* Edward Arnold, London. *viii* + 104 pp.

Pardue, J. W. 1981. Locating critical natural features information for environmental planning. *(This volume pages 69-75.)*

Perring, F. H. 1963. Data-processing for the Atlas of the British Flora. *Taxon* 12: 183-190.

_____. 1971. The British Biological Recording Network. Pages 115-121 *in* Cutbill, J. L. ed. *Data Processing in Biology and Geology.* Academic Press, London and New York.

Perring, F. H., and S. M. Walters. 1962. *Atlas of the British Flora.* Thomas Nelson, London. *xxiv* + 432 pp.

Phipps, J. B. 1975a. Kilometric distance. *Canad. J. Bot.* 53: 1116-1119.

_____. 1975b. Bestblock: optimizing grid size in biogeographic studies. *Canad. J. Bot.* 53: 1447-1452.

Prance, G. T. 1977. Floristic inventory of the tropics: where do we stand? *Ann. Missouri Bot. Gard.* 64: 659-684.

Radford, A. E., J. R. Massey, and P. D. Whitson. 1978. *Natural Heritage Classification and Information Systems.* Univ. North Carolina Student Stores, Chapel Hill, N.C. 265 pp.

Raven, P. H., B. Berlin, and D. E. Breedlove. 1971. The origins of taxonomy. *Science* 174: 1210-1213.

Reveal, J. L. 1981. The concepts of rarity and population threats in plant communities. *(This volume, pages 41-47.)*

Reveal, J. L., and C. R. Broome. 1979. Plant rarity—real and imagined. *The Nature Conservancy News* 29(2): 4-8.

Schanda, E., ed. 1976. *Remote Sensing for Environmental Sciences.* Springer-Verlag, Berlin. *vi* + 367 pp.

Seybold, S. 1972. Bericht über den Stand der Kartierung in Württemberg. *Göttinger Flor. Rundbr.* 6: 83-86.

Shetler, S. G. 1973. The Botanical Type Specimen Register. *Smithsonian Contr. Bot.* 12: 1-25.

_____. 1974. Demythologizing biological data banking. *Taxon* 23: 71-100.

Shetler, S. G., D. M. Porter, H. R. Krauss, J. H. Beaman, W. H. Lewis, J. McNeill, J. T. Mickel, A. E. Radford, P. H. Raven, R. L. Taylor, and J. H. Thomas. 1973. A guide for contributors to Flora North America (FNA) (Provisional edition). *FNA [Flora North America] Rep.* no. 65. [77 pp.]

Shreeve, D., C. Calef, and J. Nagy. 1978. *The Endangered Species Act and Energy Facility Planning: Compliance and Conflict.* Rep. BNL 50841, Brookhaven National Laboratory, Upton, N.Y.

Simberloff, D. S. 1974. Equilibrium theory of island biogeography and ecology. *Annual Rev. Ecol. Syst.* 5: 161-182.

Smith, L. C. 1977. Commentary on reports of range extensions. *Canad. Field-Naturalist* 91: 221-224.

Smith, S. D., and P. Pritchard. 1978. The National Heritage Trust Program: Information systems for a national objective. Pages 131-133 *in* Lund, H. G., *et al.,* eds. *Integrated Inventories of Renewable Natural Resources: Proceedings of the Workshop.* U.S. For. Serv. Gen. Techn. Rep. RM-55.

Soulé, M. E., and B. A. Wilcox, eds. 1980. *Conservation Biology: An Evolutionary-Ecological Perspective.* Sinauer Associates, Sunderland, Mass.

Suominen, J. 1974. Atlas Flora Europaeae: preparation and relationship to Flora Europaea. *Bol. Soc. Brot.* 47 (suppl.): 29-35.

Voss, E. G. 1967. The status of some reports of vascular plants from Michigan. *Michigan Bot.* 6: 13-24.

Webb, D. A. 1963. The treatment of alien species in Flora Europaea. *Webbia* 18: 27-34.

_____. 1978. Flora Europaea—A retrospect. *Taxon* 27: 3-14.

Weber, S. G., and R. P. Gregory. 1975. *Universal Map Code (A 7-1/2-Minute Quadrangle Identification System)* Div. Forestry, Fisheries, and Wildlife, Tennessee Valley Authority, Norris, Tenn. 12 pp.

White, P. S. 1979. Pattern, process, and natural disturbance in vegetation. *Bot. Rev.* 45: 229-299.

Whitson, P. D., and J. R. Massey. 1981. Information systems for use in studying the population status of threatened and endangered plants. *(This volume, pages 217-236.)*

Wilcott, J. C., and W. A. Gates. 1978. A review of existing geographical information systems and some recommendations for future systems. Pages 553-568 *in* Marmelstein, A. ed. *Classification, Inventory, and Analysis of Fish and Wildlife Habitat.* Rep. FWS/OBS-78/76, U.S. Fish and Wildlife Service, Washington, D.C.

Zajac, A. 1978. Atlas of Distribution of Vascular Plants in Poland (ATPOL). *Taxon* 27: 481-484.

Section 2.

Information Needs and Priorities

Rare Plant Conservation through Elements-of-Diversity Information

Robert E. Jenkins[1]

With the recent surge of interest in endangered plant species, lists have been created for the U.S. at large and most states which have been helpful in focusing conservation efforts. Of special significance has been the Nature Conservancy's initiation of State Natural Heritage programs through which we have been attempting to create dynamic atlases and data bases on the existence, characteristics, condition, numbers, status, location and distribution of the elements of natural ecological diversity, including rare plant species. The Conservancy's techniques are especially adapted to the preservation of rare plant habitats but until recently this effort was hindered by the lack of objective ecological information necessary to guide the work.

The Heritage Program methodology is outlined and some observations made about its special significance to rare plant conservation. Emphasis is given to the means by which an objective priority list of crucial land areas is derived and to the power of comparative information in facilitating protection by various means. Conservancy efforts to obtain suitable information on endangered plant localities in non-heritage states are described along with difficulties we have encountered and some specific plans for the future.

This paper is devoted to the following topics: the reasons for preserving species and other units of natural diversity; the threat against which they must be protected; the scope and character of the tasks we must undertake to accomplish this purpose; the crucial role of locality data; the relevance of The Nature Conservancy's State Natural Heritage Programs, and some of the ramifications and expectations of these programs.

The Nature Conservancy, for which I work, is wholly devoted to the perpetuation of natural ecological diversity through the protection of land areas selected specifically for their representation of certain species, communities, or other ecological elements. We come to this objective both by historical origins, and from analysis of current events. We believe that the growth of society creates two fundamental ecological problems—the threat of dysfunction in degraded natural systems and the ex-

---

[1] The Nature Conservancy, 1800 North Kent St., Arlington, Va. 22209.

Based on the keynote address "Rare Plant Conservation through Elements-of-Diversity Information" by Dr. Jenkins at the Symposium on Geographical Data Organization for Rare Plant Conservation, held 15-17 November 1977 at The New York Botanical Garden.

Pages 33-40 *in:* Larry E. Morse and Mary Sue Henifin, eds., *Rare Plant Conservation: Geographical Data Organization*, The New York Botanical Garden, Bronx, N.Y.

tirpation of specific irreplaceable system components. We find our situation analogous to that of Noah in the face of the great flood, choosing between trying to stop the flood or trying to survive it. We believe that the idea of creating an ark in this context to preserve at least the minimal number of propagules of species which might otherwise drown is a sane and rational resource management response.

The perpetuation of organic diversity and specifically of rare and endangered species is of interest for several reasons. First, there is the unknown importance of their ecological role; second, there is the potential unique utility of any species as a renewable resource for the betterment of the human condition; third, there is their scientific significance, as each may be important in answering questions not answerable by any other; fourth, there is the aesthetic and psychological significance of each species and of all species collectively in enriching the variety of the environment; and fifth, there is the partly philosophic and partly pragmatic fact that our ability to sustain a diverse and functioning environment is an indicator of our ability to live on the earth and maintain a viable human society within it. If we find ourselves powerless to prevent extinctions, how can we be confident of our ability even to sustain life support systems? The fact that all of these benefits simultaneously flow from the perpetuation of species makes this enterprise, in my estimation, one of the most under-appreciated tasks of our times.

The extinction and threat of extinction of species has at various times and instances been due to innumerable factors. These have been extensively treated elsewhere. Today and on our continent specifically, the overwhelmingly greatest threat to endangered species arises from habitat destruction through direct intensification of land uses in the critical remaining refugia. Even in those relatively few instances where direct exploitation is the important factor, the effect is still site-specifically exerted against a species having few remaining populations. Those species which are more widespread and more evenly distributed are, in most cases, only marginally endangerable and will probably manage to survive until the overall state of the landscape has become much worse than it is today. I know that exceptions can be cited to any generalization, but this one is sufficiently true for me to draw a conclusion central to this symposium's purpose: that *the successful perpetuation of endangered plant species will very largely depend upon our ability to identify individual species' localities. No matter how much we know in general about a species' niche, range, or habitat requirements, it cannot be preserved or managed unless we know the actual places in the landscape where it occurs.*

This premise has been the organizing principle of The Nature Conservancy's State Natural Heritage Programs. Through these programs, the Conservancy has been assisting individual state governments to develop a systematic continuing process for the preservation of natural ecological diversity. Endangered plant species are one of several major classes of elements which are dealt with intensively under these programs. The key to these efforts is a systematic ecological inventory, the methodology for which has been developed out of the Conservancy's experience in this field dating back to the founding of its precursor body, the special Committee for the Preservation of Natural Conditions, established under the Ecological Society in 1917.

For a number of reasons which I will not dwell upon here, these earlier inventories were largely disappointing both intrinsically and for their lack of conservation application. At least the worst of these deficiencies, I believe, have gradually been corrected in the State Natural Heritage Programs with gratifying results. Twelve such programs are now operating successfully,* and each is strengthened by the existence of, and communication with, the others. To describe generally the inventory aspect of these programs, it may be said that they are intended to create a dynamic atlas and data base on the existence, numbers, condition, protection status, location, and distribution of the occurrences of the elements of natural ecological diversity. I do not believe that we should even think in terms of such data bases ever being "completed," but as the existing ones continuously mature they are proving to be increasingly powerful tools for ecological conservation and other purposes.

Before proceeding to a brief description of these programs, I would be remiss if I did not take a moment to note that there is now a strong national initiative which has the potential of creating a federal program for the coordination and abetment of identical or similar efforts throughout all the 50 states. Part of this initiative is coming directly from the new administration. President Carter, in his environmental message in May, 1977, ordered a 120-day Department of the Interior study of the feasibility of creating a "National Heritage Trust" with the intent of protecting the elements of both our natural and cultural heritage. A task force made up of Department of the Interior personnel and volunteers from various private organizations, many also contributors to this symposium, labored on this task all summer long. This task force made its report to the President, who responded by creating the new Heritage Conservation and Recreation Service. The service, composed of the former Bureau of Outdoor Recreation with the addition of units from the National Park Service, is now proceeding to develop its program, which is expected to include separate or parallel inventory and protection programs for natural and cultural heritage resources.

Since there is already enabling legislation for a cultural resources program in the form of the National Historic Preservation Act of 1966 with a developed tradition, infrastructure, and state-federal relationship, it is likely it will continue generally along its current lines. There is no similar authority for natural heritage resources, so this program will probably take shape gradually over the next year or two and may require successful Congressional action to make it fully operational. Not coincidentally, such legislation has already been introduced in the form of National Natural Diversity bills, S. 1820 and H.R. 8650. These bills themselves have a considerable evolutionary history and emanate at least in part from the recommendations made to the Department of the Interior several years ago in the "Preservation of Natural Diversity" report (Humke et al., 1975).*

As extensive description of the State Natural Heritage Programs is available elsewhere (Jenkins, 1976, 1978; Sanders, 1978; Moyseenko, this volume; Hoose, 1980), I will try to keep the present description as brief as possible, concentrating on endangered plant species as an example and emphasizing the crucial concentration on actual locality data. The programs all being with the hiring and training of staff, the development of a classification system of the elements of ecological diversity, and the installation of the data management system and inventory apparatus. The classification is

---

* South Carolina, Mississippi, Oregon, Tennessee, West Virginia, New Mexico, Ohio, the Tennessee Valley Authority, North Carolina, Oklahoma, Washington, and Indiana. [Also, by April 1980, Arizona, Arkansas, California, Colorado, Kentucky, Maryland, Massachusetts, Michigan, Minnesota, Rhode Island, and Wyoming.—Editors]

* These bills were not passed, but by early 1980, several Natural Heritage bills (S. 1842, H.R. 6504, and H.R. 6805) had been introduced, and hearings held or scheduled.

a list of elements comprising a set of coarse and fine filters intended to capture efficiently the fullest practicable array of our diverse biological and ecological resources. For coarse filters we use rather traditional plant community typologies, and the main fine filters are lists of species of special concern. These species can be so used because there are few enough sites known to be dealt with individually and because they are the species which seem least likely to be represented on lands selected mainly as typical ecosystem examples.

In order to determine which plant species should be included we employ fairly lenient criteria. In any state program, the special plant species list is composed of all the species in the Smithsonian Institution publication, *Endangered and Threatened Plants of the United States* (Ayensu and DeFilipps, 1978), which occur in that state, any additional plant species which occur on a specific state list, and any other species for which a qualified expert makes a convincing case. In practice, of course, we give the greatest time and effort to those species generally thought to be rarest and most threatened throughout their range. However, some of the special species of interest in a particular state may nevertheless be common elsewhere.

With this foundation laid, we embark upon a cyclical process of inventory. As information is being gathered we incorporate it into a "balanced information system" employing manual and automated files organized around four main "units of record." These units are the element, the element occurrence, the managed (or protected) area, and the information source. More will be said about these elsewhere in this symposium, but I would like to concentrate here on the element occurrence.

All element occurrences are actual geographic localities, and for the endangered plant species this is nearly always a known population. Information on the existence and attributes of these occurrences is derived primarily from secondary sources at the outset. As new localities are discovered from museum collections, scientific literature, earlier inventories, or the heads and files of professional botanists and field naturalists, these individual localities are plotted on the appropriate topographic quadrangle or other base map from a complete state series, and accompanying data are incorporated into manual files. A standard "Lowest Common Denominator" abstract of this information is transcribed and encoded for the computer.

If all of this seems perfectly logical and even a bit simple-minded, a few words of amplification may be justified. Unlike many who engage in quite similar procedures, our interests are not primarily academic or even ecological though they may adapt themselves to such purposes—they are conservationist. The upshot is that as our investigations reveal one particular element, such as an individual plant species, to be relatively abundant and otherwise secure, we shift our attention away from it and toward those which will appear to be less so. Because of this we will not soon gain a comprehensive picture of the ecology or distribution of relatively more common species, since, unlike the plant taxonomist's, our interest is primarily in species in need of conservation assistance. However, in this fashion we concentrate our resources effectively on the rarest and most endangered elements in a state's flora and gain the necessary information at the earliest possible time to do something effective for their conservation. Information is accessed in cycles, during which the data base is periodically analyzed to draw our attention to the species and topics on which our ignorance is glaring. The inventory is then directed more intensively toward additional secondary sources, *de novo* searches where necessary, and field surveys as required to gain just that information which we need to plot effective conservation strategy. We are less interested in discovering even the sixth or seventh locality for a species until something is done to secure at least the first or second in protected lands.

This description, of course, is an oversimplification and as the Heritage inventory grinds on it is quite as likely to compile an exhaustive body of information as any other program. However, this sequence of data accession expresses a philosophy and purpose which may be missing from superficially similar programs. In general, what must be saved is that which is rare, and too many past conservation inventories bogged down by treating their varied subjects in essentially the same proportion as they are found in the natural landscape. Ten thousand times as much information was therefore collected about the white oaks as about rare herbs when, in fact, the former generally needs no assistance while the latter may be hanging on by their root hairs.

Although there are many additional details which are important to the successful function of these Heritage Programs, they cannot be treated at length here. The strength of these programs is their comparative information base on relative rarity and the emphasis on actual localities. Because relative rarity is a dominant factor in determining criticality, these data bases have a surprisingly broad utility at the earliest possible moment in their development. Within a year from start-up, the better programs are generating priority lists of lands for conservation, providing perspective for a wide array of siting decisions, contributing in an unprecedented fashion to environmental impact assessment, and generally serving greater or lesser purposes for a large array of users. All of these applications contribute directly to effective species conservation.

Probably the greatest interest of this audience would be to trace the process by which conservation priorities are derived. Once sufficient information has been accumulated to make realistic comparisons, analysis proceeds through a number of steps. First, element considerations are dealt with. These concern mainly the apparent status of the species throughout its range, the number of known occurrences within the state in question, and the number of these occurrences which are on already protected lands. Those elements which are most localized or rare overall, with the lowest number of populations within the state and least adequately represented on conservation lands, are given priority attention in subsequent steps.

Next, site considerations are dealt with for the occurrences of the priority elements. These site considerations include questions of viability, defensibility, quality, condition, and related site factors pertinent in the context of the individual species' biology. One site factor which is only partly ecological is the question of ensembles; that is, where an occurrence of a given plant species might be located in proximity to the occurrences of other elements with which we are also interested. This has some ecological significance because protecting a larger and more diverse land area may enhance viability and defensibility, but ensembles are also important from the point of view of cost-effectiveness.

The last set of factors to be considered, and properly so, are those affecting the feasibility of protecting the land area associated with the emerging priority occurrences. This may involve consideration of pre-existing agency interest, coinciding site values, the feasibility of raising funds for acquisition, the attitude of the current owner, etc.

A standard output from a mature Heritage Program is an "element-indexed data summary" in which each element in the classification is listed along with its status, the known occurrences and managed areas, and the priority proposed preserves for the most critical species. A prime tenet of the program is that each element in the classification system which is not already represented with relative abundance in protected areas must have at least one recommended area for preservation. I do not really see how species conservation could be dealt with any more directly and efficiently than this, and the programs promise to accomplish the objective very effectively. Even where the priority preserve recommendations cannot be immediately acted upon, lesser measures of protection are ordinarily extended by the compelling nature of the data analysis itself. Relative rarity seems to be a key ingredient in human valuation-setting, and this instinctive pricing mechanism may be the most important conservation tool we have going for us.

Though the State Heritage Programs are only a couple of years old, we can already cite innumerable conservation achievements, many of which have involved rare plant species. Almost every day, staff from many of these programs comment adversely on siting decisions that would affect populations of dwindling plant species and promote outright avoidance of such destruction. The more mature programs are now reaching the point where the Conservancy is focusing and intensifying its own conservation efforts on the priority lands, and these have begun to produce results. In South Carolina the number one priority from the inventory, Stevens Creek, was recently given to The Nature Conservancy by its private owner, The Continental Group, and conveyed in turn to the state for investment in its Natural Heritage Trust. This area contained populations of nineteen rare plants including at least one plant, *Ribes echinellum,* for which only one other locality is known. In North Carolina, six of the top ten priorities are critical endangered plant species habitats, one of which contains 23 separate plant species of special concern. All or nearly all of these areas are presently under negotiation for acquisition or other forms of protection.

One critical area, a 13,000-acre portion of the Green Swamp, was conveyed as a gift to The Nature Conservancy by Federal Paperboard. This area contains important populations of the seriously declining Venus flytrap, *Dionaea muscipula,* and other bog carnivores including *Sarracenia rubra.* The rare orchid *Habenaria integra* also occurs there. In Oregon, sites for *Lomatium minus* and *Silene douglassii* var. *oraria* have been protected. Mississippi has protected a site for *Sarracenia psittacina* and is working hard on habitats of *Apios priceana, Thalictrum debile, Aristida simplicifolia,* and *Sageretia minutiflora.* In West Virginia protection actions are being completed on sites for *Anemone minima* and *Synandra hispidula.* In each program, localities for the most endangered species have been identified and proposed for protection. In many states endangered plant species sites on public lands are being given additional protection as in *Cymophyllus fraseri, Habenaria integra,* and *Saxifraga caroliniana* populations in West Virginia.

The Nature Conservancy is attempting to begin Heritage Programs in numerous other states at this time, and as the process has proven so useful this will continue to be our primary emphasis. However, our field staff in particular are impatient to refine our conservation efforts in other states where we are active but in which there is no Heritage program or equivalent. Our conservation performance in such places has often left much to be desired as lack of information or of convincing documentation led to the protection of sub-optimal properties or the failure to act on truly critical areas. We have come to see that endangered plant species may be one area in which we can improve our performance even in non- or pre-Heritage states. On the one hand, rare plant conservation is ideally suited to protection by traditional Nature Conservancy land acquisition. On the other, it should also be possible to obtain information on obvious priorities in regard to such species simply by running down actual known localities of plant species for which endangerment is apparent. For this purpose we have recently begun to cast about among knowledgeable botanists for locality data on the most critically endangered species. In light of the gross and continuing neglect of rare plant species conservation, the Conservancy, I am proud to announce, intends to dedicate at least 20 percent of its 1978 project activity directly to endangered plant species habitat acquisition. In view of the fact that last year the Conservancy did 200 separate acquisitions, this level of activity could in a few years become the single most significant factor in plant species conservation.

This brings me to the discussion of a few problems. First, we cannot possibly concentrate on endangered plant species acquisition without the willing assistance of professional botanists in supplying us information. We all know, and others in the symposium will deal directly with, the deficiencies of the published literature, the museum specimen data, and other sources in terms of accurate locality information. In Heritage states we are equipped to deal effectively with these deficiencies, but in other states we are not, and the best information is surely locked up in the minds of the professional botanical community. However, for a variety of reasons, some admirable and some perhaps less so, botanists seem loath to part with their data. First, many of them have been burnt before by the trouble they have gone to in supplying information which was not then acted upon. We can only promise you, in this case, that we have every serious intent in the future to do our utmost to give plant species conservation the priority attention it deserves. Some botanists are tight with their information because of proprietary concerns, and on this at least I can assure you that we do not have competitive publication interests.

A more serious matter is the distressing paradox of data security and species endangerment that never ceases to amaze me. In regard to endangered plant species locality data, everyone knows that it is perfectly okay to give it to the proverbial bad guys—that is, the developmental interests—since their only concern is to avoid conflict on the identified critical sites. However, it may be exceedingly dangerous for this information to fall into the hands of certain of the supposed good guys—those with personal botanical interests. The number of cases in which known populations have been damaged or eradicated by over-collection are legion. The most recent glaring example is perhaps the Virginia roundleaf birch *(Betula uber)* in Smythe County, Virginia. Presumed extinct by the compilers of the first Smithsonian list, a single population was soon rediscovered. A year after its discovery, the population was already being transformed into a stand of pruned poles by specimen collectors (U.S. Fish and Wildlife Service, 1978).

On this last question, all I can say, besides the observation that such predatory collection is an abominable disgrace, is that we have been very careful on information redistribution in the course of our data compilation. We usually make it freely available to land use decision makers, but to others only on a selected or need-to-know basis. Under no circumstances do we encourage, or approve of, promiscuous publication of such locality information.

This said, I would like to conclude this paper on a different note by offering the services of The Nature Conservancy to any and all of you intent upon plant species conservation. We are not a bit proprietary about the Heritage Programs themselves. Most of the existing ones already have been, and all of them are intended eventually to be, transferred to a government body for continued operation and management.

During their formative stages they take most of their definition and scope, aside from the basic methodology, from the expert academic and conservation communities of the state in which they operate. Without the professional academicians, notably botanists, such a program would not even be possible. At the same time, I believe that they offer a resource of great potential to the individual scientists.

Literature Cited

Ayensu, E. S., and R. A. DeFilipps. 1978. *Endangered and Threatened Plants of the United States.* Smithsonian Institution and World Wildlife Fund, Washington, D.C.

Hoose, P. M. 1980. *Building an Ark: Tools for the Preservation of Natural Diversity.* Island Press, Covelo, Calif.

Humke, J. W., B. S. Tindall, R. E. Jenkins, H. L. Weiting, Jr., and M. S. Lukowski. 1975. *The Preservation of Natural Diversity: A Survey and Recommendations.* The Nature Conservancy, Arlington, Va.

Jenkins, R. E. 1976. "Maintenance of natural diversity: Approaches and recommendations," presentation to 41st North American Wildlife and Natural Resources Conference, Washington, D.C., 1976.

_____. 1978. Heritage classification: The elements of ecological diversity. *The Nature Conservancy News* 28(1): 24-25, 30.

Moyseenko, H. P. 1981. Limiting factors and pitfalls of environmental data management: Some considerations in developing the information system for the State Natural Heritage Programs. *(This volume, pages 237-253.)*

Sanders, R. 1978. The State Natural Heritage Programs: A partnership to preserve natural diversity. *The Nature Conservancy News* 28(1): 13-19.

U.S. Fish and Wildlife Service. 1978. Furbish Lousewort among 13 plant taxa newly listed by Service for protection. *Endangered Species Techn. Bull.* 3(5): 1, 7-8.

The Concepts of Rarity and Population Threats in Plant Communities

James L. Reveal[1]

Plant rarity is based on a combination of dynamic factors relative to actual and potential species distribution, with the area the species occupies being the operative term in establishing degrees of rarity. Rarity is a two-fold concept associated with the biology of the species and the ecology of the area. Threat is more difficult to characterize since it may be a natural consequence of biological or geological processes or be the result of past or present human activities directly or indirectly influencing the plant populations or their environment. All populations change in size and density over time, and such shifts may eventually result in the extinction of a species. It is debatable whether shifts in population size and density resulting from human activities and associated threats of extinction, are natural events in the evolution of a species, and thus should be tolerated, or unnatural events from which a species should be protected. When conservation of a species is desired, site visits as well as herbarium and library research are necessary to document the conservation status of a species.

Introduction

When a species of plant is first described it is often impossible to completely document the total range that the species occupies. One can explore adjacent sites or even search out similar habitats, but unless more plants are actually located, the given range of a species can only be assumed to be that previously understood based upon past observations. Discoverers of new species in the past did not, and could not, attempt to document the full distribution of the species, as early explorations were designed to discover plants, not to study them in detail.

Today we have a fairly good understanding of the distribution of most flowering plants in the continental United States, including Alaska, with certain areas of the nation better known than others (see Lawyer *et al.,* in press, for a state-by-state bibliography). This understanding of the flora is based upon a patch-work of local

[1]   Department of Botany, University of Maryland, College Park, Md. 20742.

Based on the presentation "Documenting Plant Rarity and Population Threats" by Dr. Reveal at the Symposium on Geographical Data Organization for Rare Plant Conservation, held 15-17 November 1977 at The New York Botanical Garden.

Pages 41-47 *in:* Larry E. Morse and Mary Sue Henifin, eds., *Rare Plant Conservation: Geographical Data Organization* , The New York Botanical Garden, Bronx, N.Y.

and regional floras, revisions, and monographs, and these various sources often use different taxonomic treatments which cannot be readily reconciled. Without a comprehensive flora and distribution atlas for the United States, we cannot precisely determine what are the *rare* plants, much less those which are endangered or threatened with extinction. Still, this current faulty overall understanding of the flora has not prevented us from attempting to determine which species are rare, and which are threatened or endangered. Kartesz and Kartesz (1977) have recently summarized the current understanding of plant rarity and threat in the United States and Canada.

In order to understand the distribution of any one species, one must explore its potential habitats. In the long run it is generally more advantageous to explore a given area for all of its plants rather than to search for a given plant throughout its suspected range. For example, who would have thought to look for *Eriogonum darrovii* in Nevada when it was known originally only from Arizona? It does take longer to collect all the plants of a given area than to collect selectively, but so much more data can be obtained. Only in this way, can one start to understand the ecological parameters of a given taxon, its interrelationship with other species, and its overall range of morphological and geographical variability.

In order to document the range of any species one must conduct field studies. In this way we shall not only find out more about the range of a given species, but perhaps find new species as well. When I found *Eriogonum mortonianum* in 1973, it was in exactly the same place Rupert C. Barneby had found *Cryptantha subglabra,* one of the candidate threatened species (Smithsonian Institution, 1975). By looking for one, the other was found, and I dare say that this type of discovery could be repeated over and over again during detailed field studies. Studies of herbarium material merely tells us where plants once occurred. It does not generally tell how common the plant was, its status in the community, nor its total distribution. Nor does it tell us if it is still there. This type of information can only come from field studies by competent individuals.

Some taxa represented by few herbarium specimens are widespread in the field and simply undercollected; Siddall and Barkley discuss such cases at length in their articles in this volume. There are several variations of this problem. A taxon with many localities documented by herbarium specimens may still be rare. Such taxa may be widely distributed with only a few plants occurring at each locality and thus known to be rare, or the plant may be highly attractive to collectors (either because of physical beauty or again because of a plant's rareness), and thus the plant is frequently gathered. Commercially attractive, widespread species may also quickly become depleted, making herbarium locality information outdated. Old records from urban or other developed areas may give the impression a taxon is much more widespread than it really is.

Rarity Versus Threat

Rarity alone does not imply endangerment or impending extinction. Rather rarity is merely the current status of an extant organism which, by any combination of biological or physical factors, is restricted either in number or area to a level that is demonstrably less than the majority of other organisms of comparable taxonomic entities. A species may be quite rare, yet reproducing and competing well in those few places where it does grow. A newly evolved species may be rare at present, but possess the potential for a greater range. Likewise, a widespread species may be locally scarce yet maintaining itself well regionally (*cf.* Dumond, 1973; Drury, 1974).

Natural extinctions are the extreme cases of natural fluctuations in number or range. A plant may be approaching extinction in several different ways due to several different causes. It may be biologically vulnerable due to natural, biological processes, or it may be unable to meet the evolutionary challenges of changes in landforms or climates, or from regional modifications in the biota. Such factors, typically working in combination, may lead to a plant's rarity or even, ultimately, to its extinction. There is little we can do to reverse such natural trends. Indeed, such extinctions are a vital part in the evolutionary development of the earth's biota.

The existence of threats and trends of threats to a plant can only be determined, and judged, by competent field workers. Although specific threats generally impact particular populations, there more often is a collective pattern of threats which may be singularly difficult to discern except by field studies over a period of time. Data gleaned from herbarium specimens are not always useful as some widespread plants are undercollected, and the label data are all too often not detailed enough. Conversely, many rare species are frequently collected *because* the species is rare, and this too fails to present a true picture of the status of the population.

Assessment of rarity must take into account the operative definition of area. When the Smithsonian report was prepared in 1974, a plant's total known range was considered in determining its status, and not just the distribution of the plant in the United States. Many species in southern Florida and in the Rocky Mountains, for example, are rare in the United States, but common elsewhere. If the concept of rarity was transferred to an operative area definition of a state, a county, or a city, then such species would indeed be rare. Many compilers of state lists take this latter view.

Plant rarity and population threat revolve around two conflicting points. If a plant is now rare, and has (in human terms) always been rare, then there must be some type or combination of types of biological and/or physical barriers preventing the species from being more common. This is the most frequent cause of rarity; a species is just naturally rare and no direct or indirect action by humans has caused this rarity to occur. A second type of rarity alludes to situations where direct or indirect human actions have so altered the habitat or the actual number of individuals from a previous condition that the plants are now rare.

If one concedes that man is a natural element in the earth's biosphere, then one must determine man's role in that environment. In my opinion, one consequence of human evolution has been man's ability to act as the "destroyer" in his ecosystem. If we accept this role for man, then the extinction of species, caused by man, is a natural event and is no different from any other predator, except that man is a generalist and will destroy both habitat and species, and is uniquely suited to render extinct a whole variety of living beings unlike any other animal of the earth. Nonetheless, because man is a rational, thinking animal, capable to evaluating his actions and determining *a priori* their consequence, one must also acknowledge that the extinction of any species caused by man for no particular reason is an unnatural stage in the evolutionary history of that plant relative to its own biological and physical parameters for here the actual or potential role of humans is paramount in the cause of the degree of rarity, and had man given some reasoned thought to his action, such an extinction would have been unnecessary.

In either case, if we wish to retain the species in the ecosystem, efforts must be made to maintain extant populations and prevent significant further deterioration of the species and its habitat.

In the first type of rarity, that unrelated to human activities, there is little we can do, or should do, to expand the geographical range and numbers of individuals of such rare species. These plants are often edaphically restricted or habitat-specific, and it is exceedingly difficult to manufacture a new, suitable habitat where such plants can survive over long periods of time without continuous human intervention. There are some notable exceptions which have done remarkably well in cultivation, such as *Franklinia, Ginkgo,* and *Metasequoia.* Nonetheless, just because a few exceptions can be cited, this does not mean that we should deplete natural populations while trying to "save" the species from extinction. Too, one should add, such examples as just cited are exceptional for their desirability as cultivated arborescent species. The vast majority of rare and endangered plant species lack such exceptional desirability and thus little public interest could be generated for their preservation.

As a biologist, I must reject the "zoo" or "garden" concept so many expound as a solution to saving endangered and threatened species. Rather, I propose that we make the effort to retain the natural habitat of such species, for by the preservation of the habitat, the species can be protected.

In the second type of rarity, that involving past or present human activities, there is much we can do. Habitat alteration can be halted so as to prevent further losses, and efforts can be made to return previously altered habitats back to their original conditions. Controlled burning, reforestation, grazing, or other types of habitat reconstruction or control for the strict purpose of providing suitable areas for the survival of rare species might be required. We have the choice of benign neglect or constructive action when it comes to altering our human strategies to take into account the long-term survival of endangered and threatened plants and animals. Unless we are able to evaluate our own record, and correct, where we can, those errors of judgment which are causing rates of extinction to increase, we will continue to lose many unique biological organisms.

Threat and Endangerment

At the federal level, we are working with definitions of "endangered" and "threatened" as determined by the Endangered Species Act of 1973. The Act defines "endangered" as "any species which is in danger of extinction throughout all or a significant portion of its range . . . " while "threatened" is defined as "any species which is likely to become an endangered species within the foreseeable future throughout all or a significant portion of its range." Section 4(a) of the Act states that evidence of threats must fall within one of five categories: (1) present or threatened destruction, modification, or curtailment of habitat or range; (2) overutilization for commercial, sporting, scientific, or educational purposes; (3) disease, predation; (4) inadequacy of existing regulatory mechanisms; or, (5) other natural or manmade factors. These are the only legal categories (endangered or threatened) or conditions (the five just cited) that can be given when proposing a species as endangered or threatened under the 1973 Act.

For a species to be legally designated endangered it must be currently subjected to an ongoing, provable threat which is actively causing the decline in numbers of individuals either through the removal of individual organisms or by the destruction or alteration of the habitat which is directly leading to the reduction in the number of individuals. Such threats are generally considered to be human-caused although the third and fifth conditions cited above would include factors other than man. This would also include threats resulting in conditions which prohibit the recovery of a species from past declines (either man-caused or not) due to modifications in adjacent habitats. Additionally, the fifth condition would cover those natural biological factors which have naturally destined a plant, in its present situation, to be exceedingly rare and thus "threatened" because any modification of its habitat could result in its immediate extinction.

The definition of a threatened species seems to leave open to consideration a whole series of potential instances in which an actual threat cannot be defined or is not present, but the potential for such a threat, as defined by the Act, can be foreseen as happening in the future. This will allow for many very rare species of plants to be listed as threatened, and thus protected, under the present Endangered Species Act. The concept of threatened status can also include species which are biologically vulnerable, and this would mean that any species whose overall distribution is limited may be considered as threatened because any natural or human factor might cause a reduction in the population which would then seriously endanger the species.

Most of the rare plants in the mainland United States are not directly and actively threatened with extinction at present, and therefore are not legally endangered. Thus, evaluation of a rare plant in terms of potential threat must be made carefully.

Given the five conditions by which threat can be legally determined, the first and fifth are the most important for plants. To determine the potential problems dealing with habitat destruction, modification, or curtailment, the past and present uses of the area must be determined. For example, the presence of an action such as grazing need not imply a threat to plant species, because the plant may not be used or the area where the plant is found may be visited only infrequently by grazing animals. The mere fact that the area is grazed is not sufficient evidence of threat. The effects of that grazing must be determined. The same holds true in evaluating threats from other activities such as mining, lumbering, recreational uses of the area, and so forth. The important point here is that if the site could be subjected to disruption or modification by a change in current practices sufficient to threaten a species with extinction, then the potential for threat is present, and the species would be considered "threatened." The evaluation of human or non-human factors which might threaten a species must be stated within the confines of reasonableness. The concept I have outlined here is broad and subject to many interpretations. It cannot be abused by misuse.

Threat due to rarity revolves around the dynamics of the population and that population's interactions with other physical and biological factors, including those resulting from human activities. As human observers, we conceive rarity in relationship to the distribution and numbers of a species relative to some preconceived concept of what constitutes "rare," "common" or "abundant" (cf. Preston, 1948; May, 1975). There is no magical numerical mark at which a plant shifts from the "rare" category to a higher category. This is a subjective decision and is therefore subject to differences of opinion. I have conceptually thought of extreme rarity in terms of a human factor: "If an operator of a bulldozer, working for four hours, could cause the extinction of a taxon—disregarding the time necessary to physically transport the machine from one population to the next—the species is endangered." The reasoning behind this concept is that many extinctions which occur, particularly in the western United States, come as a result of the unfortunate actions of one individual acting without knowledge of what is being destroyed.

Conclusions

Determining if a species is endangered or threatened, especially when it is also rare, is a matter of judgment. The evaluation of status should take into account the dynamics of the species and the human-orientated activities associated with the habitat. This does not mean that all rare species are or should be considered legally threatened or endangered. A separate case must be made for the placement of each species on the federal endangered and threatened species list.

Evaluating plant rarity, its causes, and the potential for the species to be legally endangered or threatened, requires field work, taxonomic knowledge of the species, and an understanding of the biology of the organism in its present environment; such means are discussed by Radford, by Whitson and Massey, and by Henifin *et al.* in this volume. The use of good judgment and sound reason will make the Endangered Species Act of 1973 work for the benefit of humanity, and for those few plants of this nation faced with extinction.

It can be debated whether or not man-caused extinction is a natural phenomenon; after all *Homo sapiens* is merely one more animal species on earth, with actions just as "natural" as those of any other species. Yet, we are thinking animals capable of making objective decisions. By an act of Congress, the people of the United States have stated a desire to protect and maintain our native plants and animals. Following this reasoned decision, the forced or deliberate extinction of a species can no longer be tolerated. Man, the thinking animal, has so decided. It is now up to this same thinking animal to make that decision work.

Acknowledgments

I wish to thank Bruce MacBryde, Larry E. Morse, Mary Sue Henifin, and C. Rose Broome for their many comments on this subject. Some work on this was supported by NSF grant BMS73-13063. Publication support has been provided by the University of Maryland Agriculture Experiment Station (Scientific Article No. A2450, Contribution No. 5479).

Literature Cited

Drury, W. H. 1974. Rare species. *Biol. Conserv.* 6: 162-169.

Dumond, D. M. 1973. A guide for the selection of rare, unique, and endangered plants. *Castanea* 38: 387-395.

Henifin, M. S., L. E. Morse, J. L. Reveal, B. MacBryde, and J. I. Lawyer. 1981. Guidelines for the preparation of status reports on rare or endangered plant species. *(This volume, pages 261-282.)*

Kartesz, J. T., and R. Kartesz. 1977. *Rare Plants.* Biota North America Committee, McKeesport, Pa.

Lawyer, J. I., A. M. Miller, L. E. Morse, and J. T. Kartesz. *In press.* A guide to selected current literature on vascular plant floristics for the contiguous United States, Alaska, Canada, Greenland, and the U.S. Caribbean and Pacific Islands. *Mem. Torrey Bot. Club.*

May, R. M. 1975. Patterns of species abundance and diversity. Pages 81-120 *in* M. L. Cody and J. M. Diamond, eds. *Ecology and Evolution of Communities.* Belknap Press of Harvard Univ. Press, Cambridge, Mass.

Preston, F. W. 1948. The commonness and rarity of species. *Ecology* 29: 254-283.

Radford, A. E. 1981. Introduction to a System for Ecological Diversity Classification. *(This volume, pages 199-205.)*

Smithsonian Institution. 1975. *Report on Endangered and Threatened Plant Species of the United States.* House Document 94-51 (Serial no. 94-A), U.S. Government Printing Office, Washington, D.C.

Whitson, P. D., and J. R. Massey. 1981. Information systems for use in studying the population status of threatened and endangered species. *(This volume, pages 217-236.)*

Information Needed to Use the Endangered Species Act for Plant Conservation

Bruce MacBryde[1]

The kinds of comprehensive data needed to evaluate and conserve endangered plants are not well recognized. Legal and administrative needs for this information suggest that status reports prepared according to complete guidelines would facilitate the plant conservation effort by promoting thorough, well-organized data collection and documented syntheses.

The Endangered Species Act of 1973 (PL 93-205) and the Endangered Species Act Amendments of 1978 (PL 95-632) reflect the public desire, through Congressional legislation, to protect species in their natural habitats from unintentional human threats which may cause their extinction. These laws define the terms plant, species, endangered species, threatened species, critical habitat, and conservation, and they provide certain specified protective measures against some of the threatening factors which can jeopardize the survival of a species in its ecosystem. Plant conservation is complex because of the large number of imperiled species, their different characteristics and life histories and our frequent ignorance of them, and because of the pervasive effects of humankind on the natural environment. Efficient development and organization of the diverse information pertinent to plant conservation is essential if the growing number of people both in private and in state and federal government capacities are to gain maximum benefits for plants through the various means available. The limited number of specialists in plant conservation also indicates that their effective communication with all concerned is imperative.

The basic contents of the Act, and some of its differences for plants and animals, have been reviewed elsewhere (Williams, 1976; Baker and MacBryde, 1976; Smith, 1980, and in press). The new Amendments have been summarized (U.S. Fish and Wildlife Service, 1978b and this volume), but their full implications are still being evaluated. The general scope and structure of plant aspects of the federal En-

---

[1]   Office of Endangered Species, U.S. Fish and Wildlife Service, Washington, D.C. 20240.

Based on the presentation "Requirements for Listing Rare and Endangered Species and Documenting their Critical Habitats under the Endangered Species Act of 1973" by Dr. MacBryde at the Symposium on Geographical Data Organization for Rare Plant Conservation, held 15-17 November 1977 at The New York Botanical Garden.
Pages 49-55 in: Larry E. Morse and Mary Sue Henifin, eds., *Rare Plant Conservation: Geographical Data Organization,* The New York Botanical Garden, Bronx, N.Y.

dangered Species Program have been presented (MacBryde, 1977, 1979a, 1979b; Miller, 1977; U.S. Fish and Wildlife Service, 1978a; Smith, in press) in order for those concerned to achieve extensive understanding and participation in their growth. Additionally, Zeedyk et al. (1978) give suggestions to land managers for evaluating and managing plants in danger of extinction. In two of these articles (MacBryde, 1977; Zeedyk et al., 1978) an outline flowchart indicates the general procedures required for listing a plant as endangered or threatened and/or for determining its critical habitat. The Amendments require a more elaborate process including requirements to communicate with all concerned (Smith, in press), and rules have been published to indicate the amended procedures (U.S. Fish and Wildlife Service and National Oceanic and Atmospheric Administration, 1980).

Needs for Organized Status Information

Previously, there has been no detailed presentation of the kinds of data needed to evaluate a species, its habitat, and the threats to its survival, nor of the other kinds of informatioin which might be useful for its conservation. The status report guidelines developed by Henifin et al. (this volume) are designed for effective communication of the knowledge needed by the Endangered Species Program of the U.S. Fish and Wildlife Service in recognizing and conserving plants in danger of extinction.* These guidelines are also intended to apply to and complement related plant conservation efforts, such as those conducted by other federal agencies or by state conservation organizations. Because of the increasing number of people involved in developing, evaluating, and using such data and information, the more objective and standard approach offered by use of such guidelines should be mutually beneficial.

It must be emphasized that the status report guidelines represent a comprehensive ideal for desired knowledge, applicable to a broad range of species and situations. Early completion of data for all categories and topics is not essential (indeed probably impossible) for each endangered and threatened plant, but all are considered generally useful. For example, early gathering of detailed data on the population biology of a particular plant species is not of primary concern if the immediate threat is habitat destruction, but such data might become very valuable later for management of its critical habitat. In another species, early knowledge of aspects of its population ecology may become crucial, for example when an insecticide is destroying an essential pollinator. Thus the objectives in preparing a status report and pertinence of the data to them should be kept in mind. Blanks may be left as necessary, after due consideration. Reduced or partial versions of these guidelines can be prepared to serve particular purposes, while still maintaining the value of a consistent approach. For example, a simplified format better organizes the new information obtained from recent visits to well-studied populations (Henifin, Morse, Grif-

---

* The 1978 Amendments to the Endangered Species Act became law on 10 November 1978, too late to be given complete consideration in these guidelines or this paper. This paper reflects my experience and opinions; it is not necessarily final agency policy. While there is no clearly official federal format for status reports for plants (or animals), these guidelines are in general use by the Service in contracting for candidate plants in the Endangered Species Program. At a Service-wide botany meeting in September 1979, it was formally agreed to require usage of these guidelines for such contracting purposes, and they were also formally recommended as the format for plant information in the Sensitive Wildlife Information System described by Rekas (1976).

fith, and Hohn, this volume). As another example, it should be recalled that the Secretary of Interior (or Commerce) is required by section 4(b)(1) of the Act to make his determinations of endangered and threatened species and their critical habitats "on the basis of the best scientific and commercial data available to him." The status report guidelines seek the best information possible, in an organized format, not only for listing, but also for protective management and recovery. The reduced kinds and depth of data minimally adequate for listing are still matters of administrative and scientific judgment; the authors of the guidelines have indicated their opinion by highlighting with asterisks the appropriate categories and topics (*cf.* MacBryde, 1979*b*, p. 10).

An explanation of some of the purposes and uses of information requested in the guidelines will demonstrate their value and the need for such detail. Certain items in the status report are required for legal purposes, to comply with the Endangered Species Act and its amendments, and also with the National Environmental Policy Act of 1969 (NEPA) (PL 91-190; *cf.* Council on Environmental Quality, 1978). Other requirements for data and information are to satisfy either formal or practical administrative needs. Basically, the Act requires an evaluation of the extinction potential of a species: what is it, where is it, and what threats affect it (*cf.* Reveal, this volume; Reveal and Broome, 1979; and MacBryde, 1979*a*). This evaluation requires consideration of both the natural vulnerability of a species (which is usually related to its rarity) and the threats to its population(s), as shown diagramatically in Figure 1. Currently the threats are usually caused by humankind, and may be grouped under the five comprehensive factors in section 4(a)(1) of the Act (see category eleven in the guidelines for examples).

It must be noted that the socioeconomic realm is generally not covered in the status report guidelines. However, such information is now needed because the new section 4(b)(4) of the Act (established by the 1978 Amendments) requires an analysis of economic and other impacts prior to critical habitat determination, and because of President Carter's Executive Order 12044 (President and Office of Management and Budget, 1978) and subsequent final rules (Department of the Interior, 1978), which require a determination of the significance (basically, the extent of socioeconomic consequences) of each species' listing or critical habitat determination, as well as a regulatory analysis for any proposed rule which would be significant.** Bishop (1978) and Miller (1978) offer some insight into the effects of economic evaluation of endangered species.

Examples of Information Requirements

Four examples will suggest the reasoning behind and demonstrate some of the purposes and uses of the data which are requested in the status report guidelines. As a first example, no description or illustration of the plant of concern is legally required, but both are requested as part of the status report. Such information promotes concern for the species, and may be used by the Service in public education, such as an article in the *Endangered Species Technical Bulletin,* or in fulfilling often urgent requests from news media or others. This information may also be necessary

---

** The Office of Endangered Species is preparing guidelines for the collection of economic data necessary to list threatened or endangered species.

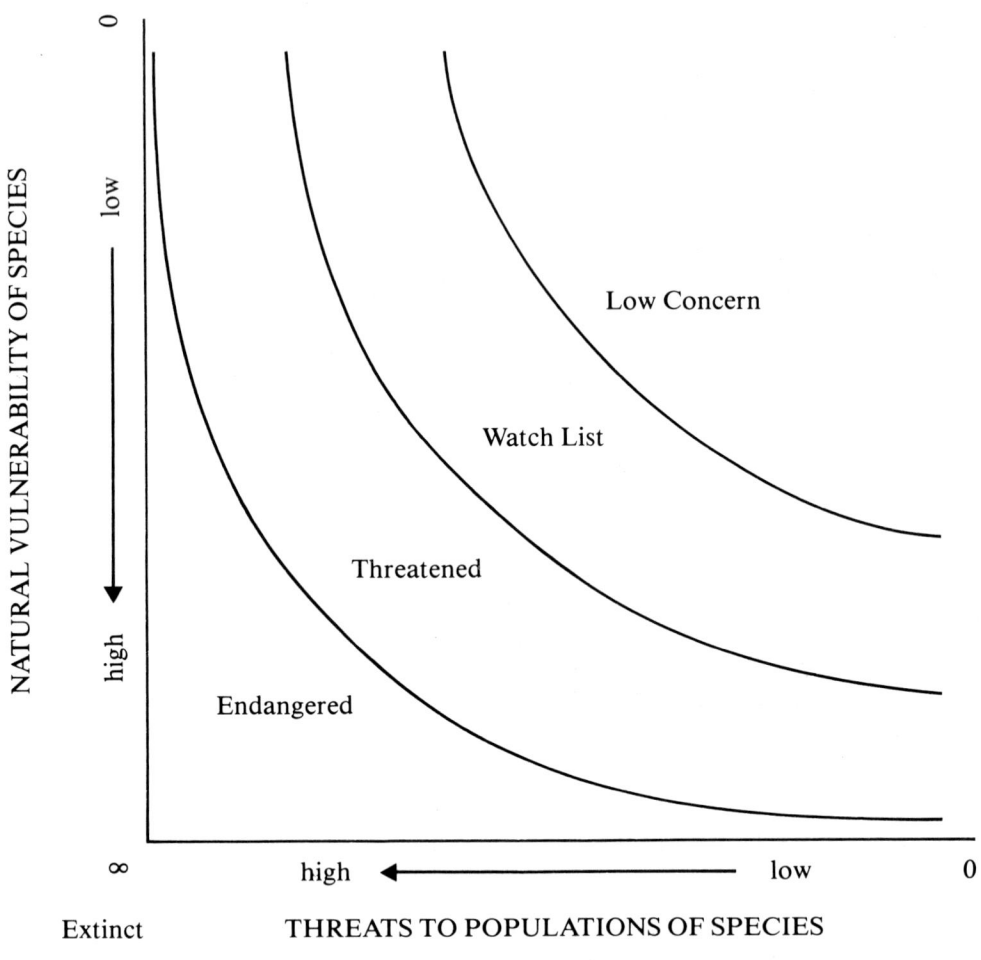

Figure 1. Extinction potential of a species. (With acknowledgments to R.F. Altevogt, M.S. Henifin, and L.E. Morse.)

for management of the plant in its habitat or enforcement of trade restrictions. As another example, the Act requires that a listed species be in danger of extinction throughout all or a significant portion of its range. To determine whether or not this requirement can be met, the guidelines request complete data on historical and present geographical distribution of a species. It is generally not sufficient to document that a species is in danger at one locality, especially when it occurs elsewhere. Frequently overlooked information is also necessary, such as sites where the plant has been erroneously reported, sites where populations of the species have been extirpated, or suspected locales for the plant which have not been verified by recent field studies. Third, the environment of the plant must be described to comply with the National Environmental Policy Act (NEPA); this information is also useful in managing its habitat. As a final example, knowledge of the land ownership of the species is essential in evaluating responsibilities for the welfare of the plant and influences the Act might have. Because the protective measures of the Act for plants are greater where there is federal involvement, some of the methods for conserving a species on private land may differ from those for a plant restricted to federal land. For examples of these extremes, see the listings of *Sagittaria fasciculata* E.O. Beal, the bunched arrowhead, found only on private land and recently vandalized, and *Coryphantha sneedii* (Britton & Rose) Berger var. *leei* (Bödeker) L. Benson, the Lee pincushion cactus, endemic to a National Park and at risk from collecting (U.S. Fish and Wildlife Service, 1979*c*, 1979*d*), and for recovery possibilities see Smith (in press).

Listing Mechanics

In essence, to protect a species under the Act, it must first be proposed as endangered or threatened in the *Federal Register,* and then listed there in a final rule. (Sometimes a *Federal Register* notice precedes the proposal.) Section 4(a)(1) of the amended Act requires that the species' critical habitat be proposed along with the species (to the maximum extent prudent). The Office of Endangered Species has prepared an internal working handbook called a Model Rulemaking Package, which provides samples of a notice, proposal, and final rule, and organizes the peripheral documents needed to recommend a species for listing action and/or a critical habitat for determination. Major draft items in the package are: the document to be published in the *Federal Register;* the Environmental Assessment (or potentially, Environmental Impact Statement); a determination of significance memorandum (to comply with Executive Order 12044) and (if a critical habitat is determined) an economic analysis; a press release; and letters to be sent to the appropriate federal agencies, state governors, scientific societies, etc. The status report on a plant should contain all the information needed to complete a package (with the exception of socioeconomic information), and it and/or any other pertinent data will be incorporated in the package as an appendix to the Environmental Assessment (which is required by NEPA), so that an objective basis for the listing action is available. The model package has been distributed to regional and some area offices of the U.S. Fish and Wildlife Service, where it is now being used by Service botanists to prepare packages and thus expedite the listing process.

Conservation of Listed Species

After a plant or its critical habitat has been legally designated under the Act, various conservation measures are legally possible to prohibit or minimize harmful activities, and to encourage active improvement of the species' condition. Even when a species is simply proposed (rather than officially listed), section 7(c) of the amended Act may postpone costruction activities and require a biological assessment under certain circumstances. In addition to their role in listing, the status report guidelines are intended to support many of these conservation measures and management needs. For another example, such data may be useful in preparing a biological opinion during any consultation required under section 7 of the Act (U.S. Fish and Wildlife Service and National Marine Fisheries Service, 1978); these rules are being revised based on Service experience and to conform to the 1978 and 1979*** Amendments). Since the Service is considering the computerization of these plant status reports in the Sensitive Wildlife Information System in order to aid these consultations, such a detailed and standardized format is particularly important. As a third example, the data in the status report may prove of value in preparing the recovery plan for virtually every listed species as now required by section 4(g) of the amended Act (U.S. Fish and Wildlife Service, 1979a). As a last example, if a state which has entered into a cooperative agreement for plants under section 6(c)(2) of the amended Act (cf. U.S. Fish and Wildlife Service, 1979b) chooses to use the same status report format, it will facilitate the combined state/federal conservation efforts. Several versions of the status report guidelines (they have been in preparation over two years) have been used in various Service contracts and some joint federal agency contracts, and it was this experience which led to the Service decision to use them uniformly in contracting on plant candidates.**** They have also proved of value in the organization of rare plant activities by some native plant societies and private conservation groups, which complement and assist the government programs (for example, they have been adopted for use by the Massachusetts Natural Heritage Program). University professors and students might also find the guidelines useful for the preparation of class papers or thesis projects by adopting candidate plants in need of conservation.

The Endangered Species Act of 1973 has been law for more than six years. The report of the Smithsonian Institution required by section 12 of the Act indicated that over 2,800 native plant taxa might be in danger of extinction and over 300 more might already be extinct. This information has been available for over five years, with an updated Smithsonian book published over two years ago (Ayensu and DeFilipps, 1978). Currently (April 1980) only fifty-six native (and two foreign) plant taxa are listed under the Act. Only these are eligible for the Act's protective measures, and critical habitat has been determined for only two of the fifty-six. In these years, the Endangered Species Program has matured in awareness of what must be done to effectively use the law. It is hoped that the detailed information requested in the status report guidelines will encourage greater participation in the efforts to understand and correct the problems of plants in ecological and evolutionary decline because of expanding, more demanding human populations.

Literature Cited

Ayensu, E. S. and R. A. DeFilipps. 1978. *Endangered and Threatened Plants of the United States.* Smithsonian Institution and World Wildlife Fund, Washington, D.C.

Baker, G. S. and B. MacBryde. 1976. The endangered and threatened plant program of the U.S. Fish and Wildlife Service. *ASB Bull.* 23(3): 141-144.

Bishop, R. C. 1978. Endangered species and uncertainty: The economics of a safe minimum standard. *Amer. J. Agric. Econ.* 60(1): 10-18.

*** On December 28, 1979, Congress enacted the Endangered Species Act Amendments of 1979 (PL 96-159). Among other changes, these Amendments allow emergency listing for plants, require federal agencies to confer with the Secretary of Interior (or Commerce) on *all* proposed species which they might jeopardize, and now provide full authority for international cooperation in plant conservation (*cf. Endangered Species Techn. Bull.* 5(1): 1-4, reprinted, this volume, pp. 313-322.)

**** I (and the other authors of the guidelines) would appreciate comments and suggestions for their continued improvement.

Council on Environmental Quality. 1978. Regulations for implementing the procedural provisions of the National Environmental Policy Act. *Federal Register* 43 [Nov. 29]: 55978-56007.

Dept. of Interior. 1978. Final rule on improvements in the rulemaking process [to implement Executive Order 12044]. *Federal Register* 43 (240,II) [Dec. 13]: 58291-58301.

Henifin, M. S., L. E. Morse, S. Griffith, and J. E. Hohn. 1981. Planning field work on rare or endangered plant populations. *(This volume, pages 309-312.)*

Henifin, M. S., L. E. Morse, J. L. Reveal, B. MacBryde and J. I. Lawyer. 1981. Guidelines for the preparation of status reports on rare or endangered plant species. *(This volume, pages 261-282.)*

MacBryde, B. 1977. Plant conservation in the United States Fish and Wildlife Service. Pages 62-74 *in* G. T. Prance and T. S. Elias, eds. *Extinction is Forever.* New York Botanical Garden, Bronx, N.Y.

MacBryde, B. 1979a. Plant conservation in North America: Developing structure. Pages 105-109 *in* I. Hedberg, ed. *Systematic Botany, Plant Utilization and Biosphere Conservation.* Almqvist & Wiksell International, Stockholm.

MacBryde, B. 1979b. Plants for all seasons: Conservation as if nature mattered. *Nature Conservancy News* 29(2): 9-11.

Miller, J. R. 1978. A simple economic model of endangered species preservation in the United States. *J. Environm. Econ. Management* 5: 292-300.

Miller, S. F. 1977. Wild flowers, the Endangered Species Act, and you. *Flower and Garden* (southern ed.) 21(12) [December]: 12-16.

President and Office of Management and Budget. 1978. Improving government regulations. Executive Order 12044, March 23, 1978. *Federal Register* 43(58,VI) [March 24]: 12661-12670.

Rekas, A. M. B. 1976. Computerized information systems for threatened and endangered species. *ASB Bull.* 23(3): 144-149.

Reveal, J. L. 1981. The concepts of rarity and population threats in plant communities. *(This volume, pages 41-47.)*

Reveal, J. L. and C. R. Broome. 1979. Plant rarity: Real and imagined. *Nature Conservancy News* 29(2): 4-8.

Smith, E. L. 1980. Laws and information needs for listing plants. *Rhodora* 82: 193-199.

Smith, E. L. *in press.* Plant conservation and the Endangered Species Act. *In:* Proceedings of the conference 'Dendrology in the eastern deciduous forest biome,' School of Forestry and Wildlife Resources, Virginia Polytechnic Institute and State University and Southeast Forest Experiment Station, U.S. Forest Service, Blacksburg, Virginia.

U.S. Fish and Wildlife Service. 1978a. [Final rules.] Determination that 11 plant taxa are endangered species and 2 plant taxa are threatened species [and general summary of comments on the listing proposal of June 16, 1976]. *Federal Register* 43(81,II) [April 26]: 17909-17916.

_____. 1978b. President signs endangered species amendments. *Endangered Species Techn. Bull.* 3(10): 1, 3-5, 11. [*Reprinted as* "The 1978 and 1979 Amendments to the Endangered Species Act: A Discussion," *this volume, pages 313-322.)*

_____. 1979b. [Final rules.] State cooperation agreements relating to endangered and threatened species of fish and wildlife or plants. *Federal Register* 44 (106,VIII) [May 31]: 31578-31581.

_____. 1979c. [Final rule.] Determination that *Sagittaria fasciculata* is an endangered species. *Federal Register* 44 (144,IV) [July 25]: 43699-43701.

_____. 1979d. [Final rule.] Determination that *Coryphantha sneedii* var. *leei* is a threatened species. *Federal Register* 44 (208,IV) [October 25]: 61554-61556.

_____ and National Marine Fisheries Service. 1978. [Final rules.] Interagency cooperation [under section 7 of the Endangered Species Act of 1973]. *Federal Register* 43 (2, IV) [January 4]: 869-876.

_____ and National Oceanic and Atmospheric Administration. 1980. [Final Rules.] Rules for listing endangered and threatened species, designating critical habitat, and maintaining the lists. *Federal Register* 45 (40, IV) [February 27]: 13009-13026.

Williams, J. D. 1976. A review of the Endangered Species Act of 1973. *ASB Bull.* 23(3): 138-141.

Zeedyk, W. D., R. E. Farmer, Jr., B. MacBryde and G. S. Baker. 1978. Endangered plant species and wildland management. *J. Forest.* (Washington) 76(1): 31-36.

The Conservation of Rare Flora through the National Natural Landmarks Program

Gary S. Waggoner[1]

The National Natural Landmarks Program, administered by the National Park Service, is presently the only ongoing program to identify nationally significant natural areas on a nationwide basis. While the Service actively encourages the voluntary preservation of these sites, the program is strictly one of recognition. Landmarks are designated for their ecological and/or geological significance. Emphasis on ecological sites is to identify, through natural region inventories based on physiographic provinces, an adequate number of outstanding areas representative of all the major ecosystem types occurring within the United States and its Territories. Major importance is given to endangered ecosystem types, threatened or endangered biota, and biogeographically significant areas. Natural Region Studies and follow-up onsite evaluation reports are contracted to competent scientists, primarily botanists and geomorphologists. Rare plant data is essential in determining the comparative values of similar areas. Numerous National Natural Landmarks have been established due to the presence of rare flora. Distributional information on biota is likewise essential in determining the relative importance of natural areas. This information can only be obtained through cooperation with the nationwide scientific community. Many important problems remain, especially in natural area management, and scientific research is critical to their resolution.

The National Natural Landmarks Program is an administratively created natural area preservation program managed by the National Park Service, United States Department of the Interior. Begun in 1962, the National Registry of Natural Landmarks presently lists 455 sites of biological and/or geological value which have been declared nationally significant by the Secretary of the Interior. These sites are lo-

[1] National Natural Landmarks Program, National Park Service, Denver, Colo.; now Natural Landmarks Group, Heritage Conservation and Recreation Service, Denver, Colo. 80225. Mr. Waggoner's present address is: Remote Sensing Section, Denver Service Center, National Park Service, Denver, Colo.

Based on the presentation "Rare Plant information Important to the National Natural Landmarks Program" by Mr. Waggoner at the Symposium on Geographical Data Organization for Rare Plant Conservation, held 15-17 November 1977 at The New York Botanical Garden. Since Mr. Waggoner's presentation at this conference, the National Natural Landmarks Program has been transferred to the newly created Heritage Conservation and Recreation Service, U.S. Department of the Interior. Some modifications have occurred to the process he describes; these are discussed in the *Federal Register* [44(225): 66599-66602, Nov. 20, 1979] and by C. T. Delaporte in "Progress and Prospect: A Report on the National Natural Landmarks Program" (*National Parks and Conservation Magazine* 54(3): 21-23, March 1980).

Pages 57-61 *in:* Larry E. Morse and Mary Sue Henifin, eds., *Rare Plant Conservation: Geographical Data Organization* , The New York Botanical Garden, Bronx, N.Y.

cated in 48 states, Guam, American Samoa and Puerto Rico. Ownership includes private, state and federal lands.

The National Natural Landmarks Program is a recognition program aimed at encouraging the voluntary preservation of the best remaining examples of our natural heritage. Ownership does not change with designation, as the Park Service is neither authorized nor funded to pursue acquisition. Registration of designated areas involves a voluntary agreement signed by the owner indicating his intention to conserve the nationally significant values identified at the site. This registration agreement, however, is not legally binding. Roughly one quarter of all the designated sites have never been registered for varying reasons but all 455 sites are maintained on the Registry because of their national significance.

Emphasis of the Program is centered on inventory and comparative study of sites. Briefly, the entire Nation and Trust Territories are being inventoried in 33 Natural Region Theme Studies (see Map 1) contracted to competent scientists from across the country. These studies provide (1) a description of the region; (2) a classification of ecosystems and geological phenomena important in the region; and (3) an inventory describing the best representative natural areas of the various types delineated in the classification system. Data on threatened and endangered biota are incorporated into these studies mainly in the site descriptions. While each contract specifically refers to the importance of data on rare biota, investigators vary in the relative importance they place on such data. All natural region studies, however, do contain a certain amount. At present, 13 such studies are completed, 11 are underway, and 9 remain to be contracted.

The next step in the process is to contract a separate onsite evaluation of each area recommended in the natural region study. Sites receiving positive onsite evaluations are reviewed by National Park Service staff, and then recommended to the Secretary of the Interior's Advisory Board on National Parks, Historic Sites, Buildings and Monuments. The Board's recommendation goes to the Secretary who makes the final determination of a site's national significance. Sites determined to be of national significance are entered on the National Registry of Natural Landmarks, published periodically in the *Federal Register*.

Protection of national natural landmarks is not automatically assured. Under this program, recognition provides the major tool for protection. The Park Service reviews all environmental impact statements and comments on possible impacts to both potential and existing landmarks. All efforts are made to avoid or minimize such impacts although no absolute veto of projects is possible.

Two recently enacted laws provide additional protection for landmarks. Section 9 of Public Law 94-429 provides for much firmer protection of natural landmarks when threatened by surface mining and Section 8 of Public Law 94-458 requires the Secretary to report annually to Congress with a listing of all natural landmarks threatened or endangered by any and all activities, both natural and unnatural. Because of the newness of these two laws, their full ramifications are not yet totally evident, but they both appear to add considerable assistance to our preservation efforts.

The Natural Landmarks Program has always stressed the importance of botanical data in identifying the nationally significant biological areas of our nation. Paramount importance is placed on recognizing the best remaining examples of all of the major ecosystem types. It is *not* assumed, however, that by getting the best examples of various ecosystems, one automatically insures the long term protection of all rare flora. Jenkins has discussed this distinction at length elsewhere in this volume. In-

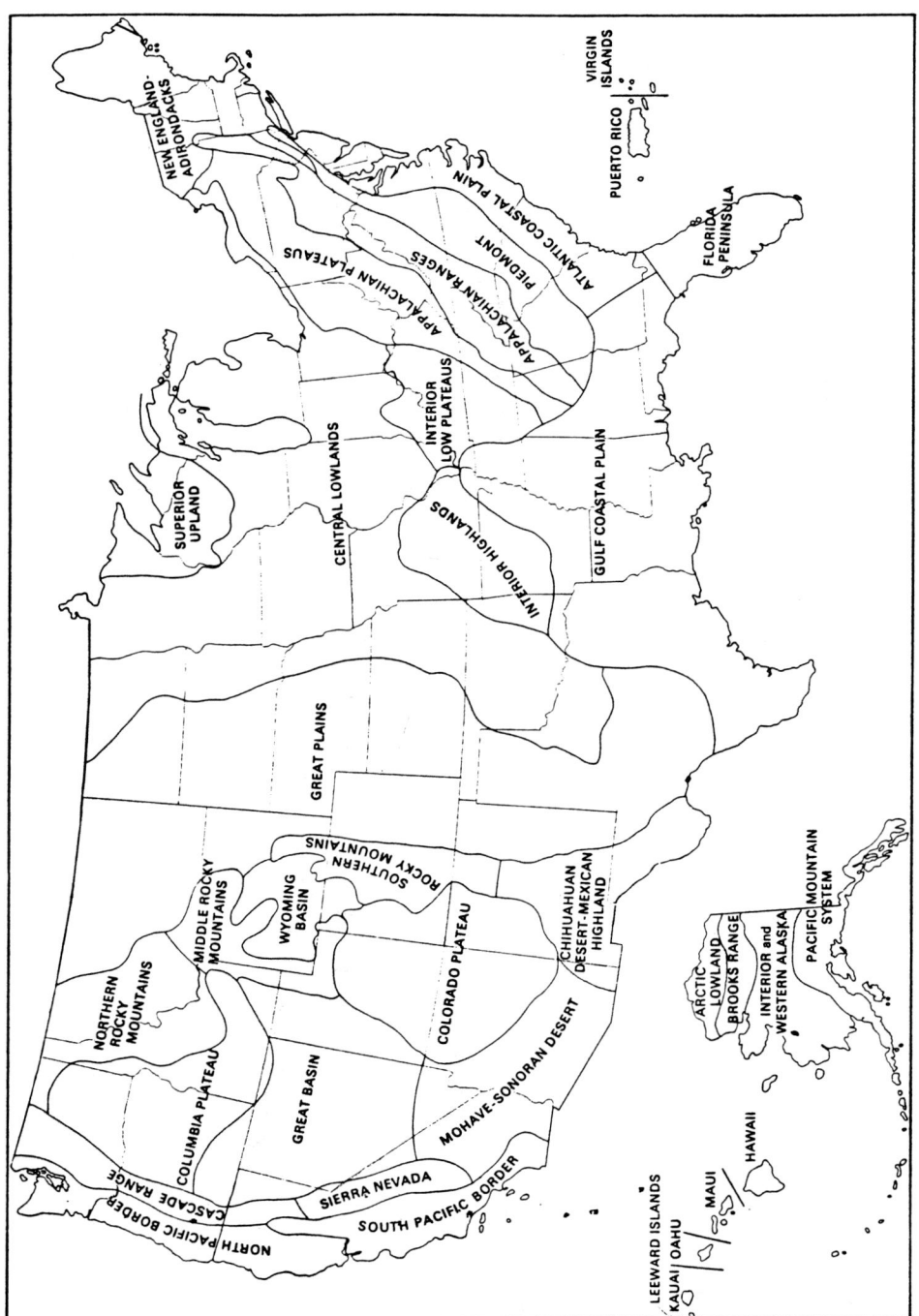

Map 1. Natural Regions of the United States, based on Fenneman's (1928) *Physiographic Divisions of the United States.*

stead, data on rare flora are often utilized in differentiating between similar areas. Of highest priority are the "best" areas representative of endangered ecosystem types such as vernal pools, cedar glades, tall grass prairie, mature longleaf pine forest, beech-magnolia forest, and the Valley Grassland of California, to name but a few. Each of these endangered ecosystem types contains characteristic endangered flora, generally several taxa. Our emphasis is to move toward protecting the best examples of these most rapidly disappearing types before the opportunities for their preservation are gone.

The determination of "best areas" is made in a variety of ways depending upon the situation. First, it must be understood that this program is recognizing only "nationally significant" sites. Therefore, the only (and thus the best) example of cypress-tupelo swamp in Delaware would not be declared a national natural landmark in and of itself, because the ecosystem type occurs much more commonly further south. If, however, this same area was determined to be one of the most pristine, old age stands in the country, there would be much greater potential for designation. Thus, there is a geographic parameter to the determination.

"Best" also reflects such factors as basic integrity or naturalness relative to other similar areas; size; successional maturity or age; lack of appreciable past disturbances, cultural intrusions or human manipulations; diversity of species and/or communities; and prospects for long-term protection. Studies are done on a comparative basis and the investigators are called on to make determinations based upon their field knowledge of similar areas within the natural region.

Other factors considered with the above include:

1. Natural areas supporting rare, threatened or endangered flora or fauna.

2. Relict ecosystems persisting from an earlier geological time.

3. Distributionally significant sites such as unusually outstanding major ecosystem type disjuncts, areas containing concentrations of endemics, areas harboring unusually high numbers of taxa occurring at the edge of their ranges, or areas containing unusually high numbers of taxa with very different geographical affinities.

The primary criterion for all natural landmarks is that each represents a true, accurate, and essentially unspoiled example of our natural heritage.

Virtually all of the professional input into the designation of national natural landmarks is obtained by the Park Service through contracts with scientists. Since the Program staff does not routinely visit these areas prior to designation, reliance on contracted expertise is great. Botanists, especially plant ecologists and taxonomists, are most frequently utilized in these investigations. However, the Natural Landmarks Program staff makes valuable use of various state and regional floras and other sources of distributional data, including other scientists, in order to monitor contracted reports. Further, money is not provided for in-depth inventory of biota present at each site so published and unpublished reports often provide the backbone of data to determine significance of an area. Where data are lacking, the Service gathers what information is available and makes a determination based essentially on the contracted investigations.

Since our office relies so heavily on comparative judgments of similar sites, a single source of botanical data, similar to that envisioned by the Flora of North America project, would greatly improve the efficiency of our work, both in speeding up the study process and in improving the accuracy of data used in the designation of natural landmarks. I heartily endorse the efforts to produce such a flora.

President Carter, in his Environmental Message to the Nation (May 23, 1977), expressed his strong intention to establish a National Heritage Trust. This program would greatly increase the federal effort in identifying and preserving natural areas largely through matching grants provided to states for inventory, acquisition, and management. Today, we know little about the environmental limitations of biota in general, much less of rare or endangered flora. Much more autecological research must be done so that proper management techniques can be identified and employed to insure the long term viability of populations. Equally important, we need to determine if the lessons of island biogeography are indeed applicable to the natural area preservation problem and the long term preservation of natural diversity as some have suggested (Cody and Diamond, 1975; Diamond et al., 1976; May, 1975; and Willis, 1974). Only further research can tell us more about (1) adequate sizes of preserves; (2) how many preserves are necessary; (3) how best can these preserves be distributed geographically; and (4) whether the shapes of preserves are of any consequence. All of these questions need to be examined in relation to the assurance of maximum genetic diversity as well as long term population viability. Further, we need to assess whether global pollution problems will negate our preservation efforts aimed at maintaining viable populations of particular taxa. As natural area preservation efforts become more and more popular, there will be much more interest focused on management and the answers to these and many other questions will be essential. While horticulture may prove to be a useful tool in the conservation of rare flora, the only real answer to the preservation of natural diversity and our life support system in general is through maintenance of ecosystem structure and function, possible only through natural area preservation.

Literature Cited

Cody, M. L. and J. M. Diamond (eds). 1975. *Ecology and Evolution of Communities.* Belknap Press of Harvard University, Cambridge, Mass.

Diamond, J. M., J. Terborgh, R. F. Whitcomb, J. F. Lynch, P. A. Opler, and C. S. Robbins. 1976. Island biogeography and conservation: strategy and limitations. *Science* 193: 1027-1032.

Jenkins, R. E. 1981. Rare plant conservation through elements-of-diversity information. *(This volume, pages 33-40.)*

May, R. M. 1975. Island biogeography and the design of wildlife preserves. *Nature* 254: 177-178.

Willis, E. O. 1974. Populations and local extinctions of birds on Barro Island, Panama. *Ecol. Monogr.* 44: 153-169.

Information Storage and Population Monitoring within Great Smoky Mountains National Park

Susan Power Bratton[1]

This paper considers efforts within Great Smoky Mountains National Park in constructing and maintaining plant inventories, including studies of rare and endangered species. National parks may or may not have a professional botanical staff or convenient access to a computer system. Parks also vary greatly in size, structure, past history, and human use. They are public and partially political entities whose management is often carefully watched by various interest groups. The parks' problems, therefore, often do not fit into "academic" categories.

Mandates and Structure

Although one mandate of the National Park Service is "to preserve and protect the native flora and fauna" (U.S. Department of the Interior, National Park Service 1970), accurate species checklists, vegetation maps, or an up-to-date herbarium may not exist. A park administrator will probably be in a great deal more trouble if a government truck disappears than if a plant population is lost. Despite these problems, most large parks have vascular species checklists and many have herbaria and vegetation maps. Some have also developed close relationships with nearby academic centers. Biological surveys are often conducted as new parks are established or in revising management plans for older parks. The quality and quantity of information varies greatly from area to area, and it is common to find that collections have not been cared for and recordkeeping has not been systematic.

Several processes are related to these patterns.

1. *Staff turnover is rapid in most parks.* In a university an herbarium curator may (for better or worse) remain for 30 years. Few, if any, parks have a permanent curatorial staff.

---

[1] Department of the Interior National Park Service Southeast Region Uplands Field Research Laboratory, Great Smoky Mountains National Park, Twin Creeks Area, Gatlinburg, Tenn. 37738.

Based on the presentation "Information Storage and Population Monitoring within National Park Service Areas" by Dr. Bratton at the Symposium on Geographical Data Organization for Rare Plant Conservation, held 15-17 November 1977 at The New York Botanical Garden.

Pages 63-68 *in:* Larry E. Morse and Mary Sue Henifin, eds., *Rare Plant Conservation: Geographical Data Organization*, The New York Botanical Garden, Bronx, N.Y.

2. *As park administrations and national policy change, the research emphasis within a park changes.* In the Great Smoky Mountains National Park there was an early attempt to collect species records, establish biological files, and map the resources of the park (Miller 1938, 1941; Jennison 1935, 1938, 1939a, 1939b; Sharp 1942a, 1942b, 1957; Stupka, 1935-1962, 1964). The park then went through a period when university-based investigators (e.g., Whittaker 1956, 1961, 1966; Crandall 1958, Schofield 1960, Oh 1964, Golden 1974, Bratton 1975b, 1976a, 1976b) were conducting community-oriented studies, but the park staff were largely involved in preserving historic structures and enlarging facilities for visitor use. Many of the park collections were given away during this period and the archives fell into disorder. An accumulation of resources management problems then stimulated the park to become involved in botanical research again. This new program is largely oriented towards exotic species and disturbed areas (Baron *et al.* 1975, Bratton 1974, 1975a, 1978; Howe and Bratton 1976; Bratton *et al.* 1978, Lindsay 1977). The recent designation of the park as a Biosphere Reserve (Johnson and Bratton 1978) and the initiation of park-wide planning efforts have created a new interest in species inventories and maps. Presently, the emphasis is not only on listing, but is also on population monitoring.

3. *Management priorities are often based on the most immediate problems or those most in the public view.* With staff and funding almost invariably in short supply, long term projects or those with no immediate obvious usefulness take a low priority. Botanical recordkeeping is more likely to be funded if threatened resources are involved.

4. *There is often very little structure to botanical recordkeeping.* Staff in many areas with good collections and accurate checklists have interacted extensively with local universities.

Any effort to maintain botanical data, therefore, must consider certain administrative safeguards. Permanent filing systems need to be independent of the presence or absence of a particular staff member. The mandate to "protect native flora" is too vague to insure accurate data gathering and recordkeeping in every case. General standards need to be established and adhered to, depending on the size and biological importance of the area, for the types of records that should be kept.

Technical Aspects

Most parks do not have ready computer access at the present time and many of the smaller areas do not have research staff. The individual areas are therefore partially dependent on technology developed at universities, by large scale National Park Service research programs, or by other agencies.

The development of more sophisticated data systems that can be utilized at a local level requires that systems developed on a "large scale" basis be locally applicable. The interface is often difficult, and programs developed for one park may or may not apply to another.

A basic problem also exists in the type of information base available. If park records are inaccurate or inadequate, improved technology may make little difference until information gathering and utilization are improved at the local level.

Inventory in Great Smoky Mountains National Park

The Great Smoky Mountains National Park (GRSM) research staff is presently working intensely to develop a better system of vegetation inventory and monitoring. GRSM, like most National Park Service natural areas, has a published vascular plant check list (Hoffman 1962), available lists for bryophyte and fungi (Peterson, in prep.), a vegetation map (compiled during the 1930's), and an herbarium in the park, as well as a master collection of voucher specimens for the park at the University of Tennessee. Records on individual rare species have been collected over the past three years, but the park lacks an organized program for monitoring communities or individual species. GRSM does not yet have the sophisticated fire modeling programs of some of the western parks, like Glacier (Kessell 1976), but potentially needs similar data for a variety of management purposes.

The present plan is to integrate part of the data collection on rare and endangered species into a much more general vegetation sampling scheme. Part of the reasoning behind this approach is the importance of understanding ecological processes and the dynamics of various impacts such as feral and exotic animals and human disturbance.

The fate of many populations, both rare and common, is integrally related to factors such as fire, human trampling, and invasion of exotic species. It is necessary not only to know "where" and "how many" for plant species but also what events favor or disturb populations. This kind of information can only be maintained through population monitoring.

In developing a vegetation inventory for GRSM, a number of processes and data sets must be considered. The general structure of the plant communities within the park needs to be described and defined. Virgin or mature forest stands should be located and mapped. The major processes and types of plant succession have to be investigated in order to understand what changes have taken place and how stands are likely to change in the future. The most important successional sequences include recovery after human-caused disturbances such as cultivation, grazing, logging, and burning; and recovery after natural disturbances such as lightning-caused fire, pine bark beetle kill, windfall, and landslides. With accurate information, these systems and processes can be modeled.

On-going disturbances are still present and these should be monitored. Two areas remain under agricultural management, which provides a reservoir for exotic plants and encourages reproduction in the deer herd; this in turn impacts the surrounding native plant communities. Human use of the park includes the maintenance of roads, trails, and campsites, all of which impact the native vegetation. Introduced animal species, such as the European wild boar and the Balsam woolly aphid, continue to modify plant community structure and remove populations of native plant species from individual sites (Bratton 1974, 1975; Howe and Bratton 1976; Hay et al. 1977). All of these on-going disturbances are a potential threat to the rarer plant species and need to be monitored closely.

Some communities also need to be described in terms of their phenological changes. Temporal monitoring provides insight into the vernal aspect of the park and includes species neglected by summer sampling.

In some cases the location of individual plant populations will be necessary to determine their status. Exotic plants have already been surveyed separately (Baron *et al.* 1975). Rare and endangered species from the national, state, and park lists require special plots and survey data. Endemic species and record size trees may also warrant individual attention. The park has been developing a system of special protection areas for unique and unusual habitats which could be easily modified by minor human disturbance.

All of this information needs to be collected under a coordinated system. How does one approach this?

In GRSM the first effort towards a general data gathering system was to standardize and computerize the park vascular plant checklist. A six-letter code (three for the genus name and three for the species epithet) was developed which can be sorted either alphabetically or by family. All field data are now taken using the standard codes. The computer list is easily updated and reprinted and other information is included, such as exotic or endangered status. Additional information can be easily added.

The second part of the park effort is the development of a set of permanent sampling plots with a standardized information base. The plots are marked in the field and recorded on master maps. Some information, like Universal Transverse Mercator Coordinates, is recorded in all cases, although special subgroups of plots may vary in size or other characteristics. All data are placed on standard computer-coded forms. Except for the six-letter species codes, all the variables are numeric and most represent a gradient or a hierarchical structure. At present, experimentation with the quantification of topographic and disturbance gradients is being conducted.

An important element of this effort is the attempt at integrating separate projects on such topics as endangered plants, wild hog disturbance, and fire history into a common data recording system. If this proves practical, it will greatly facilitate management use of the data, which can then be accessed by standard computer programs. It also should eventually produce a new and carefully standardized vegetation map of the park.

The immediate management problems requiring detailed vegetation data include: (1) fire management in one of the drier parts of the park; (2) grazing damage and disturbance of special protection areas in Cades Cove, an agricultural area; (3) accumulation of special protection descriptions; (4) monitoring of wild boar impact; (5) modeling and mapping for wildlife management; (6) assessment of human impacts; and (7) ecological analysis of disturbed "natural" areas. One of the most important elements in present park programs is the emphasis on the utility of the information to park management. This goes past the examples listed into such broad areas as fire modeling, visitor management, park or regional planning, or preparing environmental assessments.

For these uses, the data need consistent codes, accurate geographical locations, and a representative distribution through the park. They also must be accurate on a small scale (fires and campsites may be less than one hectare). To be useful to the park staff, outside consultants and planners, and visiting scientists, the data must be easily accessible, dependable, and up-to-date.

Literature Cited

Baron, J., C. Dombrowski, and S. P. Bratton. 1975. The status of five exotic woody plants in the Tennessee district, Great Smoky Mountains National Park, Manage. Rep. 2, U.S. Dept. Interior, National Park Service, Uplands Field Research Lab., Great Smoky Mountains Natl. Park, Tenn.

Bratton, S. P. 1974. The effect of the European wild boar *(Sus scrofa)* on the high elevation vernal flora in Great Smoky Mountains National Park. *Bull. Torrey Bot. Club* 101: 198-206.

_____. 1975*a*. The effect of the European wild boar *(Sus scrofa)* on gray beech forest in the Great Smoky Mountains. *Ecology* 56: 1356-1366.

_____. 1975*b*. A comparison of beta diversity functions of the overstory and the understory of a deciduous forest. *Bull. Torrey Bot. Club* 102: 55-60.

_____. 1976*a*. The response of understory herbs to soil depth gradients in high and low diversity communities. *Bull. Torrey Bot. Club* 103: 165-172.

_____. 1976*b*. Resource division in an understory herb community: responses to temporal and micro-topographic gradients. *Amer. Naturalist* 110: 679-693.

_____. 1978. Rare plant status report (In prep.) Rep. to Superintendent, Great Smoky Mountains Natl. Park. U.S. Dept. Interior, National Park Service, Uplands Field Research Lab., Great Smoky Mountains Natl. Park, Tenn. 44 pp.

Bratton, S. P., M. G. Hickler, and James H. Graves. 1978. Trail and campsite erosion survey. Manage. Rep. No. 16, Vols. I-IV. U.S. Dept. Interior, National Park Service, Uplands Field Research Lab., Great Smoky Mountains Natl. Park, Tenn.

Crandall, D. L. 1958. Ground vegetation patterns of the spruce-fir area of the Great Smoky Mountains National Park. *Ecol. Monogr.* 28: 337-360.

Golden, M. S. 1974. Forest vegetation and site relationships in the central portion of the Great Smoky Mountains National Park. Ph.D. Thesis, Univ. Tennessee, Knoxville. 275 pp.

Hay, R. L., C. C. Eagar, and K. D. Johnson. 1977. Status of the balsam woolly aphid in the Great Smoky Mountains National Park—1976. Manage. Rep. 20. U.S. Dept. Interior, National Park Service, Uplands Field Research Lab., Great Smoky Mountains Natl. Park, Tenn. 25 pp.

Hoffman, H. L. 1962. Check list of vascular plants of the Great Smoky Mountains. *Castanea* 29: 1-45.

Jennison, H. M. 1935. Trees of the Great Smoky Mountains Park. Special Rep. accompanying his November 1935 monthly rep. to the Superintendent, Great Smoky Mountains Natl. Park, Tenn. 12 pp. (typewritten).

_____. 1938. A classified list of the trees of Great Smoky Mountains National Park. 26 pp.

_____. 1939*a*. A preliminary catalog of the flowering plants and ferns of the Great Smoky Mountains National Park. Typewritten rep. to Great Smoky Mountains National Park. 249 pp.

_____. 1939*b*. Flora of the Great Smokies. *J. Tennessee Acad. Sci.* 14: 266-298.

Johnson, C. J. and S. P. Bratton. 1978. Biological monitoring in UNESCO Biosphere Reserves with special reference to the Great Smoky Mountains Natl. Park. *Biol. Conserv.* 13: 105-115.

Kessell, S. R. 1976. Gradient modeling: A new approach to fire modeling and wilderness resource management. *Envir. Manage.* 1: 39-48.

Lindsay, M. 1977. Management of grassy balds in Great Smoky Mountains National Park. Manage. Rep. 17, U.S. Dept. Interior, National Park Service, Uplands Field Research Lab., Great Smoky Mountains Natl. Park, Tenn. 67 pp.

Miller, F. H. 1938. Brief narrative descriptions of the vegetative types in the Great Smoky Mountains National Park. Rep. to the Superintendent, Great Smoky Mountains Natl. Park. 17 pp. (typewritten).

_____. 1941. Vegetation Type Map, Great Smoky Mountains National Park. In color: 27 x 60 inches. Civilian Conservation Corps Project (1935-1938).

Peterson, R. (In prep.) (Ed.) Checklist of fungi of the Great Smoky Mountains National Park. Manage. Rep., U.S. Dept. Interior, National Park Service, Uplands Field Research Lab., Great Smoky Mountains Natl. Park, Tenn.

Sharp, A. J. 1942*a*. A preliminary checklist of the trees in the Great Smoky Mountains National Park (Revised by F. H. Arnold in 1945). Typewritten rep. to Great Smoky Mountains National Park. 39 pp.

_____. 1942*b*. A preliminary list of the woody and semi-woody shrubs and vines occurring in Great Smoky Mountains National Park. (Revised by F. H. Arnold in 1945.) Typewritten rep. to Great Smoky Mountains National Park. 17 pp.

_____. 1957. Vascular epiphytes in the Great Smoky Mountains. *Ecology* 38: 654-655.

Schofield, W. B. 1960. The ecotone between deciduous forest and spruce-fir area of the Great Smoky Mountains. Ph.D. Thesis, Duke University, Durham, N.C. 176 pp.

Stupka, Arthur. 1935-62. Nature Journal, Great Smoky Mountains National Park, 28 vols. (years), each with index (typewritten).

_____. 1964. Trees, shrubs, and woody vines of Great Smoky Mountains National Park. Univ. Tennessee Press, Knoxville. 186 pp.

U.S. Dept. of the Interior, National Park Service. 1970. *Compilation of the Administrative Policies for the National Parks and National Monuments of Scientific Significance (Natural Area Category)*. U.S. Government Printing Office, Washington, D.C. 147 pp.

Whittaker, R. H. 1956. Vegetation of the Great Smoky Mountains. *Ecol. Monogr.* 26: 1-80.

_____. 1961. Estimation of net primary production of forest shrub communities. *Ecol. Monogr.* 31: 157-188.

_____. 1966. Forest dimensions and production in the Great Smoky Mountains. *Ecology* 47: 103-121.

Locating Critical Natural Features Information for Environmental Planning[1]

Joe W. Pardue, Richard J. Olson, and Robert L. Burgess[2]

Information on critical natural features (such as rare species, wilderness areas, critical habitats, and endangered ecosystems) needs to be readily available to planners early in their activities to help identify and mitigate potential environmental impacts. Much of this information is dispersed among a confusing variety of agencies, organizations, groups, and individuals, and is not readily available. Needed facts about the natural features of an area are often discovered only after considerable effort has been expended on planning, design, and construction.

Oak Ridge National Laboratory (ORNL) is attempting to facilitate this information transfer by developing an interface between prime sources and those who legitimately need the information. With the planned interactive computer system, a user could request information concerning critical natural features of a region, county, or sub-county unit and receive a list of brief descriptions of these features coupled with a list of contacts for more detailed information. References will also be supplied from a growing bibliographic data file on nature conservation. Natural features information would be used to amplify natural resource information that has been compiled at the county level by the Geoecology Project at ORNL for use in integrated regional assessments of energy facilities.

All too frequently, the destruction of rare examples of nature occurs through a lack of knowledge, often exacerbated by a paucity of workable mechanisms for transferring data concerning critical natural features from the many and varied sources who hold it to the environmental planners and decision makers who need it. This paper outlines a system under development at Oak Ridge National Laboratory aimed at facilitating the transfer of existing information and, thus, assisting in conserving nature. The rare plant geographic data discussed at this symposium is an example of the type of information necessary to make our system workable.

Stories are legion of biologists who have returned to the site of a species or ecosystem which should have been preserved, only to find it in the process of being bull-

[1] Research supported by the Department of Energy under contract with Union Carbide Corporation. Publication No. 1448, Environmental Sciences Division, Oak Ridge National Laboratory.

[2] Environmental Sciences Division, Oak Ridge National Laboratory, Oak Ridge, Tenn. 37830.

Based on the presentation "Locating Critical Nature Features Information for Environmental Planning" by Mr. Pardue at the Symposium on Geographical Data Organization for Rare Plant Conservation, held 15-17 November 1977 at The New York Botanical Garden.

Pages 69-75 *in:* Larry E. Morse and Mary Sue Henifin, eds., *Rare Plant Conservation: Geographical Data Organization*, The New York Botanical Garden, Bronx, N.Y.

dozed by a development of which they were totally unaware. Telling the builders at that late date is usually fruitless. Development plans take on a momentum with time and money spent. After a certain point, the developmental juggernaut becomes virtually impossible to sway. Sites are selected, resources allocated, people hired, and project managers would generally rather go to court than make major changes because of what they may perceive as a few useless wildflowers. In these instances, the key to protecting natural features is to provide information about those areas in need of preservation to the planners before reaching that point of no return.

Modern development can conflict with nature, often to the detriment of both nature and humans. A number of recent activities and processes may help alleviate some of these conflicts. Among these are the environmental impact statement process and the efforts of long-range planners in state and federal agencies. These activities generally suffer from similar weaknesses, primarily the requirement for masses of information from diverse fields of knowledge which must be found and digested on tight deadlines by people who are trained in narrow specialties. Quick literature surveys and superficial site analyses often are required on a short time schedule; almost without exception, this has been a frustrating and difficult task. Rapid access to good information is a prerequisite for accomplishment of these tasks, yet much useful information is neglected because the planners are not able to locate it with standard search techniques or simply cannot get it within prescribed time limits. Oak Ridge National Laboratory has many years of experience in aiding transfer of this type of information and is acutely aware of the need to make use of new methods available in current computer technology.

Definitions

"Critical natural features" refers to a suite of environmentally related entities variously defined by many classifications and a diversity of epithets recognized by federal, state and private agencies as well as the interested and concerned public. For present purposes, "critical natural features" includes (but is not limited to) wilderness areas, wild areas, primitive areas, natural areas, Research Natural Areas, Biosphere Reserves (Franklin, 1977), protected areas, National Parks, National Monuments, National Recreation Areas, National Seashores, natural preserves, bird and wildflower sanctuaries, wildlife refuges, scenic areas, rare, unique or interesting geologic structures, Experimental Ecological Reserves (TIE, 1977), and National Environmental Research Parks (College of Idaho, 1974). We also include biotic resources such as rare, theatened, and endangered species, critical habitats, wild and scenic rivers, and endangered ecosystems.

The Information

The potential sources of information are many and diverse. The published literature is often easy to access, but represents only a small fraction of the information available. More difficult to locate is information, locational as well as substantive, that is dispersed among a bewildering array of agencies, organizations, groups, and individuals. Many federal agencies have extensive files relating to their particular area of concern, including several discussed in other chapters in this volume. The Nature

Conservancy is interested in preserving unique habitats. The National Park Service maintains parks, monuments, and historic sites under its control and has voluminous data relating to its National Natural Landmarks program. The Corp of Engineers has developed a sensitive wildlife data base (Addor and LaGarde, 1974). The Office of Endangered Species has many files on endangered plants and animals. The Institute of Ecology and the National Science Foundation have proposed experimental ecological reserves (The Institute of Ecology, 1977). Many states have "Heritage" or similar programs aimed at identifying critical areas.

At present no single group has comprehensive data on the locations of critical features. There is no unified directory, inventory, or "clearing house" for these related types of information. Nor is there any way to find such data except on a piecemeal basis with no guarantee that one has located all pertinent information. These problems are further discussed by Kartesz (this volume).

The Proposal

We are proposing to use the facilities, capabilities, and experience of Oak Ridge National Laboratory to assist in and ease the transfer of existing natural features information. The geographic annotation of existing, identified critical features and the appropriate data sources differ substantially from other, already existing species/area locator systems. Our intent is not to duplicate the data collection efforts of others, nor do we have an interest in the designation of critical features. These matters have been legislatively mandated to other agencies.

We have identified several types of data sources of use in a critical natural features information locator system. Examples of published primary and secondary literature include compilations or reports by The Nature Conservancy (1974), National Park Service (1975), Humke et al. (1975), Lindsay et al. (1969), Lindsay and Escobar (1976), and Waggoner (1975). The "gray" literature (such as government reports) contains a wealth of data and reviews of the current status of research, yet are usually distributed on a very limited basis and require special effort to locate and acquire. A third information source rests in computerized data sets. These contain a wide array of useful data, but present unique problems in their location, acquisition and utilization. The subject expert (such as the field biologist), who frequently possesses valuable data that is accessible only through personal contact, constitutes another source. Finally, information systems, such as specialized information analysis centers, special libraries, herbaria, zoological collections, and Heritage Programs which provide assistance in the location and acquisition of certain specialized types of data, can be of immense benefit to a comprehensive locator systems. We intend to access as many of these potential sources as possible in order to provide a central source for locating and acquiring the needed data.

The system under development is modeled on the information analysis center concept which provides comprehensive, in-depth location, acquisition and analysis of data for special user groups (Ulrikson et al. 1975). A unique feature will be the annotation of sources with geographic descriptors (such as county name) so that a user can acquire a list of information for any area of concern. The system will provide the type of source (such as journal, report, numeric data set, agency office, museum collection, academic specialist, etc.), its location (citation or address) and other descriptive annotations. Thus, the system can rapidly provide a list of features, addresses of information sources for a given area, and a list of references for a particular feature. Collectively, these constitute a good starting point for more detailed studies.

We are using standard automated and manual search techniques to obtain journal and report literature. The other four sources would be tabulated by contacting all conceivable sources and explaining the purpose and utility of our effort. These groups often offer leads to other sources. This "college of peers" approach has been proven effective in a survey of the southeastern states. The cooperation demonstrated to date has been gratifying and is certainly essential to the completion of a workable information locator system.

The information sources will be machine processed using the Oak Ridge Computerized Hierarchical Information System (ORCHIS) (Brooks, 1974). This is a comprehensive, integrated information system developed for the research community for processing, retrieval, and analysis of alphanumeric data sets. It uses a flexible generalized hierarchical method for storing data elements which allows selection of as many fields as necessary to adequately characterize a given informaion source. These fields include standard bibliographic elements such as author, publication description, and date, and also permit additional unique features to be extracted from the sources (e.g., taxon, location, and key terms). The retrieval capabilities of the system allow Boolean search strategies to be developed which can yield custom subsets of the main data base. These listings can then be manipulated to produce various indices and tabular displays. The process can be extended to include computer-produced camera-ready copy in a variety of formats. Use of geographic description elements will allow generation of maps with line printers, plotters, or display screens.

We intend to make the data base available on the U.S.D.O.E. RECON computer network which will allow user interactive searching via display screen terminals at a number of Department of Energy facilities across the country. In addition to providing an easy access point for locating natural features information, bibliographies and information locator guidebooks will be published. For example, a comprehensive nature conservation bibliography and a guidebook for locating natural features in the southeast are currently in progress.

Related Activities

Our developing system is closely linked with several related environmental information analysis activities at ORNL. Among these are the Ecological Sciences Information Center, the Environmental Response Center, and the Environmental Sciences Data Resources Group which provide a variety of information services. Of special interest is the Geoecology Project in the Environmental Sciences Division, where a computerized Geoecology Data Base characterizes the spatial and temporal distribution of environmental resources at county-level resolution. It provides readily available data on several themes on a common spatial scale and facilitates energy-related assessment and planning at regional and national scales. Over 700 variables have been accumulated for counties in 16 southeastern states. Many of these variables are also available for the 37 eastern states, and some for the entire nation. Subject themes currently include agriculture, land use, population, climate, wildlife, terrain, forestry resources, air quality, and energy production and consumption. Files under development include water resources, natural areas, and endangered species. Data are obtained from such dispersed sources as national surveys, monitoring networks, maps, and literature. In addition to impact assessment, the data base is being used for biotic productivity studies and pollution effects studies. Output from the data base consists of tables and maps of geographic attributes and/or files for input into models and statistical analysis routines. The Geoecology Data Base is currently stored on disks for batch or interactive retrievals using the data management capabilities of the Statistical Analysis System (Barr *et al.*, 1976).

Through the Geoecology Project we have access to many data sets which permit generation of various maps and tabular displays of natural features. Among these are the International Biological Program Inventory of Natural Areas (Darnell, 1974), The Institute of Ecology Experimental Ecological Reserves (TIE, 1977), the Biogeographic Information System for Bird and Mammal Species in the Southeastern United States (Kitchings, 1976), Rare, Endangered and Endemic Taxa and Habitats of the East Tennessee Development District (Goff *et al.*, 1975), and others (Olson *et al.*, 1977).

At present, attempts to access and collate many other such data sets is the most difficult aspect of the data information acquisition problem. Many excellent data sets exist, and though they may be limited in scope or geographic coverage, their access is essential to a thorough coverage of all possible data sources. Our accumulation of data sets and provision of access to them for planners can provide a useful interim information locator source until some activity such as the National Natural Diversity Inventory (Metcalf, 1977) is completed.

An early test of the proposed locator system uses county level data in conjunction with efforts to identify candidate counties for future energy development. Studies have already located counties based on energy demand, water availability, and other criteria which potentially would support an energy facility. Currently, critical natural features often are not considered because of a lack of readily available data. We anticipate refining predictions of local and regional environmental impact through use of the data made available through the natural features information locator. Lists of counties containing critical features could then be studied at a finer scale of resolution prior to the time when a power company or public utility might explore tracts potentially available. Long range planning of this nature would identify potential conflicts in time to allow resolution or mitigation.

Concerns of Field Biologists

Earlier we mentioned the use of field biologists as potential information sources. This presents at least three concerns that have been expressed by professional biologists whose information we need.

First of all, the biologist involved in a long term project may gather data for several years before they are ready to be published. Fears that the data may be misused prompts the investigator to protect the data before they are in print. Secondly releasing locations of populations or areas may lead directly to their destruction, either by speculators (or others) with an interest in development of the land, or by specimen collectors. A third concern is that the biologists would not have the time to respond to inquiries if listed as an information source about a particular natural feature.

Each of these concerns is both valid and difficult to resolve. Each case would be unique and would require care on the part of the participating biologist not to compromise the proprietary nature of the information. There is a moral obligation, however, not to allow a population to be extirpated because data were unavailable. One approach inherent in our system would identify, at the county level, those areas for which an individual had information. If more detailed location information was needed to resove a potential conflict, then the individual could be contacted. As far as resource time to help planners is concerned, each individual would have to make that decision based on perception of the need.

We believe there is an established need for a national center for acquiring, storing, servicing, and displaying locator information concerning critical natural features. Reasons for this need have been given, as have a series of examples of how such information can be handled and how such a source would be useful to a broad spectrum of planners and environmental scientists, organizations, and concerned citizens.

The Department of Energy, in conjunction with on-going activities in environmental research, analysis, and assessment, is supporting our present work. We recommend that other agencies examine their role in the delicate mix of environmental problems, and hopefully take bold and affirmative steps to help implement a national critical features locator data base.

Literature Cited

Addor, G. E., and V. E. LaGarde. 1974. *A User-Accessed Computer System for Environmentally Sensitive Wildlife*. Techn. Rep. M-74-6. U.S. Army Engineer Waterways Exp. Sta., Vicksburg, Miss. 3 vols.

Barr, A. J., J. H. Goodnight, J. P. Sall, and J. T. Helwig. 1976. *A User's Guide to SAS 76*. SAS Inst., Inc., Raleigh, N.C. 329 pp.

Brooks, A. A. 1974. Oak Ridge Computerized Hierarchical Information System (ORCHIS) status report—July 1973. Rep. ORNL/4929, Oak Ridge Natl. Lab., Oak Ridge, Tenn.

College of Idaho. 1974. *Proceedings of the National Environmental Research Park Symposium, Idaho Falls, Idaho, October 22, 1974*. Snake River Regional Studies Center, College of Idaho, Caldwell, Idaho. 62 pp.

Darnell, R. M., P. C. Lemon, J. M. Neuhold, and G. C. Ray. 1974. *Natural Areas and their Role in Land and Water Resource Preservation*. Final report, Conservation of Ecosystems Program, US/IBP. American Institute of Biological Sciences, Arlington, Va. 286 pp.

Goff, F. G., R. L. Stephenson, and D. Lewis. 1975. Rare, endangered, and endemic taxa and habitats of the East Tennessee Development District. Rep. EDFB-IBP 75-2, Oak Ridge Natl. Lab., Oak Ridge, Tenn. 32 pp.

Franklin, J. F. 1977. The Biosphere Reserve Program in the United States. *Science* 195: 262-267.

Humke, J. W., B. S. Tindall, R. E. Jenkins, H. L. Weiting, Jr., and M. S. Lukowski. 1975. *The Preservation of Natural Diversity: A Survey and Recommendations*. The Nature Conservancy, Arlington, Va. 324 pp.

Kartesz, J. T. 1981. Maintaining awareness of state rare plant lists and projects. *(This volume, pages 103-107.)*

Kitchings, J. T., S. Anderson, and R. J. Olson. 1976. Biogeographic information system for animal species in the Southeastern United States. *ASB Bull.* 23(3): 149-154.

Lindsey, A. A., and L. K. Escobar. 1974. *Eastern Deciduous Forest. Vol. 2. Beech-Maple Region*. Natural History Theme Studies no. 3, National Park Service, Washington, D.C.

Lindsey, A. A., D. V. Schmelz, and S. A. Nichols. 1969. *Natural Areas in Indiana and their Preservation*. Indiana Natural Areas Survey, Dept. Biol. Sci., Purdue Univ., W. Lafayette, Ind. 594 pp.

Metcalf, L. 1977. *Congressional Record* 123(115): S 11446-Si, July 1.

National Park Service. 1975. *Index of the National Park System and Affiliated Areas as of January 1, 1975*. National Park Service, Washington, D.C. 136 pp.

Olson, R. J., J. A. Watts, D. B. Shonka, A. S. Loebl, M. L. Johnson, M. C. Ogle, N. S. Malthouse, D. G. Madewell, and J. F. Hull. 1977. Spatial data on energy, environment, and socioeconomic themes at Oak Ridge National Laboratory. Rep. ORNL/TM-5746, Oak Ridge Natl. Lab., Oak Ridge, Tenn. 39 pp.

The Institute of Ecology. 1977. *Experimental Ecological Reserves: A Proposed National Network.* The Institute of Ecology, Washington, D.C. 40 pp. + map.

The Nature Conservancy. 1974. The Nature Conservancy Preserve Directory. The Nature Conservancy, Arlington, Va. 154 pp. + maps.

Ulrikson, G. U., G. M. Caton, M. P. Guthrie, and H. F. McDuffie. 1975. An environmental and energy information system. *Environmental Letters* 9(4): 431-442.

Waggoner, G. S. 1975. *Eastern Deciduous Forest. Vol. 1. Southeastern Evergreen and Oak-Pine Region.* Natural History Theme Studies, no. 1. National Park Service, Washington, D.C. 206 pp.

Section 3.

INFORMATION SOURCES

Use and Abuse of Specimen Labels in Distribution Mapping[1]

T. M. Barkley[2]

The Great Plains Flora Association, a consortium of eleven Great Plains botanists, recently published an Atlas of the Flora of the Great Plains. This Atlas provides destribution maps and data for the nearly 3000 vascular plant entities of the region. Preparation of this Atlas focused attention on numerous factors of importance in assimilating herbarium label data. Among the instrinsic factors are: erroneous or incomplete locality data; inaccurate identification; obsolete nomenclature; etc. Extrinsic factors include: distribution maps reflect the vigor of collecting programs; species distributions change over the course of time, etc.

The Great Plains Flora Association (GPFA) is a consortium of eleven botanists at nine Great Plains institutions organized to produce a Flora of the Great Plains. An *Atlas to the Flora of the Great Plains* has been compiled and published (GPFA, 1977). It presents distribution maps for nearly all Great Plains taxa and county distribution statements for the infrequent entities. Preparation of the Atlas focused attention on the difficulties of using herbarium specimen labels for distributional data, and publication of the Atlas provided the reason for my invitation to address this meeting.

The need for a new floristic study of the Great Plains has been documented elsewhere (Barkley & Luehring, 1975), but a prime scientific reason is that the only treatment for the region (Rydberg, 1932) is some 45 years old and it was based on a rather scanty number of specimens. Shortly after World War II, Dr. R. L. McGregor of the University of Kansas conceived the idea of a new floristic study of the Great Plains. He soon realized that the specimen resources were utterly inadequate for such a study. Consequently he, his students, and colleagues undertook a truly monumental collecting effort that has filled the herbarium of the University of Kansas

[1] Contribution No. 78-419-A, Division of Biology, Kansas Agricultural Experiment Station, Manhattan, Kansas.
[2] Editor, Great Plains Flora Association; Herbarium, Division of Biology, Kansas State University, Manhattan, Kans. 66506.

Based on the presentation "Use and Abuse of Specimen Labels in Distribution Mapping" by Dr. Barkley at the Symposium on Geographical Data Organization for Rare Plant Conservation, held 15-17 November 1977 at The New York Botanical Garden.
Pages 79-82 *in:* Larry E. Morse and Mary Sue Henifin, eds., *Rare Plant Conservation: Geographical Data Organization*, The New York Botanical Garden, Bronx, N.Y.

with several hundred thousand specimens. The Atlas was based upon this massive amount of recently collected materials, plus the materials in the other regional herbaria, especially Kansas State University, Emporia (Kansas) State University, University of Nebraska-Lincoln, University of Nebraska at Omaha, Kearney (Nebraska) State College, Chadron (Nebraska) State College, the University of South Dakota, and North Dakota State University, all of which are member institutions in the GPFA. The distributional data were accumulated by the GPFA members, their colleagues and students, and cooperating curators.

Examination of specimens during the preparation of the Atlas revealed several interpretation problems which may be attributed to intrinsic and extrinsic factors. Intrinsic factors relate to the individual specimen, and extrinsic factors relate to interpretations of data from numerous specimens.

INTRINSIC FACTORS include the following:

1. *Old specimens with imprecise locality information.* Early exploration in the Great Plains has left a trail of specimens with labels such as "Platte River," "Louisiana," "Mandan country," or "among the Sioux." Plants collected by early-day explorers often came into the hands of botanists and, in turn, became type specimens for new taxa. Many of these collections have been the subject of bibliographic studies, with itineraries published for the collectors. Much of this early exploration is summarized by McKelvey (1953). Finding the precise localities for some of the old specimens is often possible, although the process is time-consuming and tedious. It is worthwhile to document the exact locations of old specimens only when the specimens represent unusual distribution patterns or are of nomenclatural significance.

2. *Poor or misleading data.* Specimens are sometimes deposited in herbaria to document research of cytologists, plant breeders, horticulturists, etc. Unfortunately, such voucher specimens may be well and recently prepared, but possess scant label data. Too many otherwise-good specimens bear labels simply stating "Northeast New Mexico," "Western Kansas," etc.

Certain political units in the Great Plains have had their boundaries and names altered since settlement, and specimens collected in the early years of settlement may be labelled as coming from now nonexistent localities. For example, Davis County, Kansas, was changed to Geary County at the time of the Civil War; Garfield County was absorbed into Finney County, and several subsequent counties have been carved out of what was once all Butler County, Kansas. Even earlier, much of the Great Plains region was part of "Louisiana." Obviously, an accurate plotting of distributional localities requires some familiarity with the political history of the region.

3. *Identification and Nomenclature.* At least a cursory check of the identification of each specimen was necessary for accurate plotting of distributional data. Entities reputed to be rare in a region and known from a limited number of collections naturally attracted the most effort in verifying identifications. Correct nomenclature is a more difficult problem, for virtually no one is completely current on the correct nomenclature for the whole flora. Therefore, distributional data were sometimes plotted on maps only to be re-plotted as editorial work made alteration necessary. Two examples follow. *Tragia urticifolia* Michx. has been known as a single species in the Great Plains for a long time, and many herbaria still have it filed that way. However, Miller and Webster (1967) have made a convincing case for treating Great Plains material formerly referred to as *T. urticifolia* as two distinct species, *T. betonicifolia* Nutt., and *T. ramosa* Torr. Raw distributional data taken from herbari-

um specimens, some which have been annotated to reflect this recent treatment, and others, which remain unannotated, produced three distribution maps, one each of *T. urticifolia, T. betonicifolia,* and *T. ramosa,* hence the need for careful handling of the data in the preparation of the Atlas so that two distribution maps reflecting the modern interpretation of the group were produced. The other example is the treatment of *Oenothera.* Studies by Peter Raven and his students show that the genus is best treated by segregating several genera from it, such as *Calylophus* (Towner, 1977). In some herbaria, *Oenothera* is filed as a large, inclusive genus, and this needed to be accountetd for in accumulating the distributional data.

EXTRINSIC FACTORS may cause misinterpretations on distributional data through inference. The prominent extrinsic factors encounteréd by the GPFA follow:

1. *Erroneous distributional impressions.* These may be conveyed simply by insufficient collecting. A casual user of the Atlas may conclude that there is a fundamental biological boundary associated with the Kansas-Oklahoma state line, with Kansas possessing a rich and varied flora, while Oklahoma has a depauperate flora. The cause of that impression is simply that Kansas has been collected far better than has Oklahoma, so Kansas maps contain many more distribution spots than Oklahoma maps.

2. *Intentional alterations in distribution.* These are often the result of human activities, and are not readily apparent unless collection dates are incorporated into the distributional data. Weedy plants change their distribution patterns and relative abundance with amazing rapidity. For example, *Carduus nutans* L. (or *C. thoermeri* Weinm.), the nodding thistle or musk thistle, was uncommon and rather restricted in the Great Plains until about twenty years ago; now it is an aggressive, noxious weed spread throughout much of the central prairies and plains region. Another plant with a distribution affected by human activity is *Senecio riddellii* T. & G. (Riddell ragwort). Numerous old collections attest to its natural range, including the High Plains of western Kansas, eastern Colorado, Texas, and Oklahoma. However, it is poisonous to horses, so settlers in the region spent much effort eradicating it. It was rarely collected until tractors replaced horses for agricultural work, making control efforts no longer necessary.

A similar extrinsic factor is the great reduction in range by destruction of plants' natural habitats. Two examples are *Epilobium leptophyllum* Raf. (narrow-leaved willow herb) and *Tripsacum dactyloides* Linn. (gama grass). Both are widely represented by collections made before and at the time of settlement, and both occurred in the deepest and best soils of the prairies and plains region. Both species were severely affected by settlers turning the habitat into productive croplands. *Epilobium leptophyllum* is now rare and restricted in occurrence, although the distribution maps indicate it is scattered throughout the prairies and plains. *Tripsacum dactyloides* is now scattered throughout the region in small populations in relict prairies and areas protected from grazing.

To conclude, herbaria are the only repository for exact, provable distributional information, and there are some difficulties in using herbarium label data. Problems derive from inaccurate data and erroneous interpretation of the accumulated data. To minimize error and improve the usefulness of specimen data, all field botanists should provide the most nearly accurate locality and habitat data possible. Some guidelines for preparing herbarium specimen labels are appended.

Literature Cited

Barkley, T. M., and J. Luehring. 1975. A new Flora of the Great Plains. *Pl. Sci. Bull.* 21: 51-53.

Great Plains Flora Association. 1977. *Atlas of the Flora of the Great Plains.* Iowa State Univ. Press, Ames. 600 pp.

McKelvey, S. D. 1955. *Botanical Exploration of the Trans-Mississippi West, 1790-1850.* Arnold Arboretum of Harvard University, Jamaica Plain, Mass.

Miller, K. I., and G. L. Webster. 1967. A preliminary revision of *Tragia* (Euphorbiaceae) in the United States. *Rhodora* 69: 241-305.

Rydberg, P. A. 1932. *Flora of the Prairies and Plains of Central North America.* The New York Botanical Garden, Bronx, N.Y. 969 pp.

Towner, H. F. 1977. The biosystematics of *Calylophus* (Onagraceae). *Ann. Missouri Bot. Gard.* 64: 48-120.

Appendix I: Herbarium Label Data*

Better herbarium-label data are required for better utility of herbarium specimens in providing information about the natural environment, including threatened and endangered species. Highly detailed herbarium labels are very useful, but preparation of such labels is beyond the resources of most collectors. However, we urge collectors to be particularly precise in citing localities and describing habitats.

We recommend that localities be stated as Latitude/Longitude to one minute or better resolution, and to include so far as possible other references such as Town/Range/Section County, Map Reference (*e.g.,* USGS) and/or landmark reference (*e.g.,* "12 airline mi sw of Bovineburg"), and to include altitude. Further, we recommend that habitat reference be provided in accordance with a standard scheme for biogeographic description (*e.g.,* Küchler), and that relative abundance should be carefully noted.

Standard abbreviations such as compass directions, months, etc., may be given in the editorial style of *Brittonia* (*e.g.,* 15 Nov 1977).

A minimally sufficient label should therefore include:

1. Taxon name
2. Collector's name and collection number
3. Collection date (style of *Brittonia*)
4. Determiner's name, if other than the collector
5. Collection locality
    a. Latitude/Longitude to one-minute resolution or better
    b. Supplemented with Township/Range/Section, County, Landmark, etc.
    c. Altitude, when appropriate
6. Habitat, described with reference to standard scheme, e.g. Küchler
7. Relative abundance

---

* Based on recommendations developed informally at a workshop on specimen label data held 15 November 1977, preceding the Symposium on Geographical Data Organization for Rare Plant Conservation, as recorded by John Ballman.

The Literature as a Rare Plant Information Resource

Theodore J. Crovello[1]

As botanists our obligation is to produce literature that can be used, and used efficiently. With this as our guiding philosophy, we may ask how well the literature is serving as an information resource for rare plant information, and how it can serve even better. To answer these questions we must consider the following topics: basic concepts of literature and information; the types of literature relevant to rare plant studies; methods to become and to stay informed of the relevant literature, including both the retrospective and current literature, as well as future literature (or at least the data behind it); actual and potential uses of plant information from the literature; problems with the literature as a data source; and finally, a set of recommended actions to enhance the value of literature.

Some Basic Concepts

Since rare plant data appear in more than just the formal scientific literature, it is useful to discuss all sources of rare plant information. Rare plant workers want to locate and use effectively any relevant information, regardless of its source.

The useful length of life of information varies. In some fields, anything more than one or two years old is of little value to advancement of the field. In other fields, particularly in taxonomy and in endangered species work, often the exact opposite is true. Older information is at least as valuable as the very latest. For example, we may want to know what the flora of the Lake Michigan sand dunes in Indiana was since 1900. Such old information is extremely useful as a "control" of the experiment and question: how have the flora and plant communities changed between 1900 and the present, and at what rate?

Ironically, the expanding taxonomic information base of literature (and herbarium specimens), may be considered an expanding albatross because of its very long half-life. The more data and specimens we accumulate, the more time we must devote to curating it, be it in the library or the herbarium. Once introduced, information can exist in the literature indefinitely. If it is out of date, it may become an obstacle to gaining better answers to our questions. Retrieval of desired data from

[1] Department of Biology, University of Notre Dame, Notre Dame, Ind. 46556.

Based on the presentation "The Scientific Literature as a Rare Plant Information Resource" by Dr. Crovello at the Symposium on Geographical Data Organization for Rare Plant Conservation, held 15-17 November 1977 at The New York Botanical Garden.

Pages 83-93 *in:* Larry E. Morse and Mary Sue Henifin, eds., *Rare Plant Conservation: Geographical Data Organization*, The New York Botanical Garden, Bronx, N.Y.

the literature is affected by many things, but especially by the amount of literature that must be searched.

For endangered species work, we may ask what the optimum level of information is, *below* which there is not enough to make reliable decisions about rare plants, and *above* which there is too much information, which just might muddy the waters and make our conclusions harder. Furthermore, acquiring surplus information involves more time and money; resources which might be utilized better in other areas of rare plant work.

Several *literature information gaps* exist with respect to the literature's role in providing rare plant data. The first information gap is the *publication time gap,* often lasting up to two years. This is the gap between the date of submission of a manuscript by an author and the date of actual publication of the contribution. Endangered species could become extinct in this time when the information is known only to the author and editor. This is an excellent reason to encourage or even possibly require the regular deposition of research data in a (computerized?) rare plant data bank and clearinghouse.

The second literature information gap is the *raw data gap.* This is caused by the absence in the literature of raw data, that is, the absence of the data used to make the summary descriptions or conclusions found in a published paper. The raw data gap is caused by high publication costs and physical size problems that journals and libraries would encounter if all measurements and collections were published for each paper.

The *literature retrieval gap* is the third literature information gap. It is the difference in relevant information between what actually is in the literature and the amount that a particular person obtains. The literature retrieval gap has many causes, including: characteristics of the published information; characteristics of the publication field (for example, tens of thousands of publications, without a central place of storage of articles that might be of value to rare plant studies); characteristics of the methods of information retrieval, either manual or computer; and idiosyncracies of the literature searchers themselves.

Computers are helping us overcome our problems with the literature, but there is danger in total reliance on computer searching. Persons obtain only those references for which they have key words. Alternatively, if a "backended" search is made (*e.g.,* via the ASCA Search Service of Current Contents), only that subset of relevant articles will be retrieved which cite at least one of the references provided to the computer by the searcher. In neither case can users be certain that they have located all relevant references.

Some Basic Definitions

We now address an even more basic concept—the definition of literature itself. In science perhaps the most acceptable concept would be relevant books and articles in professional journals. If these stereotyped examples of literature were the only kinds, there would be no problem with a working definition of literature. Unfortunately, when we include government reports, environmental impact statements, and now the increasing amount of relevant information stored in computer data banks, the exact boundaries of literature become vague. The only reason for raising this problem is because much rare plant data is contained in literature forms such as those just mentioned, and we cannot ignore them.

Literature must be defined to include these current diverse forms of publication and of storage information. I define literature as information on a particular subject that is (or can be) made available in many copies, and in one or several media. This definition approaches the broader concept of an information source, but is not quite as broad since it also demands that more than one copy be available.

Two kinds of publications constitute the majority of the rare plant literature. A *formal publication* is a publication printed and made readily available to the public. Public does not mean only government workers or members of a particular organization or society. By public I mean all serious workers no matter what their background. A *semiformal publication* is a publication possibly printed, but not made readily available to the public. Semiformal publications often may not be supported by copyright. Examples include environmental impact statements, government reports, and state park checklists available only at the park. Obviously, the distinction between formal and semiformal publications in practice may not be easy. We should imagine a continuum ranging from formal publication to several examples of semiformal publications which hardly might be considered publications at all. I distinguish these two types of publications only to assure that we do not overlook this increasing source of dispersed information in our rare plant studies. Once sensitized to their existence, we should devote time to retrieving relevant information from them.

Is output from a computerized data bank a form of publication? It should be considered so, because computers can produce as many copies as the public demands, and at least some are available to a wide variety of users. This paper will consider data banks and their output as literature. My reasons include the fact that the general usefulness of rare plant data banks have not otherwise been discussed at this symposium. Furthermore, data banks currently do not fall into the conventional categories of sources of taxonomic and ecological information such as museum collections, field information, laboratory information, and literature in the narrow sense. Finally, at both the commercial and individual botanist levels, much literature in the narrow sense, and its information, is being deposited in computerized data banks.

One final term needs definition at this point. We define *supporting data* as those data not included in a publication, but accumulated and used in preparation of the summary information or conclusions that are presented in the publication. Supporting data include not only quantitative measurements but also items such as field and laboratory notes, photographs, and illustrations. These should be preserved, and every publication should state how readers may gain access to them.

Types of Literature Relevant to Rare Plant Studies

Given the concept of formal and semiformal publications, the specific types of literature relevant to rare plant studies either are obvious or are not so obvious. The obvious types of literature include: taxonomic monographs; floras, including those at the county and local levels; checklists at all levels of geographic organization; phytogeographic atlases; and theses. These obvious types of literature relevant to rare plant studies are common enough such that examples are not necessary. But even with these obvious types, problems arise. For example, how many checklists would be considered unpublished in the narrow sense of the world? For example, at this symposium we received a copy of Jane Lawyer's draft manuscript, "Guide to the United States' State Lists of Rare and Endangered Taxa.'* The number of entries described as "unpublished manuscript" is high. I myself have two that deal with Indiana's rare plants.

Less obvious types of literature relevant to rare plant studies include information that might be found in biosystematic and evolutionary research papers. In addition, the zoological systematics literature may contain information on rare plants. For example, if a species of spider occurs on only one species of plant, perhaps the localities of the spiders described in the zoological paper may help to locate the host plant populations. In addition, ecological books and papers including vegetation studies, herbivore studies with gut or fecal analyses, as well as plant pathology studies that involve host plants, might contain information of value to rare plant studies. The unanswered question is whether or not it ever will be efficient to search for endangered species data in the voluminous and diverse types of literature such as these.

Environmental impact statements have become a major potential source of information on rare plant data. Setting aside the problem of reliability of identification of specimens, the increase in taxonomic and geographic information in the body of environmental impact statements should not be overlooked in rare plant studies. What is missing are one or more clearinghouses that could provide workers with details of environmental impact statements down to the species level. Information Resources Press (2100 M Street, N.W., Washington, D.C.) does publish a journal (EIS: Key to Environmental Impact Statements) which contains a 300-500 word hardcopy summary of each of approximately 1400 environmental impact statements issued per year. These summaries also are available on magnetic tape. But no species level information is in this computer bank. Additional not so obvious rare plant data sources include the *Federal Register,* the *Endangered Species Technical Bulletin,* cosmopolitan newspapers such as the *New York Times,* and more local ones.

Reference to newspapers underscores a major problem with rare plant data studies that is not encountered in routine taxonomic work. Because of legal as well as social concern for rare plants, more and more people with diverse backgrounds have begun to study them. The problem is that their findings often are difficult to locate, because either they publish in journals that are not in the narrow disciplines of taxonomy and ecology, or they appear in other media, for example, as an article imbedded in the *New York Times.* The problem does have a bright aspect however. While this additional information may be hard to locate, it *is* additional information, which once located should help us to make sounder decisions about rare plants.

Methods to Become and Stay Informed on the Relevant Literature

It is not enough to publish information. Equally important is the ability to become aware and to stay informed of information of value to answer a given question.

---

* In part the basis of "A Guide to Selected Current Literature on Vascular Plant Floristics for the Contiguous United States, Alaska, Canada, Greenland, and the U.S. Caribbean and Pacific Islands" (By Jane I. Lawyer *et al.,* in press, *Mem. Torrey Bot. Club*).—The Editors

Relevant information can be in the old, retrospective literature as well as in current literature. Other information will be in the form of support data and manuscripts which are in the process of publication.

Keeping abreast of relevant literature in formal publications can be done in several ways. The easiest way is to consult a relevant bibliography, such as the one on rare plant species being prepared by the Library of the New York Botanical Garden (Miasek and Long, 1978), or that of Wood (1977). The latter contains 1149 references obtained from several sources, including *Biological Abstracts, Bioresearch Index,* and *Wildlife Review.* Manual searching of the obvious literature sources (journals, books) until recently has been the only widely available way. In the 1960's, various abstracting services, particularly Biosis and then also The Institute for Scientific Information, made available hardcopy products that were based on searches and inversions of their computerized literature files.

Today online computer searching of the literature by individuals is even possible. Someone interested in rare plant data can carry out a computer search by using the services of a commercial retrieval service that makes available via remote terminals the literature data tapes of organizations such as Biological Abstracts, The Institute for Scientific Information, and the National Library of Agriculture. Another method is to contract with an information broker. Rare plant workers convey the nature and diversity of their information needs to brokers. They then carry out online searches of the relevant data banks. Online computer searching of the full text of journal articles still is much in the future, and it is still debatable whether it ever will be worth it.

Becoming and staying informed of relevant semiformal publications is similar to the above, but frequently involves an added step and problem. With computers *and* an unlimited budget, this added step *ideally* would not be necessary. The added step is the need to search for individual documents or series of them which are of possible value to rare plant studies. For example, what government publications on endangered species or rare plants exist for federal lands, regardless of the agency that administers them? Once a person knows that an agency issues a germane series of reports, then he or she simply will check that agency or the publication series on a regular basis. The added step and challenge is to become aware of the semiformal publications in the first place.

As to becoming aware of and gaining access to the voluminous support data on which relevant publications are based, we alluded earlier to the need for registers and clearinghouses for taxonomic, environmental and other data. A small step in this direction is the International Register of Computer Projects in Systematics maintained by Crovello. Projects like those by the Indiana Biological Survey (see paper by Crovello and Keller in this volume) and that of the California Native Plant Society (Powell, Duncan, and Howard; this volume) also should prove valuable.

Active and Potential Uses of Rare Plant Information from the Literature

Crovello (1977) described three stages in the process of endangered species studies:
1. Determine what species are endangered.
2. Communicate the status of endangered species to others, including biologists, politicians, and all other citizens.

3. Continue monitoring and communication activities that will assure the continued preservation of endangered species.

These three stages are both concurrent and cyclic in nature. For example, to determine what species are endangered requires both the accumulation of new raw data as well as the summary of accumulated data over time. Collectively, these provide dynamic snapshots about the status of an endangered species. These data must include geographic and environmental information as well as taxonomic information. Decisions must be made from the available data, including where to search for endangered species and what to conclude are, in fact, endangered species. Once such decisions have been made about what species are endangered, communication and documentation of conclusions must be effected, along with publicity to the interested public as well as to immediate decisionmakers. Simultaneously, monitoring and communication must be continued to assure that further harm to a species does not occur. All stages occur concurrently, and all stages can utilize rare plant information from the literature.

The literature and supporting data can provide information for endangered species studies at low levels of resolution, be it geographical, environmental or taxonomic resolution. Perhaps this is most obvious for geography. Publications provide information on the global distribution of a species, for example, documenting presence or absence of a species at the state or county level or above. But the literature, including the semipublished literature, also can provide high resolution specific locality level data. This is possible via information from supporting specimens or from descriptions of a site at which detailed environmental and taxonomic inventories were made.

As we consider examples of how the literature and its underlying data may be used to provide valuable rare plant information, it will be useful to remember that the literature and its data may be of value at three stages: the *prepublication* stage, the *publication* stage, and the *postpublication* stage. Since examples are somewhat more obvious from the publication and postpublication stage, we will concentrate on the prepublication stage, that is before the monograph or summary paper is formally published and distributed. In so doing, we emphasize the special value of support data if it can be made available in readily retrievable form.

The first example answers a very specific question. Just this month Notre Dame received a request from a forestry aide in the Bureau of Land Management, Idaho. She is involved in an endangered and threatened plant inventory for one of the districts of the Bureau. In an herbarium search she noticed one of my annotation labels which indicated I was involved with quantitative studies in the Brassicaceae. The label also stated that information on the specimen she observed as well as others could be had from Notre Dame. Specifically, she was interested in the endemic *Cardamine constancei,* described from Idaho and reported from only one county in the familiar *Vascular Plants of the Pacific Northwest* (Hitchcock *et al.,* 1964). In preparation for a monographic study of the genus, I had accumulated herbarium specimens from 75 museums. After checking the identification of each specimen, I input label data, annotation, and state of the specimen information into Notre Dame's computer. This data base of support data for a monograph has proved useful to many people involved in rare plant studies, including this forestry aide. A computer search of about 20,000 specimens of the genus resulted in a printout listing 57 specimens of *Cardamine constancei,* representing the holdings of these 75 museums.

In addition to providing detailed label data, we also were able to confirm a range extension of the species from the one known county to four. We also provided a printout of information on each specimen and a table which summarized the distribution of these specimens not just over geographic space, but also over time. This table can alert rare plant workers to the relative urgency to assure that the species still grows in certain counties. The actual cost to run the above search was about $5.00 for computer time. An interesting question arises: who should pay? Should I charge the federal agency for at least what it cost Notre Dame? What about my time, and the time of many other taxonomists who are involved in providing rare plant services to government agencies? In the long run perhaps the best solution, particularly if funds are in short supply, will be some reciprocal agreement whereby federal or state agencies provide some form of additional information in lieu of monetary payment.

A second example which utilizes information at the prepublication stage arises from the general need to determine local and global geographic distribution of a taxon. The question is, to assure finding all county herbarium records of rare species, is it sufficient to look only in the largest herbaria, or is it perhaps better to look only in those herbaria in the relevant states? Even though my monographic study of *Cardamine* and other genera is not yet complete, analysis of the computerized support data clearly documents that the ten largest herbaria may have fewer than 25% of county level records in a particular state, be it for an endemic or a widely distributed species.

Another use of support data involved the need to obtain better locality descriptions for a given collection, particularly of a rare plant, or of a type locality. Once label data are stored in a machine, as Crovello (1972) pointed out, it is a straightforward matter to request the computer to search for all collections of a given person, then arrange these collections, regardless of taxa involved, in chronological order. Then the detailed locality information and date of each collection are printed in such a way that, while information on any one label may be insufficient to pinpoint a location accurately, combined information from collections of diverse taxa at the same place and time may do so.

As a final example of the use of rare plant information from the literature, consider one involving the postpublication stage. From a normally published flora, without using a computer, it has been difficult to answer questions like: in what geographic area might most endangered or rare species be found; what families or tribes, etc. have the highest percentage of rare species; what type of information is most lacking? Not only have such questions usually been very difficult to answer at all, but it was impossible to answer them quickly. Computer data banks of both rare taxa and of the flora in general promise to change this. As an example, on one occasion I needed information on endemic and rare species of the Brassicaceae in the Soviet Union as quickly as possible. Particularly, I wanted to know where the largest percentage of endemic species were in the country, which species were endemic to a particular region, and in what narrower geographic areas and ecological habitats they were found. The solution was to capture nomenclatural, geographic, and ecological information on the approximately 725 species of 125 genera contained in Volume 8 of the *Flora of the USSR* published in 1939. For each species the information input into the computer was its binomial, species number, year published, distribution over the 51 phytogeographic regions of the Soviet Union, world distribution, habitat, and months of flowering. Only a few results will be presented here. At the level of the entire country, it was found and documented quantitatively that the extreme geographical regions had by far the largest percentage of geographically restricted species. The number of species restricted to the Arctic superregion, in the Causasus, and in Soviet Central Asia, all were 45% or more of the Brassicaceae flora of the superregion. The other areas of the Soviet Union, specifically the European, Siberian, and the Far East superregions, all had geographically restricted species constituting 26% or less of the flora of the superregion. At the lower geographic level, detailed lists describing each species, or simply listing them, were provided for the species restricted to each of the 51 particular floristic regions. These and other results quickly and accurately provided an overview impression of floristic information, of rare species, and of their specific ecological habitats.

Data banks based on support data or on published data are just beginning to show their potential to answer questions of great value to rare plant researchers!

Problems With the Literature as a Rare Plant Data Source

The literature as a rare plant data source, with or without computing, has many problems. Some of the more important ones are:

1. Relevant rare plant information is scattered in many journals and monographs.

2. Much valuable rare plant information appears in nontaxonomic, nonecological publications.

3. An information explosion exists. While we all welcome increased data accumulation and synthesizing activity, we are faced with the problem of being mentally buried by the geometrically increasing amounts of information.

4. Information gaps exist. Several types were described earlier, including the information that never gets published, as well as the published, relevant information in articles that never gets retrieved.

5. Information delays exist. The worst involves the delay between manuscript submission and actual publication.

6. Information decay exists, including incomplete and erroneous data. Sometimes these inaccuracies are never realized, and decisions are made on such misleading information.

7. Problems of data reliability exist. This is especially so when semipublications are considered. Frequently they do not involve any formal peer review, or substantiation by herbarium specimen checking.

8. Problems of taxonomic synonomy exist. Unless one is aware of all active and retroactive synonyms, a search of the literature will be incomplete and frustrating.

9. No accepted, uniform standards exist for describing plant data in the literature. The Flora North America Project came closest with its detailed Morphology-Ecology Vocabulary (Porter et al., 1973), but few people, even taxonomists, are even aware of it. Furthermore, there exist no accepted standards at all for describing special information on rare or endangered species.

10. Even when located, many data are difficult to extract from published literature sources, including support data and that contained in taxonomic keys.

11. Difficulty exists in determining whether too much information about a rare species is published. All of the ten previous problems prevent people from gaining access to more information that would be helpful to conserving rare species. In contrast, problem eleven states that too much public information may be dangerous if it were to fall into the hands of those who do not care about rare species conservation

but do care about private gain. Consequently, many organizations have adopted the policy of publicizing only general information, such as providing geographic information only to the county level. Exact locality data are kept in restricted manual or computerized files. The relationship of this problem to the recent United States Freedom of Information Act should be explored.

Actions to Enhance the Value of Literature as a Rare Plant Data Source

It is meaningless to complain about the somewhat low value of literature as a rare plant data source. We must individually and collectively do something. I present some general needs, and then suggest what we may do in our different roles as people interested in rare plant preservation.

Our general needs are simple to state but difficult to accomplish! We need more rare plant data in the literature. That is, we must increase the supply of rare plant data. We need more standardization of rare plant data in the literature. That is, we must increase its readability, transferability (to computer data banks) and its compatability. Finally, we need more easy retrieval and summarization of rare plant data in the literature. That is, we must computerize our data and make these data available through efficient and understandable data base management systems.

As *individual botanists,* either in academic or applied positions, we can do much to enhance the value of literature as a rare plant data resource. We should plan to standardize the collection of our data, even if we ourselves will not computerize it. We should document and deposit supporting data in proper repositories which permit easy, subsequent use by others. We should indicate in our publications which species are rare, threatened, or worse. If no species fall into these categories, then we should state that, too. We should send our publications involving rare plants to endangered species organizations, and perhaps even better, send preprint copies of our manuscripts to such organizations. This is especially important if long delays exist before actual publication. We should both indicate in our publications and inform endangered species organizations directly where the underlying support data are available. Finally, we should no longer think only of formal publications as the means to publicize results. Semiformal publications are a fact of life, as are computer data banks. We must form the habit of using these, and perhaps even contributing to them.

As *editors* or as individuals who can influence editors through our societies, we should request that every monograph and floristic study include information on synonymy, phytogeography, its change over time, and relative or average population sizes. Editors should also suggest that authors deposit documented supporting data in proper repositories. Editors should discourage the use of expensive printed pages for specimen citations, *or* for new county records, *but only if* an acceptable procedure is available to assure their availability at one or more data centers (for example, an herbarium so designated for a particular state or taxon). There are two reasons for this suggestion. Such a policy would decrease publication costs and at the same time increase the rate of acceptance of the concept and use of taxonomic information centers. Editors further should insist that among the keywords at the end of the abstract of an article, the binomials of any endangered or threatened species should be listed, and their status indicated. Editors should adopt a practice of printing the titles of articles accepted but not scheduled for publication until later, along with names and addresses of authors. This would help endangered species workers to locate possible valuable information in time.

Abstracting services such as Biosis, The Bibliography of Agriculture, and the Institute for Scientific Information, should establish rare and endangered plant subject categories, analogous to those for new species taxon lists which Biosis now does regularly.

Rare plant agencies or subdivisions such as the Office of Endangered Species, or the Division of Nature Preserves in the Indiana Department of Natural Resources, should create a list of the most needed literature items which are valuable to endangered species studies. Furthermore, they should request support to satisfy such needs. Such agencies also should take the lead and issue suggested guidelines for the accumulation, preparation, and standardization of valuable information to publish on rare plants. Finally, they should urge editors, government officials overseeing semipublications, and authors to put such guidelines into effect.

Botanical societies should permit their journals and newsletters to carry information and articles on rare plants. In fact, they should be given priority. Such societies should establish annual awards for the best endangered species article published each year. The officers and membership of such societies should urge adoption of the suggestions given above for editors.

Textbook writers and *teachers* should incorporate endangered species information and concepts into their texts, be they at the grade, high school, or university level. My daughter was given a week's unit on endangered species in her fifth grade science class, but how many university students, even biology majors, receive no information at all?

Finally, *everyone* should recognize the need for data standards and act through their various societies as well as individuals to bring them about. In choosing taxa to study, we should consider giving priority to those taxa which might involve rare plants. There is a need for articles written by people familiar with plant conservation for both professional and popular audiences. The goal of such articles should be to educate and to motivate, whether the motivation produces increased study of rare plants in the field, increased ballot box support of politicians who favor rare plants, or keeps citizens sensitized to the status of rare plants. Perhaps there is a need for a new publication which would serve as the central organ for information on rare plant data, be it geographic, environmental, taxonomic, legal, economic, or whatever. We must support the creation and maintenance of computerized data banks and clearinghouses to store and disseminate information of value to people studying rare plants. Everyone should expand their concept of literature beyond that of scientific literature in the narrow sense and beyond that of the typical journal publication to accept the fact that rare plant data appear in publications and semipublications of many disciplines.

Most importantly, we must remember that the literature has been and will continue to be an important source of rare plant data. But many actions must be taken to enhance its value. We must not just think about it. We must do something, and we must urge others to do something.

Literature Cited

Crovello, T. J. 1972. Computerization of the Edward Lee Greene Herbarium (ND-G) at Notre Dame. *Brittonia* 24: 131-141.

_____. 1977. Computers as an aid to solving endangered species problems. Pages 337-346 *in* Prance, G. T., and T. S. Elias, eds. *Extinction is Forever.* The New York Botanical Garden, Bronx, N.Y.

Crovello, T. J., and C. Keller. 1974. Use of computerized floristic data of Indiana for plant geography. *Proc. Indiana Acad. Sci.* 83: 399-406.

Hitchcock, C. L., A. Cronquist, M. Ownbey, and J. W. Thompson. 1964. *Vascular Plants of the Pacific Northwest.* Part 2. Univ. Washington Press, Seattle. 597 pp.

Miasek, M. A., and C. R. Long. 1978. Endangered plant species of the world and their endangered habitats: A selected biography. The New York Botanical Garden, Bronx, N.Y. 12 pp.

Porter, D. M., *et al.* 1973. A guide for contributors to Flora North America. II. An outline and glossary of terms for morphological and habitat description (Provisional edition). *Fl. N. Amer. Rep.* 66. [Dept. Botany, Smithsonian Institution, Washington, D.C.] 152 pp.

Powell, W. R., T. Duncan, and A. Q. Howard. 1981. The California Native Plant Society Rare Plant Project. *(This volume, pages 193-198.)*

Wood, D. A. 1977. *A Bibliography of the World's Rare, Endangered, and Recently Extinct Wildlife and Plants.* Environmental Institute, Oklahoma State University, Stillwater, Okla. 85 pp.

The Role of the Volunteer in Gathering Rare Plant Information

Jean L. Siddall[1]

The project to compile the list of rare, threatened, and endangered plants in Oregon has been almost entirely a volunteer effort, one which began in 1973 with two volunteers, Kenton L. Chambers and Jean L. Siddall, and now involves over 250 amateur and professional botanists throughout Oregon, and a 12-member taskforce of representatives from state, federal, and private agencies and organizations. Throughout the project, the primary source of data has been field information submitted by amateur and professional botanists, based upon their personal knowledge of the plants, or gathered through recent field-checking of the species. This method of gathering information is discussed and compared with information from other sources such as herbarium records and the literature.

Scope and History of the Oregon Project

In Oregon there are 96,981 square miles of land, many parts of which have never been botanized. In the East, this geographic area would accommodate Maine, Vermont, New Hampshire, Connecticut, Massachusetts, and Rhode Island, and have enough room left over for three-fifths of New York State. Or, put another way, Oregon is two-thirds the size of California and one and one-half times larger than Washington.

Oregon varies in elevation from sea level to 11,000 feet (in the Cascade Mountains), and in habitat from the coastal rain forests to alpine fell-fields to the arid high desert grasslands covering much of the eastern two-thirds of the state. Based upon the distribution of the rare and endangered plants, there are 17 physiographic provinces and seven areas of significant endemism within the State. Botanically, Oregon is second only to California in plant species diversity.

The first efforts to compile a list of the rare, threatened and endangered plants in Oregon began early in 1973 when Kenton L. Chambers, Professor of Botany and Curator of the Herbarium at Oregon State University, began searching the herbarium and the literature for species rarely collected. I joined this effort late in 1973, when as Chairman of the Oregon Chapter of The Nature Conservancy and a mem-

[1] Oregon Rare and Endangered Plant Species Taskforce, 535 Atwater Road, Lake Oswego, Ore. 97034.

Based on the presentation "The Role of the Volunteer in Gathering Field Information" by Ms. Siddall at the Symposium on Geographical Data Organization for Rare Plant Conservation, held 15-17 November 1977 at The New York Botanical Garden.

Pages 95-102 *in:* Larry E. Morse and Mary Sue Henifin, eds., *Rare Plant Conservation: Geographical Data Organization,* The New York Botanical Garden, Bronx, N.Y.

ber of the State Natural Area Preserves Advisory Committee to the State Land Board, I was asked to participate in the conference to determine research natural area needs in the Pacific Northwest, and along with Arthur R. Kruckeberg, Professor of Botany at the University of Washington, was assigned to the rare and endangered species committee. While there were good lists of the rare and endangered fish, reptiles and amphibians, birds, and mammals, there were none for plants, so we began at the beginning. I assumed that Dr. Kruckeberg would compile the Washington *and* Oregon lists, but on January 2, 1974, I got a call asking that I do the list for Oregon, and submit it by January 18! I spent the first week on the telephone asking botanists all over the state what they thought was rare in their area and the second week typing, and did turn in a review draft by the 18th.

The Oregon Project differs from those in most other states in two significant ways:

1. The Project has been almost entirely a volunteer effort, (one which began with two of us in 1973 and now involves over 250 amateur and professional botanists throughout Oregon, and a 12-member Taskforce of representatives from state, federal and private land-managing agencies); and

2. Our primary source of data throughout most of the Project to date has been field information submitted by botanists based on their personal knowledge of the species or gathered through field checking, rather than herbarium records and the literature, which is the case in most states.

This approach was born more of necessity and circumstance at the time than of deliberate project design, but it was right for Oregon and if I were to start all over, I would not change it.

Because an Oregon list was needed, but there were no funds available, it became a volunteer effort; because no herbarium search was possible in the initial time frame allowed, the information was gathered from people who knew their area of the state and its flora; and because there are only about enough professional taxonomists in Oregon to fill one station wagon, the Project included amateurs as well as professional botanists.

In the process, however, many people all over the state became involved in the Project, greatly increasing the awareness of the need to protect our native species. I found it was primarily the amateur botanists who get out and roam the hills, know which plants are rare and where they can be found. Furthermore they were submitting information about areas never before collected and therefore not available in any herbarium or the literature. This source of information usually goes untapped in most states.

One question I am continually asked, however, is "But how do you know the information from amateurs is valid?" First of all, the word "amateur" is relative. Many of these people had been studying a specific species or area or the state most of their lives. Each person I asked for information was either known to me personally or had been recommended by someone—in retrospect, an important point. Too often a conference is held and the public invited.

Secondly, I spoke only Latin, and those who did not have the level of expertise to keep up with the scientific nomenclature dropped out of their own accord. Lastly, only about one per cent of the sites reported by participating botanists have been questioned as misidentifications, as compared with an error factor of about 40 per cent for herbarium records, and when the person was sent back to collect a specimen, many of these turned out to be significant range extensions.

Since the main funding available was my own, records were kept on 4 x 6 cards in a shoebox on my dining room table, with the "known" information from floras on the face of the card and the specific site information learned from botanists recorded on the back—a very simple, effective system of data storage which is retrievable day or night at no cost.

I found there were definite advantages to being a volunteer. I could cut red tape and do whatever needed to be done when it needed to be done without waiting for funding or approval. I could plow new ground without first writing a manual of procedure. I could expand or contract my workforce with a telephone call. Since I was not aligned with any agency, I was not suspect to those I was asking for information.

Through the summer of 1974, Dr. Chambers continued to research herbarium collections and the literature, and I continued to review the initial plant list with people, dropping some species, adding others, and starring those which people considered to be endangered. In the fall when we discovered we were surfacing the same species using two different methods, we combined our lists into the "Tentative List of the Rare, Threatened and Endangered Plants in Oregon." This list, with additions from the Smithsonian Institution list (1975), was published by physiographic province as "Vascular Plants of Special Interest" in the report *Research Natural Area Needs of the Pacific Northwest* (Dyrness *et al.*, 1975).

At the time the Smithsonian Institution's *Report on Endangered and Threatened Plant Species of the United States* was published in 1975, Dr. Chambers and I were still about the only people worrying about the rare and endangered plants in Oregon, as the Governor of each state was being asked by the Office of Endangered Species of the U.S. Fish and Wildlife Service for review and comment. I wrote the Governor of Oregon, telling him of our efforts. I also asked if he wished to appoint a taskforce to continue this work, or whether he wished me to assemble an informal taskforce of the people working on the project and those who were being assigned the responsibility for endangered plants for their agency. He supported the informal taskforce, with the admonition not to spend any money. Thus the 12-member taskforce of volunteers representing state, federal, and private agencies and organizations was organized in December, 1975. In Oregon it is particularly important that the land-managing agencies be directly involved since over one-half of the state is under federal management (U.S. Forest Service, Bureau of Land Management, U.S. Fish and Wildlife Service, and National Park Service), one-fourth is under various kinds of state ownership, and one-fourth is in private ownership, much of that being private timber or ranch land.

The purpose of the Oregon Rare and Endangered Plant Species Taskforce was three-fold: to gather information about the rare and endangered vascular plants in Oregon; to write status reports for the species proposed for federal listing; and to help draft enabling legislation in Oregon for the protection of threatened and endangered plants.

Methods Used to Gather Field Information

To further review the list of rare, threatened and endangered plants, the Taskforce scheduled a conference in Portland in March 1976 to bring together as many amateurs and professional botanists as possible to review the species. I had felt we would be lucky if we could find 30 people who knew the native flora well enough to know where to find the rarer plants; however most people contacted knew at least one other whom they felt knew more than they did, and before we had finished we had called 158. Only four chose not to participate; ninety-three came to the Conference in person and the rest sent information by mail. The motley crew of volunteers participating includes homemakers, professors of botany, agency field personnel, nurserymen, foresters, biology teachers, range managers and one person who collects pollen for a living. They all have two things in common, however—they know their plants and they get out and roam the hills.

Prior to the Conference, worksheets were sent to the participants to give them time to research their own field notes and collections and record the information to be turned in at the Conference. The 568 species then on the Oregon list were grouped according to the quadrant of the state in which they occurred, and the pattern of distribution determined for each species to indicate whether they were:

Ia — Narrow endemics
Ib — Regional endemics
IIa — Wide range but scattered and rarely collected
IIb — Widely disjunct populations
III — Rare in Oregon, more abundant elsewhere
IV — Unusual population

The Conference was also organized according to quadrants, so that someone who knew only the plants of the Wallowa Mountains needed to attend only the session for the Northeast Quadrant. To our surprise, however, almost every participant attended all four sessions.

At the Conference, the species were reviewed one by one, and as botanists reported specific sites where they had seen the species, the locations were recorded on newsprint at the head of the room for all to see, and "pinned" on a large map so participants could watch the areas of endemics and concentrations of rare species develop as the Conference progressed. By the end of the Conference, there were many pins in the Wallowa Mountains, in the Illinois Valley and Siskiyou Mountains, and along the Columbia River, but very few in the ranchlands of eastern Oregon which have been plowed and grazed for 150 years. The genetic reservoirs here are in old cemeteries and small pockets of land where cows and plows cannot reach. Nor were there many pins in the Willamette Valley. Species endemic to these grasslands cannot compete with farming and urban development, and now survive primarily along roadsides and fencerows, which, with the ban on field burning, are now being mowed or sprayed to keep down the "weeds." Perhaps the least known part of Oregon botanically is the steep and rugged Coast Range. Here the list reflects where the botanists have been rather than the plants which are there.

Of the 568 species reviewed at the Conference, 186 (32%) had been seen by no one; and 137 (24%) were reported by only one person. It is perhaps more amazing, however, that the remaining 145 species (44%) had been seen recently by two or more persons.

In transcribing the information from the Conference and from the many worksheets, we realized that both methods of gathering information are needed, since botanists reported sites on the worksheets which they did not report at the Conference, and, with the stimulation of others, remembered sites at the Conference not recorded on the Worksheets.

Sighting Report forms were given to each of the participating botanists on which to report species found during the coming field season. In addition, 115 of the species not reported at the Conference were assigned for field-checking. In retrospect, the most important parts of both the Sighting Report and Field-Checking Report forms are the lines left for "Comments" and the space on the back on which to draw a sketch map of the site.

Botanists combed the state during the summer in a gigantic "treasure hunt" which reached from the Pacific Ocean to the Snake River Canyon, and reported some interesting sightings. *Lomatium greenmanii,* known only from its type collection in The Wallowa Mountains and listed by the Smithsonian Institution as Possibly Extinct, was found at two sites. *Streptanthus howellii,* last collected in Oregon in 1913, could not be relocated at the type locality, but was found at three new sites. *Sophora leachiana,* thought to be a rare local endemic, was found to have suddenly extended its range 15 miles with the advent of clear-cutting in the area; it is jumping into each new clearcut, crowding out the reforestation seedlings. On the other hand, *Lomatium bradshawii,* known from nine old localities according to herbarium records, could not be found at any of those sites.

During 1976, 103 new species were added to the Oregon list and 100 more botanists joined the effort. In May, 1977, a Second Conference on Rare and Endangered Plants was held in Portland. The difference between the two conferences is very significant in terms of gathering information. Whereas the first Conference had tapped a reservoir of information gathered over a lifetime, the second Conference mostly represented the information gathered during one year.

In the spring of 1977, with the help of travel funds made available by agencies, information was gathered from 23 herbaria. In addition, 813 field-checking assignments were sent out to participating botanists, a) to determine whether populations were still extant at old herbarium sites; b) to verify the identification of species previously reported by field botanists, if questioned (usually the botanist who had reported the species was sent back to the site to collect a specimen—*if* the population warranted—and to take a photograph, if possible); and c) to look for species for which we still had no herbarium record or field information. Of the 186 species for which no sites were reported by botanists at the 1976 Conference, 44 were found in Oregon during the summer of 1976, and 35 more were reported at the 1977 Conference.

Since 1973 when the project began, three new species have been discovered in Oregon, four others have been described and are awaiting publication, and four more have been published, thus adding 11 species to the floras. In addition, participating botanists have reported 20 species which are first records for Oregon, and 116 sites which are substantial range extensions within the state.

In summary, field information has been gathered from the participating botanists in the following ways:

- by individual communication (this works best but is very time consuming)
- on Worksheets (completed from field notes)
- at two Rare and Endangered Plant Conferences
- on Sighting Reports
- from Field-Checking Reports
- from area studies, as land-managing agencies are beginning to hire botanists to inventory specific areas on their lands.

Comparison of Field Information With Other Data Sources

As the Oregon Project is apparently the only state which began its data collection with field information and then later searched herbarium records and the literature, perhaps a comparison of the relative merits of each as a source of information is in order. I find it ironic, however, to look back in retrospect and realize that because we have done our project "backward," we apparently are now "ahead."

The herbarium records, floras and the literature are all considered to be good *additional* information, and we have used all three as information against which to compare the field data submitted by botanists; however, I would not want to compile a list or determine the status of rare or endangered plants, at least in Oregon, on the basis of any one alone or even all three together. The field information is vital to the project.

Herbarium records. Information has now been gathered from 23 herbaria, both in Oregon and out-of-state. In comparing the data taken from herbarium sheets with the field information submitted by participating botanists, we are finding that:

1. The overall error factor in herbarium specimen identification is much higher, often approaching 40 per cent. Some of these are straight misidentifications; far more often the genus has been reworked and the names changed, but the herbarium sheets have not been brought up to date. With 671 species to review for the Oregon project, however, there is no time to ask that each collection be annotated by an authority. By comparison, of the 2300 specific sites reported by botanists in 1976, the identification was questioned for only 28 (about one per cent), and in most instances the botanist was asked to field-check the species and to collect a specimen, if the population warranted.

Whereas specimens are frequently submitted to herbaria *to be identified,* people are generally not reporting sites to the Project unless they feel certain of their identification. Furthermore, when questions do arise, the person involved is still alive to ask—an advantage one does not have with most herbarium collections.

2. The information received from botanists participating in the project is current. By comparison, of the data collected from 23 herbaria thus far, 56 percent of the collections were made before 1940, with 12 percent of them before 1900. Only 5 percent have been collected since 1970, the year we have arbitrarily chosen as the dividing line between "current" and "historical" information. The other 95 percent of the herbarium reports need to be verified on the ground to determine if they are still extant.

3. The specific site information on herbarium labels is often vague, particularly on older collections. For example, "Stony swales, Eastern Oregon" could be anywhere in 60,000 square miles and is a little difficult to map, or to computerize, or to consider in land management decisions.

In other instances a collector used local names commonly accepted at the time, such as "3 miles east of Humphrey's Ranch," which never appeared on any map; or what is worse, used code names for places. We now know Suksdorf's "Falcon Valley" is near Glenwood, Washington; however we are still looking for "Keystone Canyon" in the Wallowas, the site of many Cusick collections.* On older collec-

---

* Geographic names on Oregon herbarium labels are interesting: Tin-Cup Pass, Pistol River, $8 Mountain, or Rough and Ready Creek. Then there is Fossil, Free and Easy Trail or Whorehouse Meadow. (Recent efforts have been made to change this to Naughty Girl Meadow on the maps, but they have not succeeded locally.)

tions, even the county name should be questioned as county boundaries have changed dramatically in Oregon since the 1850s. Anyone punching herbarium label data onto computer cards without first locating the site on a map could inadvertently be adding an error factor of several hundred miles to the range of a species.

4. Another limitation to herbarium labels is that there is rarely any status or abundance data included, and the information is limited to the site of that collection. Other information, such as additional sites nearby, threat to the plant or the site, habitat data, and so forth—needed to determine the status of a species—is generally not available from herbarium labels.

We are finding that herbarium records have their greatest value in determining the probable historical range of the species, against which to measure present range, and as records of places the species once grew, which can be used as starting points in field checking.

Herbaria are also serving as an invaluable repository for specimens now being collected by field botanists. Dr. David Wagner, Curator of the University of Oregon Herbarium, is now red-flagging the folders for species on the Oregon list; as new sheets are filed, a sighting report is first filled out and the information sent to the Oregon Project.

Floras and Other Literature

Floras are invaluable for identification, for determining the habitat of the species, and as a quick reference in determining their general range, although there are some significant discrepancies between them in this regard. Floras are, however, generally based upon herbarium collections and therefore are only as accurate or as complete as the herbarium information.

Although excellent monographs have been written on some genera, and there are excellent journal articles on some species, many are not covered in the taxonomic literature. Where such information is available, it has been added to the files.

A Look Forward

The next step is to get this large body of current information off my dining room table and into a "Report on the Rare and Endangered Vascular Plants in Oregon—A Summary of Current Knowledge and a Field Guide." To broaden the base of the Project, we are about to establish an Oregon Rare and Endangered Plant Study Center, to continue working with the field volunteers in gathering information, to work with universities on study projects, to work with land-managing agencies, and to begin the intensive study and recovery aspect for some of the species. In this regard, the Rae Selling Berry Garden, through volunteer effort and funding, is about to become the first botanical garden in Oregon. Once established, facilities will be available there for propagation and recovery programs.

Summary

In conclusion, let me summarize a part of what we have learned.

First, in gathering information about rare and endangered plants, we need all four data sources—current field information, herbarium records, the floras, and the taxonomic literature. None stands alone; none should be omitted.

Botanists throughout the state, both amateur and professional, are a most valuable source of information about the native flora, a resource as yet untapped in most states. In most instances, their information represents a lifetime of study.

Fieldchecking is best done by people who live in the area to be checked, and already know the area and the plants in it. Given the name of the species and its habitat, they know where to search.

The cooperative effort between the agencies and the volunteer effort made possible through the Taskforce has been very important. Each could provide things the other could not.

From the standpoint of land-managing agencies, three things are especially important: the status of the species, the specific sites where it has been found, and the ownership of the land it is on. For this reason we have divided the 36 counties in Oregon into 174 subunits based first on river drainages then on land ownership, and are compiling lists according to the subunit so that these species can then be field-checked also by the land-managing agency involved.

Several things are especially important when working with volunteers: The project needs to be well-organized, and the work must be divided into bite-size pieces which have a beginning and an end, are accomplishable, and give a sense of personal accomplishment and contribution. Instructions need to be explicit, and forms need to look easy to fill out and ask only for very basic information. You can always go back for more information, but if the initial form is complicated, it will never be sent in. Someone needs to be assigned the responsibility for being the contact who is always available, knows the project, and has the authority to make decisions.

Perhaps the hardest part so far of working with the participating botanists in the Oregon Project has been in convincing them they do not have to know the whole state and all of the species in it to help. A good friend of mine remarked, "But how can I help? All I know is *Allium*." I could only tell him that what this project needs is someone who really knows their onions!

Literature Cited

Dyrness, C. T., J. F. Franklin, C. Maser, J. D. Hall, and G. Faxon. 1975. *Research Natural Area Needs in the Pacific Northwest: A Contribution to Land-use Planning.* U.S. Forest Service Gen. Techn. Rep. PNW-38. 321 pp.

Smithsonian Institution. 1975. *Report on Endangered and Threatened Plant Species of the United States.* House Document 94-51 (Serial no. 94-A), U.S. Government Printing Office, Washington, D.C.

Maintaining Awareness of State Rare-Plant Lists and Projects

John T. Kartesz[1]

The continually increasing number of state and provincial works dealing with rare-plant information has compounded the problem of maintaining awareness of such works. Perhaps the most efficient solution to this problem would be to develop and publish monthly or quarterly a rare-plant newsletter to list newly published literature. A second or additional solution could be to develop and keep current a comprehensive book such as the publication *Rare Plants*. A third possible solution might be to recognize an individual within each state or province who would be responsible for maintaining local awareness of rare-plant information; and, a final solution could be to establish a clearinghouse for rare-plant data. In any event, accomplishing the task of maintaining awareness must stem from comprehensive coordination.

The task of maintaining awareness of rare-plant listings and projects has become a full-time job. Currently, all states in the United States and provinces in Canada have developed, or are developing, rare-plant listings for their areas. Those knowledgeable in rare-plant research or remotely interested in environmental conservation should be aware of the role that rare plant lists and related works play in the total assessment of species rarity. However, for those less familiar with this subject, the following summary points are germane:

1. Localized rare plant listings and related works often provide the only real reference source available to environmentalists in assessing the degree of species rarity and are helpful in determining species distribution, habitat requirements, and other essential facts necessary for the preparation of environmental reports and impact statements. Thus, to the environmental consultant, such listings and works are of extreme importance.

2. Such works are particularly valuable in biogeographical data awareness, often suggesting endemism, range limits, relict or disjunct populations of species and other critical range data which are often not documented anywhere else in the existing literature.

3. Often such works are also useful in identifying potentially sensitive or

[1]   Biota of North America Committee, 2202 Ridge Road, McKeesport, Pa. 15135.

Based on the presentation "Maintaining Awareness of State Rare Plant Lists and Projects" by Mr. Kartesz at the Symposium on Geographical Data Organization for Rare Plant Conservation, held 15-17 November 1977 at The New York Botanical Garden.

Pages 103-107 *in:* Larry E. Morse and Mary Sue Henifin, eds., *Rare Plant Conservation: Geographical Data Organization*, The New York Botanical Garden, Bronx, N.Y.

uniquely fragile ecosystems and in suggesting critical areas, natural areas, or other areas of special concern to conservation agencies and land managers.

4. These works add significantly to our understanding of the comprehensive biogeographical range of a taxon and assist in determining the actual status of rarity of that taxon at regional and national levels. In this way, such works supply crucial information to those preparing works of greater geographical significance.

5. It is not contradictory that some taxa appropriately included within certain state lists may be rare only at the state level but common elsewhere in North America; see Fortney *et al.* (1978) for an example of a state list and the rationale behind it, and Preston (1975) for further discussion. The significance of these lists lies in the fact that attention is drawn to these taxa which are locally rare and therefore of great interest to nature groups, botanical clubs and other wildflower enthusiasts who are often responsible for initiating interest in the rare species concept and teaching its values through conservation education. Conservation of such disjunct and peripheral populations also promotes maintenance of genetic diversity in widespread species.

6. If maintained, such works will undoubtedly provide a source of long-range data for rare-plant researchers, thus serving as environmental indicators in keeping floristicians aware of changes in degree of rarity due to increased knowledge, floristic shifts, etc. It should be indicated, however, that state or equivalent listings represent only local rarity, but not necessarily regional or national rarity.

Often more than one list may exist for an area, which not only causes confusion in determining which lists are currently recognized as authoritative and current but also creates problems in assessing which list is most accurate. Consequently, a definite need exists to develop and maintain a compendium of lists and projects as a means of promoting cooperation, minimizing duplication of effort, and providing a data source which may be used and accepted universally.

Prior to the publication of the compilation *Rare Plants* (Kartesz and Kartesz, 1977), a thorough review was made to determine the availability of existing literature. It was found that fewer than 14 works before 1970 dealt specifically with rare plant information in North America. Of these, only 8 dealt with geographical distribution. From 1970 to 1976, the number of works dealing with rare-plant information had increased to 100, of which 86 dealt specifically with geography. More recently, from March 1977 to January 1978, 30 or more new works have been developed, 10 more are in press, and still others are being revised. The consequence of this continual flow of data is the obsolescence of such current bibliographical works as Ayensu and DeFilipps (1978) and Lawyer *et al.* (in press) even before they appear in print.

This dramatic increase in data again emphasizes the need to maintain awareness of newly published and existing works; however, attempting the maintenance of such data is both costly and time-consuming. Much of the time and cost comes from determining whom to contact for the information. Although many state and federal agencies are involved with rare-plant list preparation, few agencies within a state are completedly aware of all lists and projects for their area. Consequently, several agencies must be contacted within each state in order to ensure comprehensive coverage.

Perhaps the most reliable sources of information dealing with rare plants are the major universities and museums within each state. Here are housed the major herbaria from which documentation and verification of data may be extracted from voucher specimens. Additionally, if university or museum staff members are not directly involved with existing projects, they are generally knowledgeable of possible contacts for that information. Other primary information sources include the Smithsonian Institution, the U.S. Fish and Wildlife Service, the Soil Conservation Service, The Nature Conservancy, the departments of natural resources (or equivalents) in the various states, various state native plant societies and similar groups, and the local chapters of the Audubon Society.

Rare plant laws, like rare plant works, are continually increasing. Several new state laws have recently been passed and additional bills are pending ratification. Information concerning legal status of plants for a particular state can most readily be obtained through the law enforcement branch of the state Department of Agriculture and more recently the state Fish and Wildlife Department or its equivalent. As with rare plant listings, however, usually several requests must be made in order to obtain the desired data.

As additional information is developed, the problem of maintaining awareness will continue to grow. Several possible solutions to this perplexing problem exist. Each has advantages and disadvantages but independently provide the necessary information for gaining awareness of existing data. Perhaps the most economical solution to the problem would be to develop a rare-plant newsletter published monthly or quarterly. The publication could indicate pertinent rare-plant data such as: new listings and relevant laws, field studies, planned explorations, sources of information, brief articles dealing with species rarity, problems concerning ongoing programs, and other similarly related topics. This work could serve as a supplemental bulletin complementing the widely circulated and informative *Endangered Species Technical Bulletin* (U.S. Fish and Wildlife Service) by providing state-specific data on rare plants. Such information would also serve as a means of conservation education about rare plants to help inform the non-technical biologist and layperson about the need to preserve these significant biotic components. Also, it might further help bridge the widely expanded gap which now exists between the conservationist and the developer. Such a work might be costly to develop and maintain, but the data could easily be obtained through contributions and distributed to anyone interested in seeking rare-plant information. A major disadvantage of this system however is that it prohibits state-by-state comparison of data dealing with taxa, laws, definitions, etc., unless the editors assume the responsibility of producing a tabular form of comparative data periodically.

A second solution might be to develop and maintain a comprehensive treatise such as the publication *Rare Plants* (Kartesz and Kartesz, 1977). This publication represents a comprehensive synthesis of existing rare-plant listings presented in tabular form and indicating state-by-state rarity for over 12,000 named taxa of vascular plants. The geographical scope of the book includes the United States, Canada, Puerto Rico and the Virgin Islands. A major appendix lists by state, province, or island the pertinent data including rare species specialists, alternative listings, and applicable laws. A second appendix presents a brief summary of equivalent names used as synonyms the text listings, but is not intended to indicate the currently accepted names. Such a work provides a more complete coverage of existing data and allows a comparison of data to be made on a state-by-state basis. It also enables rapid assessment of plant distribution by showing where such plant species are rare. The work was published with an optional binding to allow for looseleaf updates to be inserted periodically. The disadvantages of this solution are again associated with high costs of development and maintenance as well as the aging of data over prolonged periods. It would be valuable for an agency, such as the U.S. Fish and Wildlife Service or the Bureau of Land Management to maintain this work. The data presented could be easily computerized and reproduced in photocopy form and made available upon request.

A third possibility, which may merit consideration, would be to identify a key person within each state to assume the responsibility of providing state rare-plant information. This person would serve as a liaison rare-species specialist working with private, state and federal agencies and could help the U.S. Fish and Wildlife Service's regional staff botanists by supplying to them such information. This approach would facilitate the maintenance of awareness at the local level; however, it would provide little impetus for national awareness unless a specific effort were then made to coordinate all such state-oriented research. A major disadvantage of this approach is that political maneuvering might be encountered in the process of determining which organization or agency would create the specialist positions.

A fourth possible solution could be to establish a clearinghouse for rare-plant data. The clearinghouse would serve the needs of all state and governmental agencies and aid appropriate groups and individuals needing rare-plant information. It could also provide systematic storage of state, regional, and national distribution of rare plants, relevant laws, etc. This approach would enable continuous storage and updating of data. A likely candidate organization for establishing such a clearinghouse would be the U.S. Fish and Wildlife Service or another federal government agency that deals directly with rare species research and has the facilities to promote such an endeavor. The major disadvantage of this system again relates to its high cost of development and maintenance. The inaccessibility of data could also be a problem. For example, groups not having access to computer terminals would find it difficult to extract the information unless periodic reports and summaries were made freely available.

In light of the fact that floristically oriented research is gaining momentum in North America, the need to develop innovative techniques to maintain awareness of rare-plant data must simultaneously increase. The methods by which awareness may best be achieved could possibly include any of the previously mentioned solutions or various combinations thereof. It is apparent, however, that due to geographical, professional and political differences, it is often difficult to establish a workable solution to coordinate efforts in a field in which so many groups have a common interest. Nevertheless, until a permanent solution to the intensifying problem of maintaining awareness of rare plant data is reached, the intended values of the concepts relevant to plant conservation may not be fully appreciated, and substantial duplication of efforts may persist.

Literature Cited

Ayensu, E. S., and R. A. DeFilipps. 1978. *Endangered and Threatened Plants of the United States.* Smithsonian Institution and World Wildlife Fund, Washington, D.C.

Fortney, R. H., R. B. Clarkson, C. Harvey, and J. Kartesz. 1978. *Rare and Endangered Species of West Virginia, a Preliminary Report. Vol. I. Vascular Plants.* West Virginia Dept. Natural Resources Heritage Trust Program, Charleston, W. Va.

Kartesz, J. T., and R. Kartesz. 1977. *Biota of North America. Part 1. Vascular Plants. Vol. 1. Rare Plants.* Biota of North America Committee, McKeesport, Pa.

Lawyer, J. I., A. M. Miller, L. E. Morse, and J. T. Kartesz. *In press*. A guide to selected current literature on vascular plant floristics for the contiguous United States, Alaska, Canada, Greenland, and the U.S. Caribbean and Pacific Islands. *Mem. Torrey Bot. Club.*

Preston, E. J. 1975. Endangered plants. *Amer. Forests* 81(4): 8-11, 46-47.

NOTE ADDED IN PROOF

A review of state plant protection laws from the legal viewpoint was recently published—McMahan, Linda. 1980. Legal protection for rare plants. *Amer. Univ. Law Rev.* 29(3): 515–569.—The Editors.

Section 4.

REPRESENTATIVE PROJECTS

Smithsonian Institution Endangered Flora Computerized Information

Edward S. Ayensu and Robert A. DeFilipps[1]

The historical background and methods used in preparing the *Report on Endangered and Threatened Plant Species of the United States* (1975) and *Endangered and Threatened Plants of the United States* (1978) are discussed. The informational base and data computerization activities of the Endangered Flora Project are outlined, and illustrated by specimen data cards, reports, and geographical distribution maps of exploited southeastern species. Categories of data in the expanded computer file are demonstrated by a preliminary report on *Scaevola coriacea* (Hawaii).

While interest in rare plant conservation in the United States had been growing for some time before the 1970s, the degree to which the United States flora is endangered or threatened was still largely unknown when the Endangered Species Act of 1973 directed the Secretary of the Smithsonian Institution to review endangered or threatened plants and methods for their conservation, and report results and recommendations to the Congress.

With the assistance of many of the nation's botanists, the Smithsonian Institution prepared its *Report on Endangered and Threatened Plant Species of the United States* and submitted it to the Congress on January 9, 1975. This *Report* was a beginning which stimulated an avalanche of research, group investigations, conservation work, symposia, and data-gathering activities which must be continued until all facts are known regarding the 3,000 plant taxa presently under consideration as well as any necessary additions to this list.

Historical Background

Due to the *Report's* historical standing, the activities of importance in setting the stage for the actual preparation of the national list will be placed in historical per-

[1] Endangered Flora Project, Department of Botany, Smithsonian Institution, Washington, D.C. 20560; now Office of Biological Conservation, Smithsonian Institution, Washington, D.C. 20560.

Note: This paper was invited for presentation at the Symposium on Geographical Data Organization for Rare Plant Conservation, held at the New York Botanical Garden, 15-17 November, 1977. — The Editors.

Pages 111-122 *in:* Larry E. Morse and Mary Sue Henifin, eds., *Rare Plant Conservation: Geographical Data Organization*. The New York Botanical Garden, Bronx, N.Y.

spective. More recent activities and planned work will then be discussed.

1953. The Pacific Science Association initiated a project to produce a documented list of rare and endangered plants of the Pacific Basin, including Hawaii, under the initiative of Dr. F. Raymond Fosberg, Curator of Botany, Smithsonian Institution.

1970. A conference on Hawaiian endangered species was sponsored by the Pacific Tropical Botanical Garden, in conjunctioin with the Smithsonian Institution. For this conference, Dr. Fosberg prepared a list of rare or threatened Hawaiian plants from his data cards. This list drew upon his forty years' experience with the botany of the Hawaiian Islands, including twenty-two years working specifically on endangered Hawaiian species. Drs. Fosberg and Derral Herbst provided the Hawaiian list of endangered, threatened and extinct plants for the 1975 *Report.* An expanded version of this list appeared in *Allertonia* (Fosberg and Herbst, 1975).

1970. This year is a landmark for international concern regarding endangered plants, for it saw the first issues of angiosperm *Red Data Book* sheets, including the lost franklinia from Georgia *(Franklinia alatamaha),* the Georgia plume *(Elliottia racemosa),* and several Hawaiian plants. The *Red Data Book* is produced by the International Union for Conservation of Nature and Natural Resources (I.U.C.N.) of which the Threatened Plants Committee is located at the Royal Botanic Gardens, Kew, England. In 1970, the director of the Smithsonian's Ecology Program, Dr. Dale W. Jenkins, was chairman of the Regional Group for North America of the Threatened Plants Committee of the I.U.C.N.

1972. By this time there were committees, often volunteer, in approximately twenty states, such as Florida, Texas, and California (beginning in 1968), which were producing individual state lists of locally endangered plants. Some of the lists were highly documented and used extensively for the Smithsonian *Report,* although almost all used different criteria for what constitutes endangerment.

Also in 1972, the director of the Smithsonian's Ecology Program was working on a list of the endangered plants of the Chesapeake Bay Region, which has since been published (Kologiski *et al.,* 1975). He later expanded his study to include the whole Coastal Plain, and eventually again expanded the coverage to encompass the entire United States. The lists were originally regional in character, and were later reoriented to a national basis.

Thus, for approximately one year before passage of the Endangered Species Act of 1973, Smithsonian personnel were actively engaged in compiling a national list of endangered plants to be submitted to the International Union for Conservation of Nature and Natural Resources for inclusion in the angiosperm *Red Data Book* as a United States contribution.

1973. In March of this year, a conference was held in Washington, D.C. which prepared and adopted a Convention on International Trade in Endangered Species of Wild Fauna and Flora; it was later signed and ratified by the United States. Plant species to be monitored by issuance of export and import permits and documentation are listed in the Appendices to the Convention.

Continuing into late 1973, a national list of United States endangered plants was still under preparation by the director of the Smithsonian's Ecology Program.

1973. The Endangered Species Act of 1973 (Public Law 93-205, enacted December 28, 1973) authorized and directed the Secretary of the Smithsonian Institution "to review (1) species of plants that are or may become endangered or threatened and (2) methods of adequately conserving such species."

A program for undertaking this review was commenced in the National Museum of Natural History under the supervision of Dr. Edward S. Ayensu, Chairman, Department of Botany. At that time Dr. D. W. Jenkins became consultant to the newly established Endangered Flora Project upon his retirement, in order to assist with compliance to the directive in the Act. Other personnel, contributors, and advisors are listed in the pages of the *Report*.

Preparation of the 1975 Report

The lists of plants recommended as endangered, threatened, recently extinct, and commercially exploited, were made by the following process:

1. All species, subspecies and varieties having a very restricted range or rare status were listed from the best available published floras of regions, states, and various local areas, and all available data were compiled on special data cards.

2. These plants were then reviewed in the latest taxonomic monographs and revisions, to verify the correct name, status and geographical range.

3. Plant collections were checked in herbaria, particularly the United States National Herbarium, Smithsonian Institution.

4. Many knowledgeable taxonomists were canvassed and consulted for their opinions on the endangerment status of the taxa in which they specialize and to nominate other candidates for the list when appropriate. Candidates were thus tentatively identified as extinct, endangered, or threatened.

5. The lists were next compared with the approximately twenty individual state lists then extant, for additional information and confirmation at the state level.

6. The results were entered into the Smithsonian computer, and printouts of a preliminary list were made.

7. In September, 1974, a Workshop on endangered plants was held by the Smithsonian Institution under the chairmanship of Dr. Ayensu with staff support from the Office of Endangered Species and International Activities, U.S. Fish and Wildlife Service, Department of the Interior. The purpose of the Workshop was to review the preliminary list and to help make recommendations for conserving the plants. Workshop participants gathered in groups to review and evaluate, as best as possible, all taxa on the preliminary list. A great deal of unpublished information and field observations was freely exchanged.

8. A revised list, based on the Workshop scrutiny, was prepared and again circulated to taxonomic specialists for verification. This revised list was also again sent to many state list compilers for a final evaluation.

9. After making changes based on the comments that came in, a corrected final printout was submitted for computer processing and became the national list for the 1975 *Report*. The data were printed out accordingly to family, taxon name, range by state, and suggested status (endangered, threatened).

The *Report on Endangered and Threatened Plant Species of the United States* was transmitted to the United States Congress on January 9, 1975, and was subsequently published by the U.S. Government Printing Office (Smithsonian Institution, 1975).

The Director of the U.S. Fish and Wildlife Service accepted the Report as a petition within the context of the Endangered Species Act, and published on July 1, 1975, a notice of review of the status of the plants recommended in the *Report*. On

June 16, 1976, the U.S. Fish and Wildflife Service proposed endangered status for over 1,700 of the recommended plants. On August 11, 1977, after extensive review of their status, the Director of the U.S. Fish and Wildlife Service officially listed four California plants as endangered species. On April 26, 1978, 13 more plants were listed. The aforementioned steps indicate the Interior Department's intent to pursue a course of action that will lead to official determination of the status of endangered and threatened plants listed in the *Report*.

Preparation of the 1978 Revised Lists

The 1975 *Report* precipitated the arrival of a wealth of additional information to the Endangered Flora Project, resulting in significant revision, up-dating, and further documentation of the lists. The July, 1975, Notice of Review in the *Federal Register* similarly brought further information to the Fish and Wildlife Service, which has been available for reference by the Smithsonian group and others. Often research and mapping of species uncovered questions regarding their taxonomy, distribution, or rarity, and recognized authorities on the taxa were consulted. Additional contacts were made with interested botanists and state rare plant committees throughout the country. Information continued to come from the original contributors to the lists.

The welcome influx of newly published and unpublished data and expert opinion has helped the assessment of the status of our endangered, threatened, exploited, and extinct flora. A few species thought to be extinct have been rediscovered, some formerly thought to be rare have been found to be more abundant or found taxonomically untenable, and a number of previously overlooked species have been commended to our attention.

Revised national lists of endangered, threatened, extinct, and commercially exploited plant taxa are now available in the book, *Endangered and Threatened Plants of the United States* (Ayensu and DeFilipps, 1978). A treatment of Puerto Rico and the Virgin Islands is newly included, and the Hawaiian list is augmented by indicating the individual islands on which the taxa occur.

The Endangered Flora Project

The Project continues to maintain and up-date the extensive files of information which are the basis of its published national lists. The files include data cards (Figure 1), correspondence, reports compiled by the project (Appendix I), and folders of data on the individual taxa. As data is received and located it is added to the files, which are the source for the Project's distribution of information to legitimate enquiries from the general public, botanists, government agencies, conservation groups, and ecological and environmental consulting firms of all descriptions and interests.

The Project has recommended that a central registry be established to maintain, computerize and distribute information on endangered and threatened plants. This registry and coordination should include central card files and maps, a specialized library, use of a computer, and a small staff of experts. The register, as a data base, would require continual updating of information on the location, status, habitat re-

ENDANGERED FLORA PROJECT--SMITHSONIAN INSTITUTION

NAME Prunus gravesii Small	FAMILY Rosaceae

| ORIGINAL LOCALITY CONNECTICUT: New London Co. | TYPE OF HABITAT low shrubby thickets on gravelly sand ridges near Long Island Sound |

MAXIMUM KNOWN RANGE
   CONNECTICUT: New London Co.: Groton: Esker Pt. Park, Noank

PROBABLE PRESENT RANGE

WHERE AND WHEN LAST SEEN OR COLLECTED AND BY WHOM

ENDANGERED

Place of Publication: Bull. Torrey Bot. Club 24:45, pl. 292 (1897).

PROBABLE PRESENT STATUS (Underline)   INCREASING   DECREASING   STATIONARY   RARE
                    VERY RARE   UNCERTAIN   PROBABLY EXTINCT

REMARKS (Use Back If Necessary) One of rarest plants in Conn.   Immediately endangered. 15 individuals in 50 sq. meters. Moderately disturbed site. Adjacent to heavily used recreation area. Individuals healthy. No seedlings. Limited vegetative reproduction.

REPORT BY:
        Status Report by J.J. Dowhan (see Bampton letter 8-75)

SI-3058
11-22-74

Fig. 1. Example of an Endangered Flora Project data card.

quirements, reproductive behavior, population size, and commercial and private exploitation of endangered and threatened plant species, and threats to habitat.

Data Computerization Activities

A computer program to map the locality of the habitats of endangered and threatened species, by latitudinal and longitudinal coordinates, has been initiated. Distribution mapping will show eventually which species occur in areas such as national and state parks and forests, preserves and natural areas, and will also indicate which species need better protection. Mapping will also permit the designation of centers of endemism and aggregations of botanically diverse species growing in the same place, and will thus assist in conserving *critical habitats* for their protection.

For a pilot set of species, the commercially and privately exploited ones, latitudes and longitudes (coordinates) of the known localities have been prepared for the computer. Computer-drawn maps of two species produced from those data are shown (Figure 2). The locations of exploited species of the southeastern states have been similarly mapped by computer, in order to help identify the areas where different species are found aggregated together in the same habitat. The resultant map is shown as an example of what could be similarly prepared for the entire country (Figure 3). The species to which the symbols refer on the map of southeastern exploited plants are listed in the legend for Fig. 3.

The Smithsonian computer file on endangered, threatened, and extinct plants currently includes the name of the taxon, its family, the states in which it occurs, and

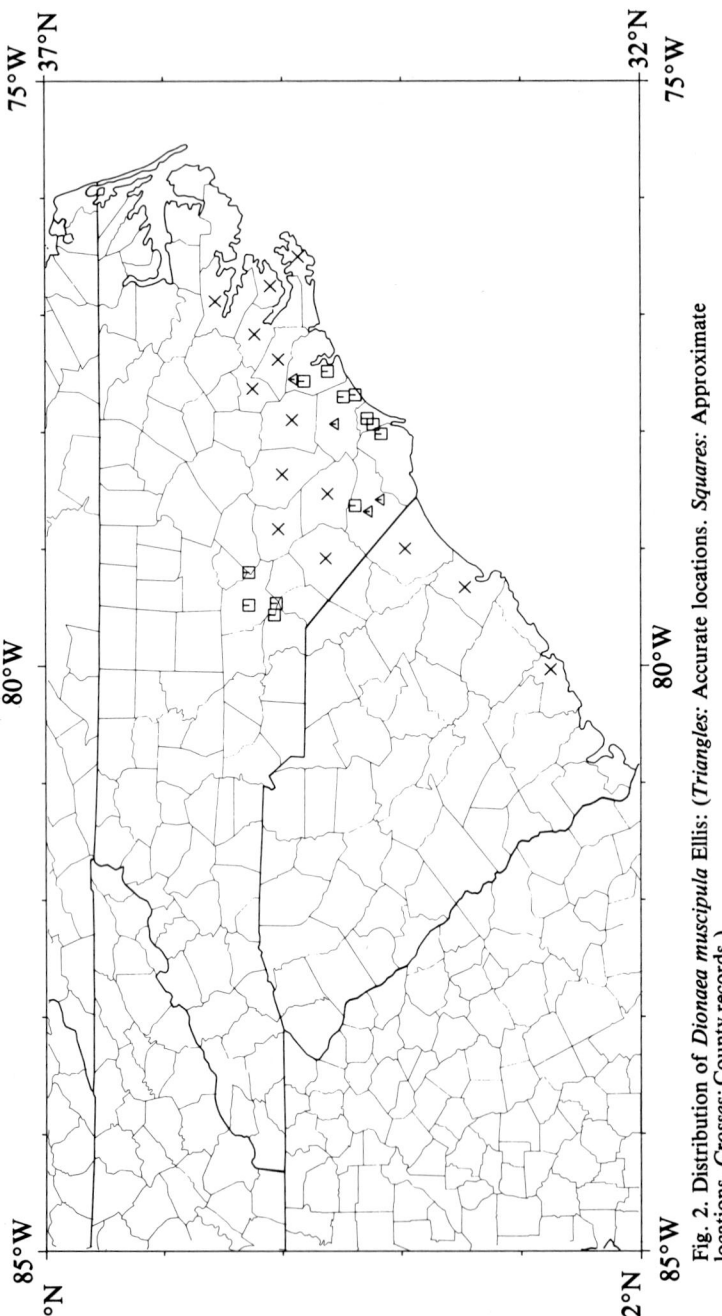

Fig. 2. Distribution of *Dionaea muscipula* Ellis: (*Triangles*: Accurate locations. *Squares*: Approximate locations. *Crosses*: County records.)

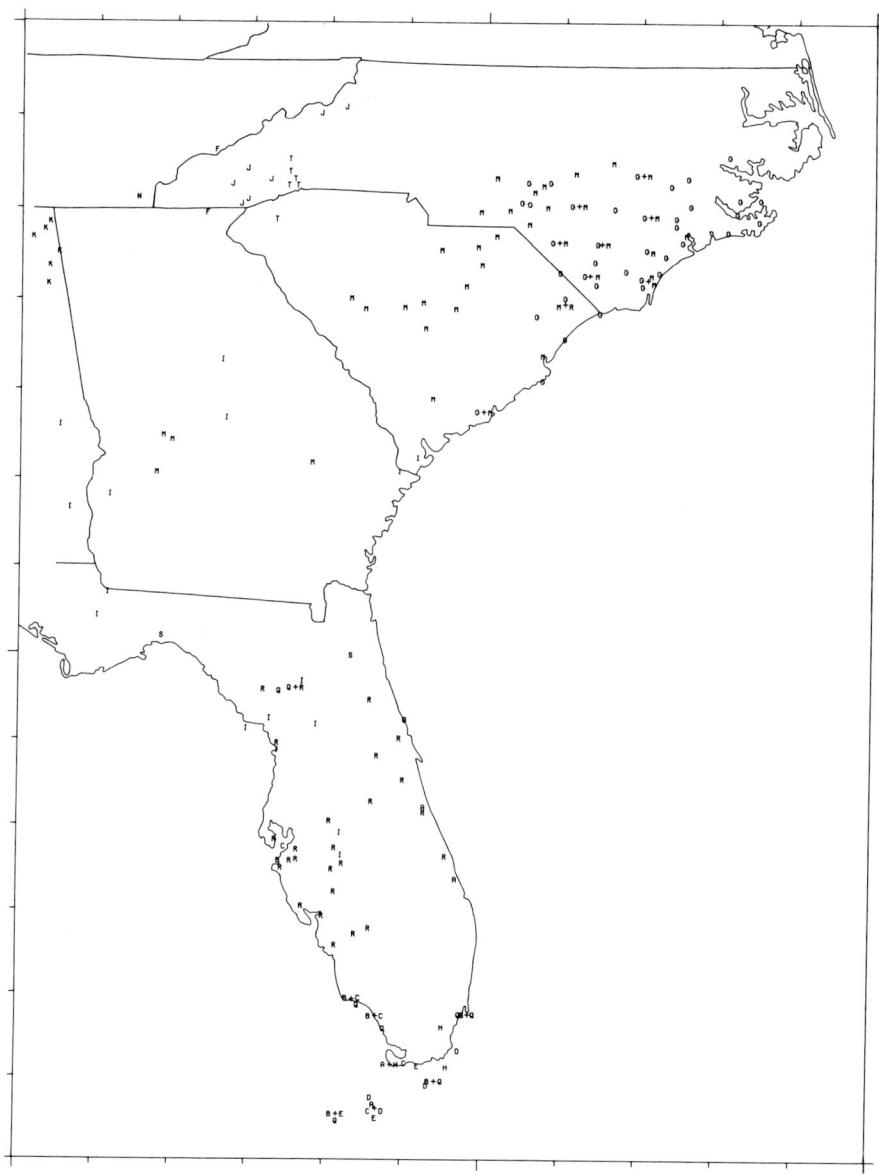

Fig. 3. Locations of exploited species in the Southeastern United States.*

*A *Cereus eriophorus* var. *fragrans*
 B *Cereus gracilis* var. *aboriginum*
 C *Cereus gracilis* var. *simpsonii*
 D *Cereus robinii* var. *deeringii*
 E *Cereus robinii* var. *robinii*
 F *Cladrastis lutea*
 G *Dionaea muscipula*
 H *Encyclia boothiana* var. *erythronioides*
 I *Rhapidophyllum hystrix*

J *Rhododendron vaseyi*
 K *Sarracenia oreophila*
 M *Sarracenia rubra*
 N *Sedum nevii*
 Q *Zamia integrifolia*
 R *Zephyranthes simpsonii*
 S *Zephyranthes treatiae*
 T *Sarracenia jonesii*

117

its status designation (endangered, threatened, extinct). All files are kept on tape in SELGEM master file format. Searches can be made on the basis of every combination of the stored data. Several kinds of standard listings can be produced, or customized reports can be provided with an existing generalized program. Data can be permuted and summarized in a variety of ways. Files can be interfaced with other systems, mapping programs, statistical packages, and so forth. We intend to computerize the *counties* in which the taxa occur next, and are also working on an expanded computer file.

To facilitate the storage of detailed information on the taxonomy, localities, population levels, habitats, threats, and reference sources for each species, and to be easily able to retrieve that information, an expanded computer file is being developed using the Smithsonian's versatile SELGEM system (Creighton *et al.,* 1972).

The file will be designed to encompass pertinent information required by botanists, government agencies, and conservation groups. This system, which is still in the planning stages, will thus be capable of printing complete data reports for each species, or listing each species whose characteristics answer to a given question asked of the computer.

Appendix II defines the data categories. A preliminary report for *Scaevola coriacea* from the Master List Program (SELLST) of SELGEM is shown in Appendix III to illustrate the extent of the file envisioned. The data elements in the new file are not yet arranged in final form, and thus do not necessarily reflect the retrievability of each element.

The types of information to be included in the various categories, which are still subject to modification, are presented as Appendix II.

Literature Cited

Ayensu, E. S., and R. A. DeFilipps. 1978. *Endangered and Threatened Plants of the United States.* Smithsonian Institution and World Wildlife Fund, Inc., Washington, D.C.

Creighton, R., P. Packard, and H. Linn. 1972. SELGEM retrieval: A general description. *Procedures Computer Sci.* 1(1). Information Systems Division, Smithsonian Institution.

Fosberg, F. R., and D. Herbst. 1975. Rare and endangered species of Hawaiian vascular plants. *Allertonia* 1(1): 1-72.

Hunt, C. B. 1967. *Physiography of the United States.* Freeman, San Francisco.

Kologiski, R. L., F. R. Hivick, C. W. Reed, and D. W. Jenkins. 1975. Rare, endangered, and endemic plants of the Chesapeake Bay region. Appendix D (pages D1-D49), *in Compendium of Natural Features Information, Volume 1.* Maryland Department of State Planning and Smithsonian Institution Center for Natural Areas.

Melville, R. 1970. *Red Data Book. Vol. 5. Angiosperms.* International Union for Conservation of Nature and Natural Resources, Morges, Switzerland.

Smithsonian Institution. 1975. *Report on Endangered and Threatened Plant Species of the United States.* House Document 94-51, (Serial Number 94-A). U.S. Government Printing Office, Washington, D.C.

U.S. Fish and Wildlife Service. 1975. *Federal Register* 40(127): 27825-27924. July 1.

U.S. Fish and Wildlife Service. 1976. *Federal Register* 41(117): 24524-24572. June 16.

U.S. Fish and Wildlife Service. 1977. *Federal Register* 42(155): 40682-40685. August 11.

U.S. Fish and Wildlife Service. 1978. *Federal Register* 43(81): 17910-17916. April 26.

Prunus gravesii Small (Rosaceae): Graves' Beach Plum

STATUS Vulnerable; it has been found only at the type locality, Groton, Connecticut. (2). This population consists of approximately 15 individuals encompassing about 50 m. on a moderately disturbed site criss-crossed by footpaths and adjacent to a heavily-used town recreational beach area. (1).

DISTRIBUTION U.S.A.; known only from Esker Point, New London County, Connecticut. (1).

HABITAT AND ECOLOGY It is restricted to coastal shrub thickets in well-drained gravelly sands bordering Long Island Sound, in association with *Lonicera japonica, Prunus serotina, Corylus americana,* and *Rubus allegheniensis.*

CONSERVATION MEASURES TAKEN In 1975 a snow fence was erected enclosing the entire stand as well as a large buffer zone around the perimeter. (1).

CONSERVATION MEASURES PROPOSED Further management recommendations must await additional information on the ecological requirements of this species. (1).

BIOLOGY AND POTENTIAL VALUE It may be of hybrid origin involving *Prunus maritima* as one parent. (2).

CULTIVATION It is in cultivation at the Morton Arboretum and Arnold Arboretum, according to the American Horticultural Society's Plant Sciences Data Center.

DESCRIPTION Unarmed shrub up to 2 m.; stems erect or ascending, much-branched with a dark, rough bark. Leaves 2-4 mm., orbicular or ovate-orbicular, serrate, sparingly pubescent or glabrate above, more pubescent beneath, especially on the veins. Stipules linear, glandular-pubescent. Pedicels 6-10 mm. long, stiff, stout, pubescent. Flowers white, 1-3 cm. in diameter, solitary or 2-3 together, scattered on the twigs near the top of the shrub. Calyx campanulate, the segments oblong, as long as the tube, pubescent. Petals suborbicular, about 5 mm. in diameter, abruptly narrowed at base. Drupe 10-15 mm. in diameter, globose, solitary, purplish-black, glaucous; stone broadly oval, acute at base, rounded at apex. Flowering late May to early June. Fruiting early September. (4).

For illustration see (3).

REFERENCES 1. Bampton, T. (1975). Personal communication to the U.S. Fish and Wildlife Service, August 13.

2. Dowhan, J. J. and Robert J. Craig. (1976). Rare and Endangered Species of Connecticut and Their Habitats. State Geological and Natural History Survey of Connecticut. The Natural Resources Center, Dept. of Environmental Protection. Report of Investigation No. 6, p. 74.

3. Small, J. K. (1897). An Apparently Undescribed Species of *Prunus* from Connecticut. Bull. Torrey Botanical Club 24: 44-5, pl. 292.

4. Wight, W. F. (1915). Native American Species of *Prunus.* USDA Bull. No. 179. Washington, D.C.

Appendix II. Provisional list of data categories for computer data file of Smithsonian Endangered Flora Project.

FAMILY.	
NAME.	Genus and species, subspecies or variety
AUTHOR, PLACE OF PUBLICATION.	Name of author, bibliographic citation
SYNONYMS.	Basionym and important synonyms, with author and place of publication
COMMON NAME.	
TYPE.	Locality, collector, date of collection
GEOGRAPHICAL SUMMARY.	Brief overview of range: type locality/collection only; name of state; series of states; "eastern United States and Canada;" including countries other than U.S.
PHYSIOGRAPHIC PROVINCE.	As described in *Physiography of the United States* (Hunt, 1967, pp. 8-9, fig. 1.1)
NATURAL REGION	A region characterized by a geological feature or natural condition of the landscape, e.g. Grand Canyon, Smoky Mountains, San Joaquin Valley, Driftless Area
ASSOCIATION.	Plant associations, e.g. spruce-fir forest, pinyon-juniper

ASSOCIATES.	Dominants and indicator plants of the community in which it occurs
HABITAT.	Key words, e.g. calcareous meadows, serpentine soils, littoral zone
ELEVATION.	Altitudinal range expressed in both feet and meters
S. I. STATUS.	Smithsonian Institution recommended status of endangered, threatened, or extinct, in 1975 and 1978 publications; record any changes of status and removals from listings
DATE LAST RECORDED.	Year of last report if extinct or not recently collected
EXPLOITATION.	Note if commercially or privately exploited; indicate facts on exploitation and source
LEGAL STATUS.	Notice of Review, *Federal Register* reference
	Proposed Rulemaking, *Federal Register* reference
	Final Rulemaking, *Federal Register* reference
	Proposed and Final Critical Habitat, *Federal Register* reference
	Convention on International Trade Appendices
OTHER LISTS.	Status and authority, e.g. state lists, Red Data Book
POPULATION DATA.	Notes on abundance, stability, rate of decline, status, or history with source
THREATS.	Natural and human causes of past or potential destruction or depletion: natural rarity, disease, succession, exploitation, strip-mining, clear-cutting, pollution, housing developments, dams
NATURAL PRESERVES.	Occurrence on lands such as national and state forests, parks, private refuges, or military reservations
CULTIVATION.	If in cultivation according to American Horticultural Society Data Bank; names of arboreta, botanical gardens, nurseries, private individuals
BIOLOGY.	Note if the species is a tree, succulent, insectivorous, parasite; specific pollinator, breeding system
POTENTIAL VALUE.	Note if the species is a mineral indicator, source of organic compound, ornamental, agricultural or weedy cogener; scientific, horticultural or ecological value
REFERENCES.	Descriptions, illustrations, treatments, letters, verbal communications with references, literature
DATE OF REPORT.	Date the sheet is printed to facilitate keeping sets up to date as subsequent sheets are printed
RANGE.	Final retrievability and hierarchy of the elements in the *RANGE* category to be determined
COUNTRY.	
STATE.	
COUNTY.	
ISLAND.	Pertaining to Hawaii, Puerto Rico and the Virgin Islands
DISTRICT.	Pertaining to Hawaii, Puerto Rico and the Virgin Islands
SITE.	Relation to towns, roads, and natural and political reference points
COORDINATES.	Latitude × Longitude expressed to the nearest minute

Appendix III

DATE: 04/18/77

```
                                    ....2....V....3....V....4....V....5....V....6....V....7....V....
```

SERIAL	CAT-DEFINITION	CATEG	LINE	
00001300	FAMILY:	010	01	GOODENIACEAE
	NAME:	020	01	SCAEVOLA CORIACEA
		021	01	
	AUTHOR, PLACE OF PUBLICATION:	030	01	NUTTALL/ 1843/ TRANS. AM. PHIL. SOC./ N. S. 8: 253.
		031	01	
	COMMON NAME:	050	01	FALSE JADETREE, NAUPAKA
		051	01	
	TYPE:	060	01	HAWAII: KAUAI, NEAR THE SEA/ NELSON/ COOK EXPEDITION
		061	01	
	GEOGRAPHICAL SUMMARY:	070	01	HAWAII: AT ONE TIME PROBABLY ON ALL OF LARGER ISLANDS, NOW
		070	02	EXTANT ONLY AT WAIHEE, MAUI.
	HABITAT:	071	01	DRY, CONSOLIDATED, LITHIFIED, COASTAL SAND, ARID LOWLANDS, LAVA
	ELEVATION:	200	01	NEAR SEA LEVEL
	S. I. STATUS:	201	01	
		210	01	
	LEGAL STATUS:	211	01	ENDANGERED
		220	01	
	OTHER LISTS:	221	01	NOTICE OF REVIEW, FEDERAL REGISTER 40: 27904. 1 JUL 1975.
		250	01	VL, VR, EN, C - FOSBERG & HERBST/ 1975/ ALLERTONIA/ 1(1): 38.
		251	01	
	POPULATION DATA:	260	01	100 PLANTS AT WAIHEE, MAUI, 1968 (LAMOUREUX).
		261	01	LESS THAN 500 WILD PLANTS IN ONLY ONE AREA.
		270	01	
	THREATS:	271	01	AREA GRAZED BY CATTLE, SUBJECT TO DEVELOPMENT IN NEAR FUTURE.
		280	01	
	BIOLOGY:	281	01	TRAILING SHRUB
		310	01	
	REFERENCES:	311	01	
		330	01	CARLQUIST, S./ 1970/ HAWAII: NATURAL HISTORY/ 155. (ILLUS.)
			02	CARLQUIST, S./ 1 APR 1970/ PERSONAL COMMUNICATION.
			03	DEGENER, O./ 1950/ FL. HAWAIIENSIS. (DESCRIP., ILLUS.)
			04	FOSBERG, F. R./ DATA CARDS.
			05	HILLEBRAND, W./ 1888/ FL. HAWAIIAN ISLANDS/ 266. (DESCRIP.)
			06	LAMOUREUX, C. H./ 28 APR 1970/ PERSONAL COMMUNICATION
			07	LAMOUREUX, C. H./ SEP 1975/ PERSONAL COMMUNICATION.
			08	ST. JOHN, H./ 1973/ LIST FL. PL. HAWAII/ 346.
	DATE OF REPORT:	331	01	
		340	01	DEC 1976
00001301	COUNTRY:	410	01	U. S. A.
	STATE:	420	01	HAWAII
	COUNTY:	430	01	HAWAII
	ISLAND:	440	01	HAWAII
00001302	COUNTRY:	410	01	U. S. A.
	STATE:	420	01	HAWAII
	COUNTY:	430	01	MAUI
	ISLAND:	440	01	MAUI
00001303	SITE:	460	01	"ISTHMUS" (DEGENER)

```
...2....V....3....V....4....V....5....V....6....V....7....V....

SERIAL    CATEG  LINE   CAT-DEFINITION

00001304    410   01    COUNTRY:          U. S. A.
            420   01    STATE:            HAWAII
            430   01    COUNTY:           MAUI
            440   01    ISLAND:           MAUI
            450   01    DISTRICT:         WAILUKU

00001305    460   01    SITE:             WAIHEE, WAIHEE GOLF COURSE
            470   01    COORDINATES:      20 55 30 N X 156 31 00 W

00001306    460   01    SITE:             KALEPOLEPO
            470   01    COORDINATES:      20 45 00 N X 156 27 30 W

00001307    460   01    SITE:             WAILUKU TO WAIHEE PT.

00001308    410   01    COUNTRY:          U. S. A.
            420   01    STATE:            HAWAII
            430   01    COUNTY:           MAUI
            440   01    ISLAND:           LANAI
            450   01    DISTRICT:         LANAI

00001309    460   01    SITE:             KEONOHAU

00001310    410   01    COUNTRY:          U. S. A.
            420   01    STATE:            HAWAII
            430   01    COUNTY:           MAUI
            440   01    ISLAND:           LANAI

00001311    460   01    SITE:             LAE WAHIA, MAHANA

00001312    410   01    COUNTRY:          U. S. A.
            420   01    STATE:            HAWAII
            430   01    COUNTY:           HONOLULU
            440   01    ISLAND:           OAHU
            450   01    DISTRICT:         WAIANAE

00001313    460   01    SITE:             KAENA PT.
            470   01    COORDINATES:      21 34 15 N X 158 16 45 W

00001314    410   01    COUNTRY:          U. S. A.
            420   01    STATE:            HAWAII
            430   01    COUNTY:           HONOLULU
            440   01    ISLAND:           OAHU
            450   01    DISTRICT:         EWA

00001315    460   01    SITE:             BARBER'S PT.
            470   01    COORDINATES:      21 17 45 N X 158 06 30 W

00001316    410   01    COUNTRY:          U. S. A.
            420   01    STATE:            HAWAII
            430   01    COUNTY:           KAUAI
            440   01    ISLAND:           KAUAI

00001317    410   01    COUNTRY:          U. S. A.
```

122

Regional Coordination of Rare Plant Information Synthesis by the New England Botanical Club

William D. Countryman,[1] Joseph J. Dowhan,[2] and Larry E. Morse[3]

The Endangered Species Committee of the New England Botanical Club has set as its current goal the preparation and publication of coordinated lists of the rare, threatened, and endangered indigenous vascular plant species of the six New England states. Inclusion of species in the various lists is to be based on uniform, explicitly defined criteria, and the lists are to be presented in a form useful to conservation agencies, legislators, and regional planners, as well as to experienced botanists. Emphasis is being placed on information concerning the current field status of these taxa, although historical information from herbarium specimens and other sources is also included. Synthesis of the six state reports into a publication on regionally endangered species is planned.

New England consists of six small but intensely developed states: Connecticut, Maine, Massachusetts, New Hampshire, Rhode Island, and Vermont. Due to its repeated glaciation in the Pleistocene, the region has few endemic species (Fernald, 1929; Schafer and Hartshorn, 1965; Jorgensen, 1971). However, the climatic and topographic diversity of New England and the rapid development of the present postglacial vegetation have resulted in many interesting patterns of range limitations and disjunctions, and many peculiar species assemblages.

Much of the New England landscape has now been developed for agriculture or forestry, and an increasing proportion of the land is presently being urbanized or devoted to rights-of-way for highways and powerlines. Nearly all the New England coastline is now committed to industrial uses or to private or public recreation; only a handful of significant natural areas remain to conserve the coastal vegetation. Rec-

[1] Dept. Biology, Norwich Univ., Northfield, Vt. 05663.
[2] Connecticut Geological and Natural History Survey, Storrs, Conn.; present address: U.S. Fish and Wildlife Service, 2800 Cottage Way, Sacramento, Calif. 95821.
[3] Cooperative Parks Study Unit, The New York Botanical Garden, Bronx, N.Y. 10458, and Gray Herbarium, Harvard University, Cambridge, Mass. 02138; present address: The Nature Conservancy, 1800 N. Kent St., Arlington, Va. 22209.

Based primarily on the presentation "Rare Plant Data Priorities of the New England Botanical Club" by Drs. Countryman and Dowhan at the Symposium on Geographical Data Organization for Rare Plant Conservation, held 15-17 November 1977 at The New York Botanical Garden.
Pages 123-131 in: Larry E. Morse and Mary Sue Henifin, eds., Rare Plant Conservation: Geographical Data Organization, The New York Botanical Garden, Bronx, N.Y.

reation is also causing significant impacts to the vegetation of New England's high mountains such as the White Mountains in New Hampshire or the Green Mountains in Vermont. Other factors, such as suppression of historical fire patterns (*cf.* Ross, 1978), have also contributed to changes in species abundance and distribution in New England. Even the vast, virtually uninhabited "North Woods" of Maine is not immune to major disturbances, as shown by the current controversy over the proposed Dickey-Lincoln School Lakes hydroelectric project, which includes in its area most of the known populations of the Furbish lousewort, *Pedicularis furbishiae,* a federally protected endangered species and one of the few species endemic to the New England area (Macior, 1978).

The escalating pressures from the consumptive use of natural resources in the remaining wild areas of New England, along with the increased demand for ecologically sound planning and development, make the determination of plant conservation priorities for the region an urgent task. We hope the work reported here will make a sizeable contribution to this effort by providing state and regional lists of plant taxa regarded as rare, threatened, endangered, or unique, with these decisions documented by specific criteria on distribution, abundance, threat, and vulnerability.

Background for Current Work

The vascular flora of New England is generally quite well known, although recent comprehensive reviews by the Club's Committee on Vascular Plant Distributions have confirmed that some areas of the region are significantly underexplored botanically (Morse *et al.,* this volume). Seymour (1969) has published a regional manual that includes many distributions to the county level, and state checklists with county distributions have been published recently for Maine (Bean, Richards, and Hyland, 1966) and Vermont (Vermont Botanical and Bird Club, 1973). Literally hundreds of county, town, and local floras and checklists exist, some quite recent and others over a century old—258 New England floristics works before 1900 are listed by Day (1899). Many herbaria also contain important documentation of the New England flora because of their substantial numbers of New England specimens; the largest regional collection is that of the New England Botanical Club (Deane, 1899), which is especially rich in collections predating 1920.

The New England Botanical Club (NEBC) was founded in 1895 by several botanists living in the Boston and Cambridge area. Then, as now, the Club was composed of both professionals and amateurs, all interested in the plants of New England. Most of the Club's activities are carried out on a volunteer basis by its members, and for more than 75 years these efforts have been channeled into two major areas, the Club's journal *(Rhodora)* and its herbarium. From its inception, the NEBC has always maintained a close relationship with Harvard University, and its herbarium is housed in a special section of the Harvard University Herbaria. Only specimens of plants from the six New England states are included in this collection, which now numbers over 400,000 sheets.

Since most active botanists in New England are members, and since the Club's library and herbarium in Cambridge are essential to the serious study of rare New England plants, the New England Botanical Club is an appropriate organization to coordinate the production of lists of rare and endangered plants of the region. The Club has a strong tradition of interest in documenting distributions and recording changes in distribution patterns; a precursor to the present project was the work of the "COCERNEP" Committee, the NEBC's Committee on Change in Environment and Range of New England Plants, active in the 1960's.

The Club's present Endangered Species Committee draws many of its ideas and several of its members from the Club's earlier Natural Areas Criteria Committee, which was formed in 1971 to prepare a report on criteria for identification and evaluation of natural areas, at the request of the New England Natural Resources Center.

The criteria developed by this committee to evaluate New England natural areas were:

1. Areas with plant communities having several *rare or endangered* species merit greater consideration than those with fewer such species.

2. Other things being equal, areas containing features fitting two or more of the listed physiographic categories (e.g., alpine areas, cliffs and ledges, bogs, etc.) merit greater consideration than areas having one such feature.

3. In general, areas in later successional stages of vegetation, such as virgin forests or pre- and post-climax stands, merit greater consideration than areas in earlier stages of succession. Consideration, however, should be given in each New England state to the preservation of typical examples of all significant successional stages. In this connection it may be necessary to recommend such management practices as burning, thinning, removal of competing species, etc., in order to maintain intermediate seral stages and thus preserve certain desirable species or plant communities.

4. The preservation of larger areas should take precedence over smaller areas of a similar nature.

5. Areas which have been studied scientifically and documented should receive special considerations.

6. Areas containing rare and/or endangered plant species that are found in close proximity to a center of human interest, such as an educational institution, should be given priority over areas of otherwise similar value at more distant places.

7. Except in special instances recommended by competent botanists, the highest value should be accorded those areas which have been subjected to the least disturbance by man.

In addition to the preparation of this report (Countryman *et al.,* 1972), the committee recognized the need for well-documented state rare and endangered species lists, well before passage of the U.S. Endangered Species Act of 1973 required the Smithsonian Institution to prepare a national list of candidate endangered and threatened plant species. The Natural Areas Criteria Committee also began preparing such a list for each of the six New England states for publication in the Club's journal, *Rhodora.* However, none of these were published, and only one list was completed, George Church's for Rhode Island. The partially completed New Hampshire list by the late Albion Hodgdon was later published in abbreviated form (Hodgdon, 1973). This committee also developed criteria for recognizing different categories of rare and endangered vascular plants; these currently being used by the various state projects, and provided a basis for the recently developed revisions presented in Appendix I.

Following completion of the Natural Areas report, several members of the Club maintained interest in Club participation in environmental impact assessment, natural areas evaluation, and endangered species research, and recognized the continuing and urgent need for a comprehensive rare and endangered plant species list for the New England region as a whole. Particular stimuli to organization of the current work were the Massachusetts research initiated in 1976 by Katharine G. Field and Jonathan Coddington, the New Hampshire studies begun about the same time by Irene M. Storks under the supervision of Garrett E. Crow, the developing series of

Rare and Endangered Plants of

SPECIES	STATUS	HABITAT	# OF KNOWN STATIONS						TOTAL	MODE OF OCCURRENCE IN N.E.
			Me.	N.H.	Vt.	Ma.	Ct.	R.I.		
ROSACEAE										
Prunus gravesii Sm.	U.S. Endangered (Proposed)	Gravelly ridge near shore of L.I. Sound					1		1	Rare and Local
EBENACEAE										
Diospyros virginiana L.	N.E. Endangered	Sandy clearing near shore of L.I. Sound					1		1	Local
							[1]		[1]	
						1*				
GRAMINEAE										
Calamagrostis nubila Louis-Marie	Presumed extinct	Shore of mountain lake	[1]						[1]	Rare and Local
ORCHIDACEAE										
Arethusa bulbosa L.	Declining	Acid sphagnous bogs	11	4	7	13	6	2	43	Local
			(6	7	5	4	3)	—	(25)	
			[5	9	4	2	2	1]	[23]	
CYPERACEAE										
Eleocharis equisetoides (Ell.) Torr.	Vulnerable	Lake shores					1	1	2	Local
							(1)	(1)	(2)	

Note: Some of the data used in this example were estimated here for the sole purpose of illustration and do not represent actual values.

Figure 1. Example of a regional summary compiled from state reports.

status reports on Maine species being prepared by Leslie M. Eastman of the Maine Audubon Society, and the publication of Dowhan and Craig's (1976) report on rare and endangered Connecticut species. Considered together with Church's earlier work on rare plants in Rhode Island, Hodgdon's in New Hampshire, and Countryman's in Vermont, it became evident that a New England synthesis was feasible. The desire for coordination, encouragement, and review of these various state projects was the major reason for the Club's organization of the Committee on Rare and Endangered Species early in 1977, chaired by Dr. Countryman.* Also significant in

* Membership in the committee has recently included George Church, Jonathan Coddington, William D. Countryman, Garrett E. Crow, Joseph J. Dowhan, Leslie M. Eastman, Katharine G. Field, Martha N. Fisher, Mary Sue Henifin, Les Mehrhoff, Larry E. Morse, and Irene M. Storks.
    The authors acknowledge their thanks to the various other members for discussions in which ideas presented here were developed, and also thank Kate Field, Martha Fisher, and Mary Sue Henifin, as well as Barbara Ertter and Jane Lawyer, for extensive comments on earlier drafts of this manuscript.

New England   Summary Sheet

OVERALL POPULATION TREND	GENERAL DISTRIBUTION (RANGE)	PRINCIPAL REASONS FOR RARITY	COMMENTS
Stable	Restricted — State Endemic — Ct. only	Extremely restricted range	Possibly of hybrid origin. (*P. maritima* × ?) Known only from type locality. Total population size: 14 shrub individuals (1977).
Approaching regional extirpation	Common in SE U.S., rare in N.E.	Peripheral (NLR); habitat destruction	Native stand of 140 trees (1820) in New Haven now reduced to 5 individuals (1977). Hurricane of '38 and construction are major factors in decline. *Mass. station not native.
Unknown (Extinct?)	Restricted — State Endemic — N.H. only	Extremely restricted range. Causes of extinction unknown.	Last collected in 1862 by Wm. Boott. Taxonomic status uncertain. Known only from Lake of the Clouds, Mt. Wash.
Long-term decline	Common outside N.E. (Canada), becoming rare in N.E.	Habitat destruction; peripheral (SLR)	Very few recent (post-1930) records for southern N.E. Many local extirpations throughout N.E. region.
Unknown	Local north of Florida	Peripheral (NLR)	Local and disjunct throughout large parts of its range; common in Gulf States.

catalyzing the current New England synthesis was the interest and encouragement of Richard W. Dyer, Endangered Species Botanist, U.S. Fish and Wildlife Service.

General Guidelines for State Projects

At its first meeting, the committee agreed that listings of taxa as rare or endangered should be based on consideration of at least the following information: the scientific name of the taxon; its usual or preferred habitat; a list of vouchered county or town distribution records for the state; the number of known stations (towns or similar-sized areas) in the state; the explicit reasons for listing; and any other pertinent knowledge. It was felt that generally all these data should be published, although discretion should be used where appropriate. Additionally, the Committee encouraged the preparers of state lists to keep on file any pertinent additional information, such as detailed locality and habitat records, dates of collection, bibliographic references, and any other information substantiating the reasons for inclusion of a taxon in their lists. State projects were also encouraged to take note of taxa declining in abundance but not presently meriting listing as rare and endangered species.

The committee defined its role as coordinator and reviewer of the several independent state projects, and left considerable discretion and independence to the compilers of the six state lists in the details of their work. For example, the feasibility of assessing the status of varieties and subspecies was left to each state project to determine. Similarly, state projects were encouraged to expedite their work, where necessary, by postponing consideration of poorly understood or inadequately collected taxonomic groups such as the complex genera *Rubus* and *Crataegus*.

The issue of site locality documentation was discussed at length. For a great number of well-recognized reasons, it was readily agreed that exact locality data for rare plants should not be routinely published or publicized. On the other hand, it was recognized that presentation of distribution information to the county level, whenever possible, greatly enhances the usefulness of the lists. The committee recommended that state projects use the greatest discretion in responding to inquiries for detailed distribution or habitat information, and also stated its willingness to provide state projects its advice on such matters on request.

Refining the Criteria

On the topic of specific criteria or reasons for listing of species, the committee developed a revision of the criteria from the 1972 Natural Areas report, presented here as Appendix I. These revised criteria are being considered for use in synthesis of information from state rare plant projects in New England, in order to make explicit the reasons for listing of taxa and to promote comparability of information in state projects in the New England region. Review of criteria used by several other groups was essential to revision of the NEBC guidelines; among the methods consulted were those of the California Native Plant Society (Powell, Duncan and Howard, this volume), South Africa's Rare and Endangered Species Survey (Hall, this volume), the Sensitive Wildlife Information System (Rekas, 1976), and the guidelines for status reports developed by Henifin *et al.* (this volume).

These NEBC criteria do not themselves define the familiar terms such as "endangered," "threatened," or "vulnerable." Instead, the criteria provide a way of organizing the information being used to set priorities and categorize the various taxa in a state. Evidence that might merit the status of "endangered" in one state may only indicate the status of "rare and vulnerable" in another, yet to be comparable this evidence should be gathered and organized in a similar way in each state.

When the same criteria have been used in several states, then the state-by-state information they contain can be readily synthesized to produce information on regional significance. A sample page illustrating this method of information synthesis is provided as Figure 1.

Project Coordination

Currently, the committee is coordinating the development and revision of the six New England lists, with much of the state-level work now supported by the U.S. Fish and Wildlife Service. The general work plan being followed in each state is to: (1) review and revise extant lists of threatened and endangered plants for the state;

(2) field-check status where both necessary and feasible to provide current information on selected species of high interest; and (3) develop status reports for each plant taxon in the state currently under review for threatened or endangered species listing by the U.S. Fish and Wildlife Service.

Coordination of this work by the NEBC committee is particularly important in avoiding duplication of effort on taxa occurring in more than one state, and to ensure that reports developed in the various states can be synthesized readily into a regional overview. The committee members are also reviewing drafts of the various state reports on request.

The committee hopes the completion of these state projects** will allow early preparation of a comprehensive publication on the rare and endangered vascular plants of New England emphasizing taxa of regional significance and documenting all listings by information organized by the recommended criteria. With support from the U.S. Fish and Wildlife Service, the NEBC held a symposium in May, 1979, on the "Rare and Endangered Plant Species of New England," with the proceedings published subsequently in *Rhodora*.

Literature Cited

Bean, R. C., C. D. Richards, and F. Hyland. 1966. Checklist of the vascular plants of Maine, rev. ed. *Bull. Josselyn Bot. Soc. Maine,* No. 8. 71 pp.

Countryman, W. D. 1977. Threatened and endangered species problems in North America: The Northeastern United States. Pages 30-35 *in* Prance, G. T., and T. S. Elias (eds.), *Extinction is Forever,* The New York Botanical Garden, Bronx, N.Y.

Countryman, W. D. (Chairman), H. E. Ahles, G. L. Church, A. R. Hodgdon, G. A. L. Mehlquist, and R. B. Pike. 1972. Guidelines and criteria for the evaluation of natural areas. (A report prepared for the New England Natural Resources Center by the Natural Areas Criteria Committee of the New England Botanical Club, Inc.) [Cambridge, Mass., 12 pp.]

Day, M. A. 1899. The local floras of New England. *Rhodora* 1: 111-120, 138-142, 158, 174-193, 194-196, 208-211.

Deane, W. 1899. The herbarium of the New England Botanical Club. *Rhodora* 1: 56-57.

Dowhan, J. J., and R. J. Craig. 1977. Rare and endangered species of Connecticut and their habitats. *Connecticut Geol. Surv. Rep. Invest.* 6. 137 pp.

Fernald, M. L. 1929. Some relationships of the floras of the Northern Hemisphere. *Proc. Int. Congr. Pl. Sci.* 2: 1487-1507.

Hall, A. V. 1981. Information handling for Southern Africa's Rare and Endangered Species Survey. *(This volume, pages 167-184.)*

Henifin, M. S., L. E. Morse, J. L. Reveal, B. MacBryde, and J. I. Lawyer. 1981 . Guidelines for the preparation of status reports on rare or endangered plant species. *(This volume, pages 261-282.)*

Hodgdon, A. R. 1973. Endangered plants of New Hampshire: A selected list of endangered species. *Forest Notes* [Soc. Protection N.H. Forests] 114: 2-6.

Jorgensen, N. 1971. *A Guide to New England's Landscape.* Barre Publ., Barre, Mass. 256 pp.

** Note: The six state reports were recently published by the U.S. Fish and Wildlife Service, Newton Corner, Mass.:

Church, G. L. 1978. Rare and Endangered Vascular Plant Species in Rhode Island.

Coddington, J., and K. G. Field. 1978. Rare and Endangered Vascular Plant species in Massachusetts.

Countryman, W. D. 1978. Rare and Endangered Vascular Plant Species in Vermont.

Eastman, L. M. 1978. Rare and Endangered Vascular Plant Species in Maine.

Mehrhoff, L. J. 1978. Rare and Endangered Vascular Plant Species in Connecticut.

Storks, I. M., and G. E. Crow. 1978. Rare and Endangered Vascular Plant Species in New Hampshire.

Macior, L. W. 1978. The pollination ecology and endemic adaptation of *Pedicularis furbishiae* S. Wats. *Bull. Torrey Bot. Club* 105: 268-277.

Morse, L. E., M. Lamson, A. F. Tryon, and R. R. Walton. 1981. Specimen locality indexing by the New England Botanical Club. *(This volume, pages 185-191.)*

Powell, W. R., T. Duncan, and A. Q. Howard. 1981. The California Native Plant Society Rare Plant Project. *(This volume, pages 193-198.)*

Rekas, A. M. B. 1976. Computerized information systems for threatened and endangered species. *ASB* [*Assoc. Southeastern Biol.*] *Bull.* 23: 144-149.

Ross, S. R. 1978. *The effects of prescribed burning on ground cover vegetation of white pine and mixed hardwood forests in southeastern New Hampshire.* M.S. Thesis, Dept. Botany and Plant Path., Univ. New Hampshire, Durham, N.H.

Schafer, J. P., and J. H. Hartshorn. 1965. The Quaternary of New England. pp. 113-128 *in* Wright, H. E., Jr., and D. G. Frey, eds. *The Quaternary of the United States.* Princeton Univ. Press, Princeton, N.J.

Seymour, F. C. 1969. *The Flora of New England.* Charles C. Tuttle, Rutland, Vt. *xvi* + 596 pp.

Vermont Bird and Botanical Club. 1973. *Checklist of Vermont Plants.* Burlington, Vt. 90 pp.

Appendix I. Revised NEBC Criteria for Listings of Rare and Endangered Vascular Plants.

I. New England Status—Degree of Threat or Endangerment
   A. United States Endangered (Proposed or listed)
   B. Regionally (New England) Endangered
   C. United States Threatened (Proposed or listed)
   D. Regionally (New England) Threatened
   E. Declining or Vulnerable (United States)
   F. Declining or Vulnerable (New England)
   G. Possibly Extinct/Extirpated
   H. Probably Extinct/Extirpated
   I. Presumed Extinct/Extirpated
   Z. Undetermined

II. Reasons for Status Reccommendation
   1. Number of Known Stations (Localities) in New England
      A. Extant Stations, recently verified in field
      B. Reported Stations, not recently verified extant
      C. Stations Extirpated, no longer extant
   2. Mode of Occurrence in New England
      A. Rare and Local
      B. Local
      C. Rare
      D. Few Collections
      Z. Unknown
   3. Overall Population Trends in New England
      A. Approaching Extinction/Extirpation
      B. Long-term Decline
      C. Recent Decline
      D. Stable
      E. Increasing
      Y. Presumed Extinct/Extirpated
      Z. Unknown
   4. General Distribution over Total Range
      A. State Endemic (Specify state)
      B. New England Endemic
      C. Regional Endemic, occurring only in an area comparable in size to New England
      D. Disjunct, with main range outside New England
      E. Widespread, but rare or local throughout most or all of range, including New England
      F. Widespread and relatively common outside New England, although rare or local within New England

5. Principal Reason(s) for Rarity or Decline in New England
   A. Extremely restricted total natural range
   B. Habitat restricted:
      1. Essential habitat both naturally scarce in New England and currently being destroyed or threatened by human activities
      2. Essential habitat currently being destroyed or threatened by human activities
      3. Essential habitat naturally scarce in New England, but not currently threatened with destruction
   C. Relict or disjunct population(s)
   D. Exploited species, commercially or privately
   E. Peripheral populations at limit of taxon's range
   F. Other ecological factors, such as disease, competition with alien species, grazing, beaver, windfalls, or erosion
   G. Other human activities, such as frequent fires, fertilization, insecticides, or herbicides
   H. Frequently overlooked species, including ephemerals and taxa rarely collected
   I. Taxonomically difficult species, including those in groups avoided by collectors, taxa similar to familiar species, and taxonomically uncertain groups
   J. Other (Explain)
   Z. Unknown

The Indiana Biological Survey and Rare Plant Data:

An Unending Synthesis

Theodore J. Crovello and Clifton Keller[1]

Basic information to make optimal long term decisions about Indiana's natural resource heritage, especially rare plant and animal species is lacking. The Indiana Biological Survey has been reactivated to contribute information and insights to help in such decision-making, as well as to make Indiana's citizens more aware and appreciative of their living heritage. The computerized Flora Indiana Project is one activity that can contribute significantly to studies of today's rare plants as well as those that may become rare in the future. Six steps are described for the creation and continuation of a plant survey—an unending synthesis. Our philosophy toward rare plants data organization involves not only data collection but also publicizing and disseminating information to the diverse types of people who require it, and providing such data, and summaries of it, in a form that the recipients can understand and use. Consequently, decisions during and after data capture are made with this general philosophy in mind. Examples are provided of how different types of computer results can help rare plant studies in these ways. Finally, we consider future developments, interactions with other organizations, problems, sponsorship, and immediate challenges—all topics of importance to anyone considering the organization, maintenance and use of rare plant data.

Every year more people become increasingly aware of the high level of interaction and dependency between human activity and the quality of our environment, particularly its living components. This awareness is increasing because plants and animals growing in more or less natural communities satisfy many physical and mental requirements of our lives. But such awareness also is due to a growing moral maturity which includes not only concern for the human species or consideration of primates or other animal species, but also a moral responsibility towards plants and of the general environment itself. Questions traceable to this widening moral awakening have revealed an embarrassing and critical gap in our knowledge of rare animal and plant data. We are in the difficult position of knowing what we should do in terms of moral decision-making, but we are unable to do it due to lack of sufficient and reliable information!

[1] Department of Biology, University of Notre Dame, Notre Dame, Ind. 46556.

Based on the presentation "The Flora Indiana Project and Rare Plant Data" by Dr. Crovello at the Symposium on Geographical Data Organization for Rare Plant Conservation, held 15-17 November 1977 at The New York Botanical Garden.

Pages 133-147 *in:* Larry E. Morse and Mary Sue Henifin, eds., *Rare Plant Conservation: Geographical Data Organization,* The New York Botanical Garden, Bronx, N.Y.

© 1981 The New York Botanical Garden

133

This paper provides insights into what is being done about rare plant data in Indiana to overcome the above problem. It also can serve as a model for the organization of floristic studies in any state on a long term basis. This must extend beyond one person's or institution's interests.

History of the Indiana Biological Survey

The Indiana Biological Survey (IBS) is the main activity of the Biological Survey Committee (BSC), a standing committee of the Indiana Academy of Sciences. The current BSC chairman (Crovello) is fortunate to have the services of six committee members, both botanists and zoologists. Formed in 1892, the BSC is almost as old as the Indiana Academy itself. As stated in the 1893 Proceedings of the Academy, its original purpose was, " . . . 1) to ascertain what has already been accomplished in the direction of making known the character and extent of the life of the State, and to this end to prepare a complete bibliography of materials bearing on the botany, zoology and palentology of Indiana, to be published by the Academy; 2) to associate the various workers throughout the State and so correlate their labors that all will work together towards a definite end, and ultimately accomplish the main purpose of the survey, namely, the making known of the entire fauna and flora of Indiana, its extent, its distribution, its biological relations, and its economic importance; 3) to stimulate the teachers of biology throughout the State to encourage in their pupils the accumulation of material, which shall make known the local extent and distribution of life forms, and thus contribute facts that will be useful in the survey and at the same time develop acute observers for continuing the study of the natural resources of the State; 4) ultimately to secure for the academy a collection that will illustrate the biology of the State."

Even though several people (Deam, 1940; Lindsey, 1966; and Lindsey et al., 1969) actively studied and surveyed the State's biota, the last several decades saw BSC activity reduced to an annual survey of the literature.

In the last few years the growing need for diverse and detailed information on the biota of the State has become evident. But potential users of such information, no longer restricted to academic biologists, now include such diverse people as writers of environmental impact statements, employees of State and federal agencies, teachers at all levels, area and regional planners, and individuals with various private interests. More importantly, federal legislation (e.g., the Endangered Species Act of 1973) now requires developers to demonstrate that their activity will not endanger certain species. This and other legislation requires a basic, detailed knowledge of the biota of Indiana. It is essential that such knowledge be accumulated and that subsections of it relevant to a specific task be made available in a useful way.

Consequently, in 1974 Crovello proposed to the Academy: "That a formal Flora Indiana Project be created whose major purpose will be the continued maintenance and dissemination of information about the kind, quantity and distribution of the plant life of the State. The information accumulated would be maintained and disseminated in ways that are both useful and efficient." The motion was adopted by the Academy in 1975 and subsequently formed the basis for an expanded revitalization of the BSC. The current goals of the BSC are the same as when the Academy was founded: to accumulate, store and make available in useful forms, information on the biota of the State. Given the magnitude of the data base of interest, and the

134

need for fast retrieval of selected pieces of it, we also considered how computers might help the BSC achieve its goals.

The BSC is organized into the following subcommittees:

1. *BSC Literature Subcommittee*—Its purpose is to maintain and to accumulate published references on the biota of the State. In the past the BSC Literature Project presented its findings in conventional printed form, but we now also are building a retrospective computer-assisted data bank. It provides updated, integrated bibliographies of particular topics in response to specific requests.

2. *BSC Endangered Species Subcommittee*—Its purpose is to develop and maintain information on the status of species that might be threatened or endangered. In progress are surveys of several large taxa, such as mammals, fish, and vascular plants. In 1977 two summary lists of rare plant species were produced from a computer data base, and mimeographed copies were distributed around the State. They are serving as a first draft for a list of the State's threatened and endangered plants, being prepared by John Bacone, Division of Nature Preserves, Indiana Department of Natural Resources, Indianapolis.

3. *BSC Flora Indiana Project* (FLIP) *Subcommittee*—FLIP's goals were stated above. Currently, computerized information is available on the presence or absence of each of the approximately 2,500 vascular plants in each of Indiana's 92 counties. Based on Deam's (1940) *Flora of Indiana,* plus 8,000 new county records as verified by herbarium specimens, this computerized information is available for easy updating, customized search requests, and the production of printed summaries. We also computerized the county-level data of Swink (1974) for eight northwestern Indiana counties. While FLIP's data deal only with plants (and at present only with vascular plants), the hope is that computer programs and other techniques developed in FLIP will be valuable to a Fauna Indiana Project.

4. *BSC Environmental Impact Statement Subcommittee*—Its purpose is to use and make more readily available the information contained in Indiana environmental impact statements; and to help achieve the purposes of the BSC. Still in the discussion stage with the Environmental Protection Agency, we expect computers to be of use. At this time IBS does not take a direct role in reviewing impact statements, but it could.

5. *BSC People Power Subcommittee*—Its purpose is to develop and maintain information on persons interested in any or all aspects of the biota of the State, including the accumulation and use of such data. People with interests in the biota of one geographic area immediately come to mind, as do people with interests in one taxon, one group of species, etc. But people who can help meet the goals of the BSC are not restricted to biologists. The expertise of geographers, geologists and archeologists, to name a few, is of great value to the BSC. A People Power Survey is made regularly to invite additional people to participate.

The vision and activities of the BSC, all of which constitute the Indiana Biology Survey (IBS), is long term as well as short term. The BSC is not just a stopgap organization intent on rushing from one endangered species emergency to the next. We also intend to build a firmer, broader, data base. Only with this long range view will endangered species and their study be served adequately, since today's common species may become tomorrow's rare ones. We also encourage more and different types of people and organizations to become involved in IBS work since this enhances its value. This involvement should not be just at the user stage, but also in areas of data accumulation and dissemination. Finally, like any taxonomic work, IBS and its activities must accept the frustration and challenge of being an unending synthesis (Constance, 1964).

Present Status of the Indiana Biological Survey

Current status of IBS and its use for rare plant data can best be approached via the major stages involved in the creation, maintenance, and use of a plant data bank, whether computerized or not. Thus, even if one is not interested in the specifics of Indiana's rare plant data, it may be useful to use the Flora Indiana Project as a source of examples in thinking about what is involved in the creation, maintenance, and use of any rare plant data bank.

While a biological survey can be described in discrete steps, readers must realize that compartmentalization of a more or less continuous process may be misleading. Surveys are unending, multifaceted assessments of the changing state of dynamic natural and human ecosystems. Many of its different activities are carried out simultaneously. Nevertheless, for convenience we describe IBS in six stages: 1) planning; 2) data accumulation; 3) data bank deposition; 4) data bank maintenance; 5) data bank output; and 6) publicity. Each step is briefly outlined below.

Stage One: Planning.   Planning has been effected mostly by BSC members, many of whom are not just academic biologists but also are active in conservation and nature preservation groups such as the Nature Conservancy and the Audubon Society. Interaction occurs also with people in several divisions of the Indiana government (especially within the Department of Natural Resources), in local government, with members of other committees of the Indiana Academy, such as the Science and Society Committee, and with other interested citizens throughout the State.

Perhaps our people power situation is similar to that of many states. Indiana has few state employees with floristics expertise of any type as part of their job description. Few professional plant taxonomists or plant ecologists live in Indiana.

During our planning step essential questions are asked including the following: (1) What are our current and future sources of support? (2) What are the types of data required? (3) What types of users are expected? (4) What types of people are available to help? (5) What types of output are in demand now and will be in demand in the future?

Stage Two: Data Accumulation.   Three substages of data accumulation exist: a) field work; leading to b) specimen preparation, typed summaries, or publication; and finally, c) data assembly. It is valuable to recognize these substages because different people frequently are involved with each substage, each can be done at different times, and each involves different procedures.

Field information can be stored in written notebooks, on photographic slides, as dried specimens, etc. Hopefully, the format of the information will be such that it easily can be deposited at a later date into a computer data bank. As far as FLIP and rare plant data, no new organized field work has been done. We receive a small amount of new county record information from workers on environmental impact statements, as well as from State Park naturalists, but this is only beginning.

Substage Two, the transition of field information and materials to more permanent or summary forms, may involve preparation of specimens on herbarium sheets, or of publications. Readers should refer to the other paper by Crovello in this volume for further information on many types of publications and "semipublications" that exist, the problems associated with their retrieval, and the use of their information (see also Crovello and Keller, 1974, and Keller and Crovello, 1973).

The final substage of data accumulation, data assembly, frequently is overlooked. Yet it is both essential and time-consuming. Data assembly involves finding the proper floras, the relevant herbarium sheets, etc., and then extracting from it the *specific* information required for a *particular* task. FLIP has assembled sources of data including the county distribution maps prepared by Deam (1940), and Swink (1974), and information from herbarium sheets at the Indiana University Herbarium.

Stage Three: Data Bank Deposition. Data bank deposition involves all activities which transfer information from herbarium sheets, publications, or data sheets into a computer bank. For FLIP we processed information from Deam (1940) and as mentioned above from the "official" copy of Deam's book at Indiana University, which includes the 8,000 new county records mentioned above. For each vascular plant we recorded: scientific name, author, map number in Deam, counties of known occurrence, and whether the species was considered weedy or non-weedy. So, unlike Crovello's current monographic work in the mustard family, and unlike his earlier computerization of the Edward Lee Greene Herbarium (Crovello, 1972), the computerized Flora Indiana Project is species-oriented, not specimen-oriented. Both kinds of data banks are useful in rare plant studies. They provide different levels of resolution in answers to taxonomic, geographic, or ecological questions about rare plants.

Stage Four: Data Bank Maintenance. Too often in this age of the Computer Revolution we mistakenly think that once data are in a computer, users can obtain *any* type of desired analysis. The computer can only do what programs have been instructed to do, and they can only operate on data that have been input and continuously updated to remain current. The computer is fast in retrieval, but many other tasks are associated with the creation and maintenance of a rare plant data bank.

Two types of computer files are important. *Data files* contain the total body of information which, when properly searched and analyzed, provide answers to our questions about rare plant data. Such rare plant data files must be kept current which involves correction of information in the data bank, addition of new information, and subtraction of information of no use. The *program file* is the second type of computer file. New questions arise at a regular interval and new programs must be written to answer these new, usually more sophisticated questions. Also the need exists periodically to adapt what was once a properly running program to new situations, even at the same computer center.

Data bank maintenance is essential. If biologists involved in rare plant surveys do not want to become involved in the mechanics of computing, then measures should be taken to assure that someone in their organization will have computing experience as part of their job descriptions.

Stage Five: Data Bank Output. The four major types of data bank output are: 1) prose printouts (or publications); 2) tabular results; 3) summary graphics; and 4) summary statistics.

Prose results or publications which have been requested from the FLIP data bank include computerized printouts of a county checklist of all plant species, as well as county checklists for only rare plant species. We have arbitrarily defined rare species as those that have been recorded in five or fewer counties in Indiana. Requests for such information come most frequently from environmental impact statement workers and park naturalists. Monographers have received prose printouts summarizing the county level distribution of each species in the family of interest. All such results were easily and cheaply obtained from the computerized FLIP data bank.

Two results from the FLIP data bank were of sufficient importance to warrant large scale mimeographic duplication but without retyping. Both involved rare plant species of Indiana. The first was a summary listing of the 485 species occurring in five or fewer counties. Printed in alphabetical order, the listing contained the species binomial, Deam's map number for easy page reference to the standard work, and the numbers of the five or fewer counties in which each species was found. The last page was a map of Indiana with an unique number for each of the 92 counties. Workers quickly and easily can determine for a given species counties in which the species has been found, and ones in which it might also be found with further study. The second computer search reproduced was a listing of the 409 species (out of the 2,500 vascular species that occur in the State) which have not had any new county records since Deam's flora was published in 1940. A third useful result would be a combination of these two searches, giving an overlapping subset of the 485 species that occur in five or fewer counties *and* the 409 which have not had new county records reported since Deam's work.

Prose results also are possible from the IBS Literature Bank. Currently, the author, date, title and citation for each article relating to the State's plants or animals that appeared in the Proceedings of the Academy since its beginning in 1891 are included. If increased support becomes available, we hope to expand the IBS Literature Bank to include an indexed summary of key words that describe the taxa, counties, and ecological units involved in each article.

Tabular Results most frequently requested include information on the number of species per county as well as the number of species per family. The scope of tabular results, like Graphic Summaries, is limited only by the creativity of the questioner.

Graphic Results may take several forms, the most common of which is the production of State distribution maps at the county level, usually one map for each species. Information from several sources is used to produce county level distribution maps. For example, users of distribution maps may want to know whether a county record for a given species is based on Deam's 1940 study or a more recent one. This can be answered rather easily if the source of each county distribution record also is stored in the computer. A species map at the county level can use one of several symbols to indicate a county record. For example, a D in a county may indicate that that county record was included in Deam, but that it was first collected in 1973. Finally, a compound such as D 73 might mean that for a specific county it was both recorded by Deam and the *latest* updated reverification was made in 1973.

Computer generated maps can provide insights, but they can also cause problems. These also occur in mapping without a computer. The first is that the map may become too complex, although a simple coding scheme such as that just given is a good compromise. A second, more serious problem is the uneven reliability of the information which comes from several souces, including published floras, monographs, herbarium specimens, ecological papers, and environmental impact statement lists. Our solution has been to provide a simple way to identify the origin of each record. Another alternative is to have the computer produce several maps for each species, one map for each major source of data.

Unfortunately, as workers all around the country are discovering, our phytogeographic distribution information is incomplete at the county level. The value of publishing an expensive atlas of distribution maps at the county level is debatable, especially before large amounts of new data are obtained. No one wants to publish something that is incomplete, but on the other hand it cannot be made complete until it exists and is available for revision! One solution would be to publish an atlas of county distribution maps (at least of the rare plant species) in a relatively inexpensive way so that existing data are put into the hands of field workers, and at the same time large resources are not committed to something that will be revised in a few years.

A second form of graphic summary is the frequency of rare species by phytogeographic or political region. For example, rare plant data workers may be interested in knowing which counties have the highest number of rare plant species. A computer-assisted search of the entire data bank of information on 2,500 species can, in our case, identify those 485 species occuring in five or fewer counties. The computer then can provide a graph which has 92 positions along the X axis, each one referring to a different county. The Y axis indicates the number of the 485 species that occur in each of the counties. By visual inspection of the graph, workers obtain a quick summary insight into geographic areas of rare plants.

A geographic frequency distribution even more valuable than the one described above can be obtained from the computer when the distribution is plotted on a geographic base. This can be done easily using sophisticated computer programs that are available from the Harvard Graphics Laboratory. *Two-dimensional graphic summaries* which take the form of differentially shaded maps are available via the program SYMAP. The area of each county is filled with a certain symbol to represent the value of the character (here the number of rare plant species it contains). A finite set of character states usually is available. This results in a computer-printed map in which several counties with approximately the same number of rare species will have the same level of differential shading. Crovello and Keller (1974) provide an example of this two-dimensional technique as well as two-dimensional contour mapping. SYMAP also can produce such contour maps automatically.

Although SYMAP is useful, for IBS and similar rare plant studies, the results of SYMVUE are even more valuable (SYMVUE and SYMAP are graphics program packages available from the Laboratory For Computer Graphics, Harvard University, Cambridge, Massachusetts). SYMVUE provides a *three-dimensional view* of a frequency distribution over a given area, here the State of Indiana. Figure 1 is an example of a SYMVUE result, summarizing the distribution of rare plant species in Indiana. The height of the "landscape" on the graph represents the number of species in each of Indiana's 92 counties that only occur in five or fewer counties in the State. The data input to SYMVUE included the XY coordinates of the center of each of the 92 counties, and also the number of species that occur in a given county but in no more than four additional ones. The basic presence/absence data for each county was based on the species maps in Deam (1940). The computer then fitted a high degree polynomial to these data to obtain the "best" fitting surface for them. The user also may specify from what viewing direction the perspective drawing should be made. In Figure 1 the observer is looking at the State from the southeast at an elevation high above the surface. The highest peaks on the graph, in the northwest cor-

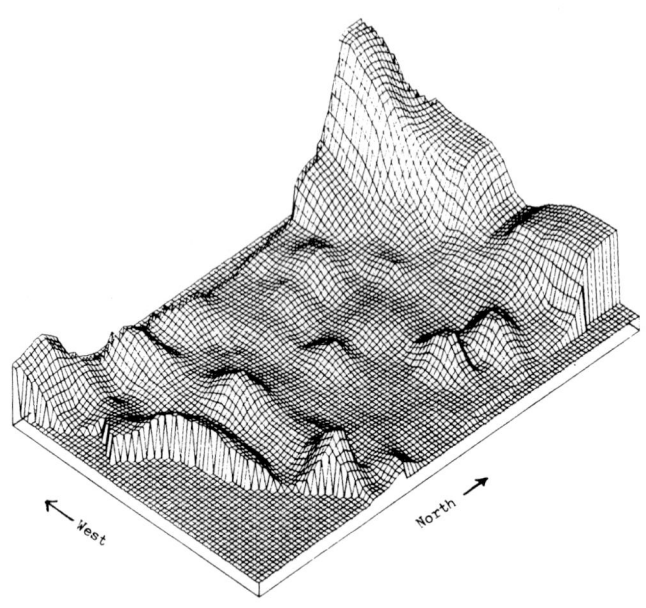

Figure 1. Rare plant species topography of Indiana, based on the number of rare plant species recorded in each of its 92 counties. View is from the southeast, and height of surface reflects number of rare species, ranging approximately from none to 130 per county.

ner, represent approximately 130 rare species per county. Some counties (such as those in the east central part of the State) have no rare species.

Figure 1 summarizes at a glance much information about the distribution of "rare" plants in Indiana, and is proof that one graph is worth a thousand data points! It presents reviewers unfamiliar with the State's flora a concise summary of the geographic areas that should be evaluated first for endangered species. The SYMVUE graph also allows easier correlation of physiographic, climatic and other geographically oriented variables with the distribution of rare plants. Graphic results provide a heuristic, suggesting meaningful questions of value to endangered species. In this context graphic results are useful at two stages of endangered species work. First, a summary of available rare plant data in the preliminary study stages of rare plants is provided to help determine where to use limited fieldwork resources. Second, graphic results are of value in creating a final floristic map of an area, which should allow more accurate, efficient fieldwork.

Summary statistics also have great potential in the development of an understanding of rare plant data and for utilizing rare plant data banks. Summary statistics can be provided for each set of taxa—for example, the average number of counties in which each species occurs. Similar statistics can be compiled to answer questions about a particular geographic unit (such as the average and variance of the number of species per county in the State), or for particular ecological units (for example, the average and variance of the number of species in a community). Bivariate analyses involving correlation statistics and perhaps also regression have, as yet, not been used in rare plant data studies.

Multivariate analysis has rarely been used in rare plant studies. Crovello and

Keller (1974) used numerical taxonomic procedures to gain insight into relationships among the counties of Indiana. County distributions of the 81 species of the Brassicaceae growing in the State were analyzed. Interesting results were obtained, but these data did not particularly involve rare plant taxa.

To summarize Stage Five, in IBS and especially the Flora Indiana Project, we believe that conventional printed output will always have value for rare plant studies. But we also believe that other kinds of computer output will do much to deepen our understanding of the past, present and future status of rare plants. Furthermore, in these days of tight budgets, routine computer use may prove cost effective.

Stage Six: Publicity.   The best rare plant data information system is completely useless if potential users are unaware of it. Consequently, every appropriate medium is used to publicize the existence of Indiana's rare plant data system. This includes the distribution of the minutes of meetings of the Biological Survey Committee and BSC People Power questionnaires at Indiana Academy of Sciences meetings and by mail as part of the Newsletter of the Indiana Academy of Sciences. Reference is made to IBS data banks in talks on various subjects throughout the country. Publications and fliers explain its existence, availability and use. The newsletter of the Indiana Public Responsibilities Network of the American Institute of Biological Sciences carries information on the IBS data banks. Members and correspondents of the BSC are urged to establish bulletin boards in their departments that provide information on BSC activity. Postcards with Figure 1 are given away and used regularly.

Future Developments.   Efforts are currently being made to interest more professional botanists, both at universities as well as in government agencies, to accumulate more data. We use every opportunity to encourage environmental impact statement workers to send new county record vouchers, species lists, etc., and point out the potential for favorable publicity for their companies should they cooperate with IBS.

An immediate goal is an updated inventory of the woody plant species of Indiana, especially trees. In addition to determining which species are becoming rarer, we are concentrating on the approximately 160 tree species because they are: a) a relatively small, easily manageable, and well circumscribed group of species; b) relatively easy to locate in the field even in winter; c) relatively easy to identify; d) relatively hard to destroy by overzealous collectors; e) often the dominant feature of a plant community, thereby able to serve as indicators of the presence of many other plant species that may be rare or endangered; and f) familiar to many people throughout the State (Scouts, 4H Clubbers, Audubon members, etc.) who have an interest in plants. Resulting publications on trees of Indiana should bring much favorable publicity to the Academy. Our goal is to carry out an extensive reverification of the distribution of all woody species of the State. This will be at both the county level and, as a completely new project, to obtain more detailed distribution information of trees below the county level. Following the lead of the RAPIC (now the PIN) system in Colorado (see Adams *et al.,* 1975, and Dittberner *et al.,* this volume) we are considering the issuance of two types of certificates: a new county record certificate; and an updated, reverified county record certificate.

It would also be sensible to inventory existing public and private nature preserves and state parks for rare plant species and the county level flora in general before embarking on a much broader survey of private property. Trained personnel already are available in nature preserves, and possible negative reaction by land owners is reduced if activity on private land is minimized. Unfortunately, and not just in Indiana, some landowners see the study and protection of endangered species as violations of their rights to do with their land as they wish, and to maximize economic gain. The role of IBS would not be to carry out actual inventories, but to integrate, maintain, and produce useful summaries of data from these and other sources.

Future data accumulation will utilize an X-Y digitizer which permits semi-automatic capture of additonal geographic distributions as well as easier creation of special maps of the State, such as overlay maps showing climatic, edaphic, and other data. A new use of the digitizer involves the capture of drawings of entire plants or specific organs of a species, which then can be used for online identification and for the efficient production of camera-ready illustrations for hard copy manuals and floras. The expansion of our data banks will incorporate the new data mentioned above as well as expansion of the IBS Literature Bank, and of vegetative data available from earth resources satellites.

Possible new developments in output include the use of a statewide computer network called INDIRS, the Indiana Information Retrieval Service, which is sponsored on an experimental basis by the Indiana University School of Business. Users throughout the State need only visit a public library to request use of its remote terminal which is connected to INDIRS. Currently, information is provided on the county level for such economic measures as unemployment rate, county tax levels, etc. We hope to add to INDIRS information on rare plant species for each county. A request also would be made to users of INDIRS to provide the Indiana Biological Survey with any updated information they may have.

An endangered species booklet for plants, mammals and fish is underway. Questions as to the best format, type of illustration, and level of exact locality data are still under discussion. As indicated above, we are also considering the use of digitized data as a basis for our illustrations.

The use of online identification programs for woody plants as well as for endangered or rare plant species is planned for the near future. We will utilize AUTO-KEY, a system of programs created by Ronald Hellenthal at Notre Dame. Current groups available for online identification include aquatic insects and trees of Indiana.

As to publicity, as Susan Bratton has described at this symposium, and as members of the Wabash College faculty have done in Indiana, we believe a critical need exists for botanists to write newspaper articles and prepare talks on rare plant species for their local, state, and national public media. Furthermore, we want to increase our interactions with primary and secondary school teachers. This best can be effected by participation in the regional grade and high school science fairs which are held throughout the country. We also plan to involve at least the woody plant project in the Indiana Junior Academy of Sciences' Achievement Program. Finally, use of our computer program called ESP, the endangered species simulator, can increase interest and understanding of rare plants at the college and high school level as well as in other organizations and events throughout the State. This program simulates the decision processes involved in rare plant conservation, showing the roles of contrasting economic, political, and conservationist arguments. These activities are designed to increase the sensitivity of the public, and of potential contributors, to the need for rare plant data as well as the need to protect plants.

Interactions of IBS with Other Organizations

Following our philosophy that the most valuable and complete work of the BSC will emerge from combined efforts of many people, both from within and outside of the Indiana Academy, continuous efforts are made to involve individuals and organizations throughout the State. An indication of current and planned involvement by other groups can be shown by reviewing the stages of IBS activity presented earlier. To help in our planning we have enlisted the aid of informed members of other relevant committees of the Academy, as well as of the membership at large. We also draw upon the State employees, particularly in the Department of Natural Resources, who have provided both information and evaluation of our planned projects. Indiana has one national forest, and its personnel have provided information about rare plants and animals. Finally, field workers involved in environmental impact statements who contact us for information are also asked to outline their views on the most urgent needs for rare plant and animal data.

In data accumulation, we have used printed matter already published by various sources, including the Department of Natural Resources and The Ford Foundation. In addition, the Department of Natural Resources maintains files of information within its various divisions which are of help in determining the distribution of rare plants at the county level and below. Naturalists engaged by the Division of State Parks also will figure prominently in data accumulation and deposition in IBS data banks. Similarly, conservation groups, such as the Audubon Society and the Isaac Walton League, include many members knowledgeable in plants and animals. Many advantages will accrue if the lands of such organizations will be inventoried early. Finally, particularly with the woody plant species, we expect that the 4-H Club organization, which is partially funded in every State by the federal government, can be utilized as a source of taxonomic, geographic and ecologic information.

No formal liasion exists between IBS and organizations in adjoining states, although we exchange information with both government and private agencies. Both Crovello and the Director of Indiana's Division of Nature Preserves serve as Indiana liaisons with the federal Office of Endangered Species.

Data evaluation has not been presented in this paper as a separate stage because it runs through every step of any survey. Nevertheless, if data evaluation had to be added to our scheme as a formal stage, it might best come between data accumulation and data bank deposition. Most decisions about data are made there, and it is also where they can be made most efficiently and economically. Once data are computerized it is more expensive to evaluate and correct them. Data evaluation will be done primarily by IBS personnel, but also by the Department of Natural Resources. Computer-generated summary statements of the data will be sent to data suppliers who will be asked to check for errors or to make updated improvements.

Data bank deposition at this point is an activity of the IBS itself. Perhaps far in the future an opportunity will exist for others to deposit data *directly* using remote terminals, but at present it does not seem economical to establish and maintain such a system. Limiting direct data deposition into computer banks to IBS personnel also will make it easier for IBS to evaluate the quality, reliability and consistency of the data in the data bank. A good example of a system where the users themselves input data from many remote terminals is the Biostoret System of the U.S. Environmental Protection Agency. However, its users have been trained to use this system, are familiar with the kinds of data being input, and just as importantly, a staff of trained computer editors exists to evaluate the data deposited in the temporary file during a previous time interval. For reasons of simplicity, data bank maintenance and data evaluation are best performed by one group. Here the most sensible group would be IBS itself.

Perhaps the largest interaction with other organizations and individuals will be in the area of data bank output. If handled correctly, the IBS data banks can serve many diverse organizations and individuals, including other committees such as the Science and Society Committee of the Indiana Academy, its individual members, and Department of Natural Resources personnel. It could serve as a data base for information on woody plants for such groups as the 4H Club, Boy Scouts, or Girl Scouts. We have yet to explore interactions with the State's Departments of Public Instruction, both at the elementary and high school level. The Environmental Protection Agency and other federal organizations will or should find geographical, taxonomic, and environmental information on the plants and animals of the State of great value for many of their projects. Indiana's Division of Fish and Wildlife can both benefit and contribute to IBS data bases in their study and management of fish and wildlife. National organizations such as Sigma Xi, the American Association for the Advancement of Science, and the American Institute of Biological Sciences, all have science and society committees. Once IBS is begun, it could be a routine task to provide such organizations with information of interest regarding species in Indiana.

The remaining stage in our cycle is publicity. In early stages most of the publicity about IBS and its activities and services must be generated by itself. But the best publicity will be generated as more and more people become aware of IBS and use its services, or contribute to its data banks. A major problem with a biological survey is that major users no longer are major contributors, and users are not simply members of one segment of an established taxonomic or biological community. Rather, major users include workers in industries such as public utilities, environmental impact organizations, civil engineers, area plan commissions, and many other agencies of the various levels of government. This makes the problem of publicity extremely difficult and challenging.

Problems

Several problems are in the narrower taxonomic sphere. What is the actual number of species present today in Indiana? This is difficult to answer because it involves a consideration of immigration, extinction, synonymy, level of collection, and the reliability and completeness of the last generally published flora of the State. The problem can be more clearly stated in a prose equation:

Number of species actually present today in Indiana equals:

the number in the last general flora (Deam, 1940)

PLUS the number of nomenclatural births since Deam (1940)

PLUS the number of new species collections since Deam (1940)

PLUS the number of immigrating and introduced species established since Deam (1940)

PLUS the number of newly evolved species (for completeness!) since Deam (1940)

MINUS the number of nomenclatural deaths since Deam (1940)

MINUS the number of actual State extinctions or extirpations since Deam (1940).

From the above equation, it should be obvious that we rarely ever will know for sure the number of species present in a given geographic area. Rollins (1953) has reached this same conclusion for Massachusetts. Again we are reminded of the valuable concept of taxonomy as an unending synthesis. This concept applies also to biogeography. Recognized in this light, the problem of the number of species in an area becomes a livable challenge rather than a consuming frustration.

Synonymy was included in the above equation, but such large amounts of effort are required to establish true synonymy that it deserves separate mention. The problem arises because nomenclatural decisions always involve a certain amount of subjectivity. This is both a weakness and also a strength of the taxonomic system. Synonymy exists because: workers always have incomplete information; the information varies from worker to worker; communication between workers is not always adequate; and the biological units being named are part of a dynamic, evolving system. How can we best solve the synonymy problem for rare plant studies, including the creation of computer data banks? Perhaps the approach of John Kartesz and his Biota of North America Committee will prove useful and expedient. Kartesz has integrated the latest available information on the plant species that grow in each state, including synonymy. In addition he analyzed recently published floras that cover several states. When Kartesz compares and evaluates these lists genus by genus and at the species level, he claims to be able to provide a high percentage of the best taxonomic decisions and synonymy possible. IBS has contracted with John Kartesz of the Biota North America Committee to update the species names in Deam's (1940) Indiana flora.

Geographic synonymy also is a problem. For example, six towns in Indiana are named Mount Pleasant. Fortunately, they all are in different counties so with just that additional piece of information the confusion can be resolved.

Another problem is the danger of making unwarranted inferences about endangered status of species from certain rare plant information. In our activities we have refrained from any claims of endangered species circumscription, because our presence/absence data currently available only can indicate rareness (frequency), and only at the level of county. Naturally our computer data bank can accommodate detailed population and other information. But IBS itself does not yet have the resources to accumulate such data. Consequently, our current data base can serve as an initial screen to the delimitation of endangered species, but not as a source of the ultimate decision.

Finally, IBS and its specific projects such as FLIP and the rare plant survey are examples of a multistage decision process (Crovello, 1970). Not only does it have many stages, but each stage requires decisions that can affect the results. The problem of choosing among alternatives is a difficult one.

Moral and Monetary Support

Crovello sought the sponsorship of the Indiana Academy of Sciences for several reasons. It permits operational freedom since it is not associated with any one institution, especially a state institution which often requires formal procedures that use time and energy. Many diverse individuals and other organizations also feel free to associate with IBS. This would not be the case if its primary existence and sponsorship depended on an agency of the state or on a particular educational institution. In addition, some firms and organizations either prefer or are unable to engage the services of a state agency, or even a state-supported educational institution. Thus Indiana Academy of Science sponsorship permits the greatest flexibility and at the same time the largest accessibility to different organizations and individuals. Coincidentally, but not to be dismissed as insignificant, the Indiana Academy of Sciences has and will continue to gain much favorable publicity from the expanded activities of its Biological Survey Committee.

Monetary support at the level of hundreds of dollars has come from the Indiana Academy of Sciences. Unfortunately, it does not seem to have the resources to do more than supply monies to meet basic costs of mailing, telephones, etc. Some support is available for other activities, but not much. The University of Notre Dame, particularly the Department of Biology and the Computer Center, have provided significant amounts of real dollar support as well as released faculty time and computer time. Without this support from Notre Dame, we would have been unable to pursue the diverse activity of the BSC even as far as we have. The Department of Natural Resources provided support by contracting to purchase 50 copies of the June 1977 list of species of plants occurring in five or fewer counties. The charge was $50, and was meant only to defray part of the real costs, as well as to explore what is involved in dealing with the State government on a financial basis.

As to the future, we hope that the Indiana Academy will be able to provide up to several thousand dollars per year to the BSC. Perhaps the State's Department of Natural Resources may find it of value for the BSC to carry out certain contract work for them. Nevertheless, it is obvious that these sources can not provide the financial support necessary to establish IBS activities at the level needed to serve all of the citizens and organizations in and out of the State. We believe that after a three or five year initiation period with strong outside support, user fees should be sufficient to provide significant amounts of financial resources necessary to continue IBS activities. Without such strong outside support for this initiation period, many of the BSC's goals will remain out of reach. Unfortunately, the adverse effect of poor outside support on our State's natural resource heritage may be severe.

Immediate Challenges

1. We need a taxonomic synonymy update as soon as possible.

2. Additional field work is imperative.

3. We must educate and convince the public about concern for nature, especially rare and endangered species and habitats.

4. Our greatest challenge is to initiate and maintain contact with all of the many diverse individuals, organizations and agencies which could both use and contribute to the activities of IBS. This is perhaps the most difficult task we face, because all of our BSC members are biologists. What the challenge demands is the exploration of every area in the State to establish contacts, and then to provide information on how IBS can serve them better and satisfy the various needs. To initiate and maintain contact with such a diverse audience requires great amounts of time, which is our most scarce resource.

Summary

To summarize the state of the art of Indiana rare plant data studies, we may say simply that they and other IBS projects are coming to life! The long range goal of the BSC is to accumulate, maintain and disseminate information about the kind, quantity, and distribution of the biota of the State. Its short range goals are to study endangered species and to recommend protection of them by first concentrating on State nature preserves, county parks, and land held by sympathetic organizations such as The Nature Conservancy and The Audubon Society. In addition, we plan to initiate a long term study of woody plants to obtain useful results about the status of these species throughout the State, and also to publicize the BSC and to educate the public, by their contributing to the study, as well as their use of its results, and simply by learning that more people care about the State's natural resource heritage.

Our philosophical point of view has always been that such surveys are an unending, multifaceted task. This task requires many people but it promises both tangible and intangible rewards to both participants and to society in general. We have found personally that IBS is a very time consuming process, and this time we could have spent in other significant activities. Nevertheless, we believe in the need for IBS and that in the long run it will serve both academic goals as well as those of Society.

Literature Cited

Adams, R. P., et al. 1975. RAPIC, the missing link? *BioScience* 25: 433-437.

Constance, L. 1964. Systematic botany—an unending synthesis. *Taxon* 13: 257-273.

Crovello, T. J. 1970. Analysis of character variation in ecology and systematics. *Annual Rev. Ecol. Syst.* 1: 55-98.

Crovello, T. J. 1972. Computerization of specimen data from the Edward Lee Greene Herbarium (ND-G) at Notre Dame. *Brittonia* 24: 131-141.

Crovello, T. J. and C. Keller. 1974. Uses of computerized floristic data of Indiana for plant geography. *Proc. Indiana Acad. Sci.* 83: 399-406.

Deam, C. C. 1940. *Flora of Indiana.* Indiana Dept. Conservation, Indianapolis. 1236 pp.

Keller, C. and T. J. Crovello. 1973. Procedures and problems in the incorporation of data from floras into a computerized data bank. *Proc. Indiana Acad. Sci.* 82: 116-122.

Lindsey, A. A. (ed.). 1966. *Natural Features of Indiana.* Indiana Academy of Sciences, Indianapolis. 597 pp.

Lindsey, A. A., D. V. Schmelz, and S. A. Nichols. 1969. *Natural Areas in Indiana and their Preservation.* Indiana Natural Areas Survey, Lafayette, Indiana. 594 pp.

Rollins, R. C. 1953. How many species of vascular plants grow without cultivation in Massachusetts? *Rhodora* 55: 361-362.

Swink, F. A. 1974. *Plants of the Chicago Region.* Morton Arboretum, Lisle, Illinois. 474 pp.

The Use of the Plant Information Network (PIN) in Rare Plant Conservation

P. L. Dittberner,[1] **G. Bryant,**[2] **and K. C. Vories**[3]

The Plant Information Network (PIN) is an information storage and retrieval system designed to provide information for management-related questions. Major uses at this time are for providing information for environmental inventories, impact assessments, and planning and design of reclamation projects. Approximately 4100 native or naturalized vascular plant species found in Colorado, Montana, and Wyoming are included in PIN. Advantages of the data bank and use of PIN are: (1) information can be curated and kept up-to-date; (2) accessibility is easy and convenient and could be nearly instantaneous with interactive terminals; (3) information may be retrieved in any combination of attributes; (4) PIN helps identify information gaps and reduces the chance of new information being "buried;" and (5) the expense of obtaining information is reduced compared to conventional methods.

The Plant Information Network (PIN) is a computer-based data bank for rapidly retrieving and organizing information on the native and naturalized vascular plants in the states of Colorado, Montana, and Wyoming. PIN is not a new concept or a new development in information systems. It is a new adaptation or modification of previous systems. The previous programs were called (1) TAXIR and (2) RAPIR. TAXIR is a program started at the University of Colorado as a means of recording, cataloging and updating herbarium information. In this system TAXIR catalogued the individual specimens collected, vouchered, and recorded in the herbarium. RAPIC, the Rapid Access Plant Information Center, using the RAPIR program software, in addition to cataloging herbarium information, was also expanded to include other taxonomic, geographic and some biological and economic information.

The Colorado State University Experimental Station funded the RAPIC program for 5 to 6 years and spent about $60,000 on its development and operation. During this time, the principals involved in the project were Dr. William Klein and Dr. Robert P. Adams of the Botany and Plant Pathology Department. In 1974, the

---

[1] U.S. Fish and Wildlife Service Western Energy and Land Use Team, Ft. Collins, Colo. 80526.
[2] Dept. Botany and Plant Pathology, Colorado State University, Ft. Collins, Colo. 80523.
[3] Dept. Range Science, Colorado State Univ., Ft. Collins, Colo. 80523.

Based on the presentation "The Plant Information Network (PIN)" by Dr. Dittberner at the Symposium on Geographical Data Organization for Rare Plant Conservation, held 15-17 November 1977 at The New York Botanical Garden.

Pages 149-165 *in:* Larry E. Morse and Mary Sue Henifin, eds., *Rare Plant Conservation: Geographical Data Organization*, The New York Botanical Garden, Bronx, N.Y.

Office of Biological Services, of the Fish and Wildlife Service, was established. Within the Office of Biological Services, four national teams were established, of which the Western Energy and Land Use Team began its initial staffing in 1975. As the Team began to plan and develop its projects the applicability of RAPIC was quickly recognized as a potential tool for land use managers and people making decisions and needing information about plant species and vegetation. A new project was initiated which included the basic concepts of the RAPIC system and greatly expanded its geographical coverage and its informational base. At this time the name of the system was changed to the Plant Information Network or PIN. The concept behind this system was to develop a tool that could provide information about plant species and vegetation to land use planners and resource managers. Much of the information these people need is often unavailable to them because of a lack of library facilities, or a lack of time to obtain the appropriate information. The PIN system is designed to provide that information in a short time to the users.

PIN is not intended to replace biologists or field inventories. It is only a means of cataloging the information to make it available to more people. It also brings together information that was previously only available from a wide range of other unavailable sources. The objective of PIN was and is to catalog information about plant species that will help resource managers make better resource management decisions. In the past, as well as the present, management decisions are often made based upon incomplete information and data. This will continue to occur but PIN will help the manager make more knowledgeable decisions.

Sources of information included in PIN are: (1) herbarium specimen labels from major herbaria in each of the three states; (2) extensive searches through scientific, professional, and popular publications pertaining to the taxonomy, geography, biology, ecology, and economics of the plants found in the three states; and (3) the unpublished judgment and experience of long-time experts and field researchers on the attributes and utility of plants in the areas of wildlife-plant relations, livestock-plant relations, soil-plant relations, and man-plant relations.

Three major uses of PIN were anticipated and are now resulting from the system. These include (1) vegetation inventories—lists of species found in any county of the three states can be queried from the system, (2) environmental assessment and impact analysis—economic and ecologic value of species are recorded in the system and can be queried and used for anticipating the effects of disturbances or land use changes, (3) reclamation planning—queries can be constructed to obtain lists of plants adapted to site conditions and having desirable characteristics to fulfill reclamation objectives.

Information Available

The basic information entry in the system is by species. The species included are the ones for which there are vouchered specimens collected and on file in four herbaria. These herbaria are the Colorado State University Herbarium, the University of Colorado Museum Herbarium, the Rocky Mountain Herbarium at the University of Wyoming and the Montana State University Herbarium in Montana. Currently there is information on approximately 4,000 plant species entered in the system. The characteristics for these species are grouped into five categories. These categories include taxonomic, geographical, biological, economic, and ecological information. The information in these categories is divided into units called descriptors. At the present time there are 239 descriptors for each plant species in the system. These descriptors include 96 general bits of information in the 5 categories, plus the 143 counties of the three states. Within each descriptor, there are classes or ranks of information called descriptor states. The descriptors and descriptor state definitions (Vories and Sims, 1977) are listed in the Appendix.

Care must be exercised in interpreting the results of a PIN query. Where possible, definitions were used that are commonly accepted in the plant ecology, range science, and revegetation fields of expertise. However, because the descriptor states *must* be mutually exclusive this was not always possible. In these cases, working definitions as noted in the appendix were used. A brief list and description of some of these descriptors follow. This description includes descriptors that will be of most interest or those which may cause some confusion or misinterpretation.

Taxonomic. The taxonomic descriptors include the division, family, genus, species, infraspecific names and common names for each one of the species in the system. Most of this taxonomic information came from local floras and botanical literature. The infraspecific names may be varieites or subspecies. The common name descriptor information came from a ranked order of references. The first reference used was Plummer (1976), the Intermountain Range Plant Symbols. If Plummer did not have a common name for the species in PIN, then Kelsey and Dayton (1942), Standardized Plant Names was used. If neither of these two sources had a common name then Beetle's (1970) Recommended Plant Names was used.

Geographic. The geographic category of descriptors in PIN includes origin, counties, and minimum and maximum elevations at which the vouchered collections of that species have been found in each of the three states. The county locational information is of two levels of reliability. A county "present" record means that the species has a vouchered specimen collected and stored in one of the four herbaria listed above. A county "reported" record means that a reliable source has reported the species' occurrence in that county, but no specimen was collected.

Biologic. The biologic category of descriptors includes anthesis, carbon dioxide fixation, habit, life cycle, reproduction and trophic status. Anthesis is the mode month in which a plant flowers or pollinates. Habit is the growth form or outward appearance of the plants, such as grasslike, trees, or shrubs.

The life cycle descriptor includes the descriptor states of perennial, biennial, annual and combinations of these for plants that are not distinctly one or the other of the three.

The reproduction descriptor includes the sexual or asexual process by which a plant generates others of the same kind. The descriptor states for reproduction include sexual, vegetative, apomictic, and combination of these three.

Ecologic and Economic. The above descriptors and descriptor states have been completed for the 4000 species, if the information was available in documented sources. Because of the immensity of the task to develop the information base of ecological and economic descriptors and descriptor states for this many species, a subset of approximately 400 "important" species was chosen having high economic, ecologic, or biological values such as importance to grazing resources; importance to wildlife populations; importance as rare, threatened or endangered plants; reclamation value, or predominance in the three state area. For this 400 species the ecological and economic descriptor information was completed and added to PIN as described below.

The ecologic categories of plant descriptors include structural relationships or the proportional influence of a particular species in 20 vegetation types. These vegetation types are taken from Küchler (1964) and are the vegetation types he mapped in the three states included in PIN. Each species is ranked as dominant, codominant, subdominant or a component for each one of these 20 vegetation types.

Also included in the ecological descriptors are disturbance indicator; edaphic indicator; growth relationships of the species on textural classes of soils; habitat or ecological conditions under which a plant grows; the optimum slope on which the plant is normally found; the optimum soil depth on which the plant normally obtains the best growth; vegetation indicator; major dispersal agent for seed dispersal; mycorrhizal relationship; nodule forming information; nitrogen fixing information; potential biomass production; *and population dynamics, including endemic species, rareness in the three states, species stability or the relative vulnerability of the species to extinction including endangered, threatened, or vulnerable species by states.* The endangered and threatened states are determined by the latest U.S.D.I. Fish and Wildlife Service lists. The vulnerable state is determined by local taxonomists.

Because of the subject and emphasis of this symposium volume, this paper will concentrate on the Population Dynamics descriptors and other descriptors directly related to population dynamics of rare plants and their protection and enhancement. The examples to follow later in the paper will give some practical field problems exemplifying PIN use.

The economic category of descriptors include allergenic; edible; culture; erosion control potential; establishment requirements; revegetation potential; weediness; cover for various classes of wildlife, food value for cattle, sheep, horses and various classes of wildlife; forage value for cattle, sheep, horses, and various classes of wildlife; and poisonous to livestock.

The establishment requirements include descriptor states of high, medium, and low and describe the relative extent of cultural practices which must be employed to insure successful planning of the species.

The revegetation potential descriptor has descriptor states of high, medium, and low. This is the ability of a plant to become established and persist on sites to which it is adapted.

The culture descriptor is a text descriptor. Text descriptors are included in PIN in a narrative format and are retrieved in narrative form. All other descriptors are recorded as bits of information of one to three words usually. These descriptor states must be mutually exclusive and that is the reason that the descriptors, the descriptor definitions, and the descriptor states and their definitions must be strictly adhered to in querying or asking the system questions and interpreting the results. This mutual exclusiveness is one of the reasons that system is so efficient in using and keeping costs of use of PIN down.

Example Uses of PIN

Below are three examples of how PIN can be used in identifying rare, threatened or endangered plants that have been found to occur in specific localities. The most specific geographical descriptor that can be queried is counties. Hence, in the first example query (Table 1) the species listed are characterized by at least one of the following descriptor states: (1) endemic to Wyoming; or, (2) rare in Wyoming and non-

PRINT, GENUS, SPECIES, INFRASPECIFIC, ENDEMIC, RARE-WY, SPECIES STABILITY-
WY, MAXIMUM ELEVATION-WY, MINIMUM ELEVATION-WY, SWEETWATER-WY FOR ALL
SPECIES WITH (ENDEMIC, WYOMING OR CO-WY OR CO-WY-MT OR WY-MT OR (RARE-
WY, YES AND WEEDINESS, NON-WEEDY OR UNKNOWN) OR SPECIES STABILITY-WY,
ENDANGERED OR THREATENED OR VULNERABLE) AND SWEETWATER-WY, PRESENT OR
REPORTED*

COLUMN DESCRIPTORS
A =   GENUS--SPECIES--INFRASPECIFIC
B =   ENDEMIC
C =   RARE-WY
D =   SPECIES STABILITY-WY
E =   MAXIMUM ELEVATION-WY
F =   MINIMUM ELEVATION-WY
G =   SWEETWATER-WY

RESPONSE: A	B	C	D	E	F	G
ABIES CONCOLOR NONE		YES				P
ASTRAGALUS PROIMANTHUS NONE	WYOMING	YES	ENDANGERED	7100	7100	P
BALSAMORHIZA HISPIDULA NONE		YES				P
BRICKELLIA SCABRA NONE		YES				P
CHAMAECHAENACTIC SCAPOSA NONE		YES				P
CHENOPODIUM LEPTOPHYLLUM LEPTOPHYLLUM		YES		6800	6700	P
CHRYSOTHAMNUS GREENI NONE		YES				R
CRYPTANTHA CAESPITOSA NONE	WYOMING		VULNERABLE	7200	5400	P
DRABA OLIGOSPERMA NONE	NO	YES	VULNERABLE	10500	6900	P
PHACELIA DEMISSA NONE		YES				R
PINUS EDULIS NONE		YES		6900	6100	P
SEDUM LANCEOLATUM NONE		YES		10900	4500	P
TANACETUM CAPITATUM NONE	WY-MT		VULNERABLE	7500	6000	P
TOWNSENDIA SPATHULATA NONE	WYOMING	YES	THREATENED	5800	5400	P

NO. OF ITEMS IN QUERY RESPONSE = 31---NO. OF ITEMS IN THE DATA BANK = 3986
PERCENTAGE OF RESPONSE/TOTAL DATA BANK = .778

Table 1. Partial printout from PIN of rare or endemic or endangered plants of Sweetwater County,
Wyoming, and some selected habitat information about them.

weedy or unknown; or, (3) endangered or threatened or of vulnerable populations; and, (4) found in Sweetwater County, Wyoming. A complete list of the descriptors, descriptor states, and descriptor and descriptor state definitions are listed in the Appendix.

The PIN system can also be queried to determine the habitat characteristics for sites that a rare, threatened, or endangered plant may be expected to be found on. The habitat characteristic descriptors for rare, threatened, and endangered are not complete in the PIN data base at this time and additional work is being concentrated in this area to update and complete that information.

Another query form is to specify the site characteristics by descriptors and descriptor states and query for the rare, threatened, or endangered species one might expect to find on those site conditions. Table 2 is only a partial query printout of the rare plants found in Larimer County, Colorado. This example shows the second of two print formats that can be used in PIN print statements. Table 1 illustrates the tabular format and Table 2 illustrates a "string" format with the descriptor and descriptor states printed on one line and separated only by commas. The complete list for the Table 2 query contained 108 species but for the example only a few of these were included. These two examples also illustrate that the PIN system is very flexible in querying abilities. Any combination of species, descriptors, and descriptor states may be queried depending on the user's needs.

In some instances queries are asked of PIN for which the entire response is negative or for which the PIN data base is incomplete in that particular information.

```
RARE PLANTS OF LARIMER COUNTY                      DATE:  78/09/0

PRINT, GENUS, SPECIES, INFRASPECIFIC, COMMON NAME, D-LARIMER-CO FOR ALL
PLANTS WITH LARIMER-CO, PRESENT OR REPORTED AND RARE-CO YES*

RESPONSE:

ABRONIA, CARLETONII, NONE, UNKNOWN, LARIMER-CO:REPORTED
ACHILLEA, MILLEFOLIUM, MILLEFOLIUM, COMMON YARROW, LARIMER-CO:PRESENT
ACORUS, CALAMUS, NONE, DRUG SWEETFLAG, LARIMER-CO:PRESENT
ADOXA, MOXCHATELLIA, NONE, MUSKROOT, LARIMER-CO:PRESENT
AGALINIS, TENUIFOLIA, NONE, UNKNOWN, LARIMER-CO:PRESENT
AGASTACHE, FOENICULUM, NONE, FENNEL GIANTHYSSOP, LARIMER-CO:PRESENT
AGRIMONIA, STRIATA, NONE, ROADSIDE AGRIMONY, LARIMER-CO:PRESENT
AGROPYRON, TRICHOPHORUM, NONE, STIFFHAIR WHEATGRASS, LARIMER-CO:PRESENT
ALETES, HUMILIS, NONE, UNKNOWN, LARIMER-CO:PRESENT
SPOROBOLUS, HETEROLEPIS, NONE PRAIRIE DROPSEED, LARIMER-CO:PRESENT
STIPA, RICHARDSONII, NONE, RICHARDSON NEEDLEGRASS, LARIMER-CO:PRESENT
STIPA, SPARTEA, NONE, PORCUPINEGRASS, LARIMER-CO:PRESENT
TELESONIX, JAMESII, NONE, UNKNOWN, LARIMER-CO:PRESENT
VERBASCUM, BLATTARIA, NONE, MOTH MULLEIN, LARIMER-CO:PRESENT
VERONICA, SCUTELLATA, NONE, MARSH SPEEDWELL, LARIMER-CO:PRESENT
VIOLA, BIFLORA, NONE, TWINFLOWER VIOLET, LARIMER-CO:PRESENT
VIOLA, SELKIRKII, NONE, WILDERNESS VIOLET, LARIMER-CO:PRESENT

NO. OF ITEMS IN QUERY RESPONSE = 179
NO. OF ITEMS IN THE DATA BANK = 3986
PERCENTAGE OF RESPONSE/TOTAL DATA BANK = 4.491
```

Table 2. A partial query listing of rare plants found in Larimer County, Colorado.

```
ENDANGERED PLANTS SPECIES OF ROSEBUD, BIG HORN OR POWDER RIVER COUNTY,
MT*  DATE:  78/09/0

PRINT, GENUS, SPECIES, INFRASPECIFIC, ROSEBUD-MT, BIG HORN-MT, POWDER
RIVER-MT FOR ALL SPECIES WITH SPECIES STABILITY-MT, ENDANGERED
AND (ROSEBUD-MT, PRESENT OR REPORTED OR BIG HORN-MT, PRESENT OR
REPORTED OR POWDER RIVER-MT, PRESENT OR REPORTED)*

RESPONSE:   NO. OF ITEMS IN QUERY RESPONSE = 0
            NO. OF ITEMS IN THE DATA BANK = 3986
            PERCENTAGE OF RESPONSE/TOTAL DATA BANK = 0.000
```

Table 3. List of endangered plants found in Rosebud, Big Horn, or Powder River Counties, Montana. List of zero plants met these conditions.

In this case a response similar to Table 3 is received; for example, no species are currently listed in PIN as being rare, endemic, or endangered in Rosebud, Big Horn, or Powder River Counties, Montana as shown by Table 3.

The above three PIN queries were chosen as examples of how the Plant Information Network (PIN) may be queried. There are numerous other types of queries that may be processed and results received from PIN. The PIN system was designed to be very flexible and efficient to operate. Any combination of descriptors may be used for querying and the selection will depend upon a user's interests and needs.

The average cost of processing PIN queries is now about $8.50 for computer time. The costs range from about $1.00 to $100.00 The more complex queries requiring more printout and complex searches are more expensive for processing. The Plant Information Network is now available for public use. Charges for the use of the system are minimal and can be obtained by contacting the PIN staff. Either call or write the PIN staff to explain your information needs and they will give you an estimate of costs. The PIN address is:

> Philip L. Dittberner
> Plant Information Network
> U.S. Fish and Wildlife Service
> Fort Collins, CO 80526
> Telephone: 303-223-2040

Literature Cited

Beetle, A. A. 1970. Recommended plant names. *Agric. Exp. Sta. Res. J. (Wyoming)* 31. 124 pp.

Kelsey, H. P., and W. A. Dayton. 1942. *Standardized Plant Names* (2nd ed.) J. Harace McGarland, Harrisburg, Pa. 675 pp.

Küchler, A. W. 1964. *Potential Natural Vegetation of the Conterminous United States.* Special Publ. 36. American Geographical Society, New York. 39 pp. + map.

Plummer, A. P., S. B. Monson, and R. Stevens. 1976. *Draft Revision: Intermountain Range Plants Symbols.* Intermountain Forest and Range Exp. Sta., Ogden, Utah.

Vories, K. C., and P. L. Sims. 1977. *The Plant Information Network: A Users Guide.* USDI Fish and Wildlife Service, Office of Biological Services Western Energy and Land Use Team, Ft. Collins, Colo. 56 pp.

Appendix I. PIN Descriptors and Descriptor State Definitions

Taxonomic

The classification of plants into the following appropriate categories.

1. DIVISION—a non-scientific category of the major vascular plant groups.
   a. *conifer*—(Gymnospermae) a member of a group of predominantly evergreen and cone-bearing trees.
   b. *dicot*—(Dicotyledoneae) one of the two major divisions of angiosperms, characterized by a pair of embryonic seed leaves.
   c. *monocot*—(Monocotyledoneae) one of the two major divisions of angiosperms, characterized by a single seed leaf.
   d. *fern allies*—vascular plants not producing seeds or true flowers, reproducing by spores (Lycopodiophyta, Equisetophyta, Polypodiophyta).

2. FAMILY—a name category ranking below an order and above a genus. This category has the ending "-aceae."

3. GENUS—a name category ranking below a family and above a species. The use of the genus followed by a Latin adjective or epithet forms the scientific name of a plant.

4. SPECIES—the lowest, most commonly used category of taxonomic classification, ranking below a genus.

5. INFRASPECIFIC—a morphologically recognizable category of classification ranking below a species. Usually a *variety* or *subspecies*.

6. COMMON NAME—the colloquial epithet in general usage and language of the inhabitants of a geographic region.

Geographic

The attributes of a plant which pertain to a specific region.

7. ORIGIN—the geographic area to which a plant is indigenous.

   a. *Africa*
   b. *Asia*
   c. *Europe*
   d. *North America*—North America takes precedence over South America or other geographic areas.
   e. *South America*
   f. *Eurasia*
   g. *native*—any plant known to be indigenous to Colorado, Montana, or Wyoming. The descriptor state, "native," takes precedence over the descriptor state, "North America."

8. COUNTIES—each county name is a descriptor state and must be asked for specifically. Each county name ends with "-CO, -MT, or -WY" to designate the appropriate state it is in—Colorado, Montana or Wyoming, respectively. Includes all counties for Colorado (63), Montana (56), and Wyoming (24).

   a. *present*—a specimen of the plant has been deposited and verified in one of the following herbaria: Colorado State University; University of Colorado; U.S. Forest Service Herbaria, Fort Collins; Montana State University; Rocky Mountain Herbarium, Laramie, Wyoming.
   b. *reported*—the plant is known to occur in that county, but there is no herbarium record of it.

9. MAXIMUM ELEVATION-CO—the highest elevation at which a plant has been observed in Colorado. Recorded in 100 foot intervals from 3,000 to 15,000 feet.

10. MAXIMUM ELEVATION-MT—the highest elevation at which a plant has been observed in Montana. Recorded in 100 foot intervals from 1,500 to 13,000 feet.

11. MAXIMUM ELEVATION-WY—the highest elevation at which a plant has been observed in Wyoming. Recorded in 100 foot intervals from 3,000 to 15,000 feet.

12. MINIMUM ELEVATION-CO—the lowest elevation at which a plant has been observed in Colorado. Recorded in 100 foot intervals from 3,000 to 15,000 feet.

13. MINIMUM ELEVATION-MT—the lowest elevation at which a plant has been observed in Montana. Recorded in 100 foot intervals from 1,500 to 13,000 feet.

14. MINIMUM ELEVATION-WY—the lowest elevation at which a plant has been observed in Wyoming. Recorded in 100 foot intervals from 3,000 to 15,000 feet.

Biologic

The attributes of a plant which pertain to its own life processes.

15. ANTHESIS—for angiosperms, this is the time of *flowering* of a plant; for gymnosperms, this is the time of *pollination* of a plant. Anthesis time is recorded by month, or by the *mode month* if it occurs in more than one month. Grasses, sedges, and rushes are generally not scored for this descriptor.

The plant flowers or pollinates during the month of: a. January, b. February, c. March, d. April, e. May, f. June, g. July, h. August, i. September, j. October, k. November and l. December.

16. CARBON DIOXIDE FIXATION—the biochemical and physiological mechanism associated with incorporation of CO_2 and its ultimate conversion into carbohydrates.

   a. C_4—the plant uses a pathway where the first step in CO_2 fixation involves the formation of four-carbon compounds, the stomata are open and CO_2 is fixed in the daylight.
   b. C_3—the plant uses a pathway where the first step in CO_2 fixation involves the formation of three-carbon compounds, the stomata are open and CO_2 is fixed in the daylight.
   c. *crassulaceous*—the plant uses a pathway where the first step in CO_2 fixation involves the formation of four-carbon compounds. The stomata are open and CO_2 is fixed in the dark.
   d. *other*—the plant uses another type of pathway.
   e. *none*—the plant does not fix carbon dioxide.

17. HABIT—the growth form, or outward appearance of a plant.

   a. *tree*—a woody plant that usually produces one main trunk or bole and a more or less distinct and elevated head.
   b. *shrub-tree*—a plant whose growth form is intermediate between that of a shrub and tree.
   c. *shrub*—a woody plant that remains low and produces shoots or trunks from the base.
   d. *vine*—any woody plant whose stem requires support, and which climbs by tendrils or other means.
   e. *forb*—a non-woody plant dying down each year that is not grass-like.
   f. *grasslike*—herbaceous plants with narrow leaves usually belonging to the grass, sedge, and rush families.

18. LIFE CYCLE—the series of stages in form and mode of life which an organism exhibits between successive recurrences of a certain primary stage.

a. *perennial*—the plant grows for three or more years duration.
b. *biennial*—the plant grows for two years duration from seed to maturity to death.
c. *annual*—the plant grows for one years duration from seed to maturity to death.
d. *perennial-biennial*—the plant has the potential for growth as either a perennial or a biennial.
e. *biennial-annual*—the plant has the potential for growth as either a biennial or an annual.
f. *perennial-annual*—the plant has the potential for growth as either a perennial or an annual.

19. REPRODUCTION—the sexual or asexual process by which a plant generates others of the same kind.

a. *sexual*—the plant reproduces by pollination and fertilization.
b. *vegetative*—all cases where structures such as bulbils, tubers, stolons, rhizomes, etc., which are normally accessory means of reproduction, take over the whole reproductive process of a plant.
c. *apomictic*—the plant has a type of reproduction which results in the formation of seeds and embryos by a non-sexual process.
d. *vegetative-sexual*—the plant reproduces sexually and vegetatively.
e. *sexual-apomictic*—the plant reproduces sexually and apomicticly.
f. *vegetative-sexual-apomictic*—the plant reproduces by all three methods.

20. TROPHIC STATUS—a plant's method of nutrient procurement.

a. *autotrophic*—the plant is capable of self-nutrition; can use carbon, nitrogen, and sulfur in inorganic combinations and obtain energy from the sunlight.
b. *parasitic*—the plant lives on and/or in other living organisms and obtains some or all of its nutrients from the host.
c. *saprophytic*—the plant lives on and/or in dead organic material and obtains nutrients from it.
d. *symbiotic*—the plant lives in close association with another plant and the symbiots derive nutritional requirements from each other.

Ecologic

The attributes of a plant which pertain to its relationship to community structure and function, environment and population dynamics.

Structural Relationships

Relative Dominance—the proportional influence of a plant within each of the vegetation zones listed below. Each zone is a descriptor and must be asked for separately. Vegetation zones as defined by Küchler (1964). Descriptors 21-40 use the following states:
a. *dominant*—a plant, which by means of its number, coverage, or size has major influence upon the environmental conditions within the vegetation type.
b. *codominant*—a plant, which by means of its number, coverage, or size, has in association with other plants a major influence upon the environmental conditions within the vegetation type.
c. *subdominant*—a plant, which by means of its number, coverage, or size, has a moderate influence upon the environmental conditions within the vegetation type.
d. *component*—a plant species existing in the vegetation type exclusive of dominant, codominant, and subdominant.

21. ALPINE MEADOWS/BARREN

22. WESTERN SPRUCE-FIR FOREST

23. SW SPRUCE-FIR FOREST (SW = Southwestern)

24. DOUGLAS FIR FOREST

25. PINE-DOUGLAS FIR FOREST

26. BLACK HILLS PINE FOREST

27. WESTERN PONDEROSA FOREST

28. EASTERN PONDEROSA FOREST

29. NORTHERN FLOODPLAIN FOREST

30. JUNIPER-PINYON WOODLAND

31. MOUNTAIN MAHOGANY-OAK SCRUB

32. GREAT BASIN SAGEBRUSH

33. SALTBUSH-GREASEWOOD

34. SAGEBRUSH STEPPE

35. WHEATGRASS-NEEDLEGRASS SHRUBSTEPPE

36. FOOTHILLS PRAIRIE

37. SANDSAGE-BLUESTEM PRAIRIE

38. WHEATGRASS-NEEDLEGRASS

39. GRAMA-WHEATGRASS-NEEDLEGRASS

40. GRAMA-BUFFALOGRASS

41. DISTURBED AREA—the relative dominance of a plant on areas of environmental disruption.

   a. *dominant*—the plant becomes a dominant on disturbed areas during one or more of the early stages in natural revegetation through processes of secondary succession.
   b. *codominant*—the plant becomes a codominant on disturbed areas during one or more of the early stages in natural revegetation through processes of secondary succession.
   c. *subdominant*—the plant becomes a subdominant or disturbed areas during one or more of the early stages in natural revegetation through processes of secondary succession.
   d. *component*—the plant becomes a component on disturbed areas during one or more of the early stages in natural revegetation through processes of secondary succession.
   e. *no*—the plant does not normally occur on disturbed areas.

Environmental Relationships

42. DISTURBANCE INDICATOR—a plant whose growth and distribution commonly indicates one of the following types of disturbance. (Indicators are only scored once in the order of precedence shown.)

   a. *erosion*—the general process of the wearing away of rocks and soils at the earth's surface by natural agencies.
   b. *mechanical*—the physical disturbance of the soil and vegetation by trampling of man or animals or machinery used in road building or agriculture.
   c. *overgrazing*—excessive feeding by domestic or wild animals.
   d. *fire*—disturbance by burning.
   e. *other*—indicates a disturbance not listed.
   f. *no*—the presence of the plant does not indicate a disturbance, nor will it grow in a disturbed area.

43. EDAPHIC INDICATOR—a plant whose growth and distribution commonly indicates the presence of one of the following unusual soil characteristics.

   a. *boron*—soils which contain boron-containing minerals such as borax, sassolite, ulexite, colemanite, boracite, tourmaline.
   b. *gypsum*—soils contain hydrous calcium sulfate.
   c. *selenium*—soils which contain selenium or selenids such as clausthalite.
   d. *serpentine*—soils which contain hydrous magnesium silicate, indication of a magnesium-calcium imbalance.
   e. *very acidic*—soil pH less than 5.
   f. *saline*—soils with a conductance of saturation extract exceeding 4 mmho/cm but with Na comprising less than 15% of the absorbed cations.
   g. *very alkaline*—soils pH greater than 8.5
   h. *saline-alkaline*—plants are indicators of either or both saline or alkaline soils.
   i. *other*—other unusual soil characteristics may be indicated. User should consult expert.
   j. *none*—plant does not indicate any unusual edaphic characteristic.

Texture and Growth Relationships

The relative ability of a plant to show the full development of all phases of its growth potential on a particular soil texture where the plant normally occurs. Descriptors 44-48 use the following states:

   a. *good*—the plant is highly adapted to growth on a particular soil texture.
   b. *fair*—the plant is moderately adapted to growth on a particular soil texture.
   c. *poor*—the plant shows little adaptability to growth on a particular soil texture.

44. GROWTH ON SAND—a soil in which the sand separates (0.05 mm and larger) make up 70% or more of the material by weight.

45. GROWTH ON SANDY LOAM—a loamy soil which is intermediate in texture between sand and loam.

46. GROWTH ON LOAM—a soil which is considered to have an ideal texture for gardening. It should contain about equal amounts of silt (0.05-0.002 mm) and sand, and less than 25% clay.

47. GROWTH ON CLAY LOAM—a loamy soil which is intermediate in texture between clay and loam.

48. GROWTH ON CLAY—a soil must have at least 35% clay separates (less than 0.002 mm) by weight.

49. HABITAT—the type of locality or set of ecological conditions under which a plant grows.

   a. *submerged aquatic*—the plant grows in a fresh water environment with its vegetative parts not rising above the water surface.
   b. *emergent aquatic*—the plant grows in a fresh water environment with its vegetative parts rising above the water surface.
   c. *wet*—the plant grows in soil that is saturated with water.
   d. *moist*—the plant grows in soil that is characterized by conditions of medium soil moisture.
   e. *dry*—the plant grows in soil that is characterized by conditions of extended periods of soil drought.
   f. *epiphytic*—a plant which germinates on other plants and grows without obtaining nutriment at the cost of the substance of the host.
   g. *phreatophyte*—a plant which derives its water supply from the water table and is more or less independent of rainfall.

50. OPTIMUM SLOPE—the slope conditon on which the plant is normally found.

   a. *level-sloping*—0-8% slope.
   b. *rolling-hilly*—9-30% slope.
   c. *steep-very steep*—31% and greater slope.

51. OPTIMUM SOIL DEPTH—depth of soil to parent material on which the plant normally obtains best growth. Measured in inches.

  a. 0-6
  b. 7-12
  c. 13-24
  d. 25-36
  e. 37-60
  f. 61-120

52. VEGETATION INDICATOR—the plant's presence must indicate exclusively one of the following vegetation zones. Zones defined by Küchler (1964).

  a. *Alpine Meadows/Barren*
  b. *Western Spruce-Fir Forest*
  c. *SW Spruce-Fir Forest* (SW = Southwestern)
  d. *Douglas Fir Forest*
  e. *Pine-Douglas Fir Forest*
  f. *Black Hills Pine Forest*
  g. *Western Ponderosa Forest*
  h. *Eastern Ponderosa Forest*
  i. *Northern Floodplain Forest*
  j. *Juniper-Pinyon Woodland*
  k. *Mountain Mahogany-Oak Scrub*
  l. *Great Basin Sagebrush*
  m. *Saltbush-Greasewood*
  n. *Sagebrush Steppe*
  o. *Wheatgrass-Needlegrass*
  p. *Foothills Prairie*
  q. *Sandsage-Bluestem Prairie*
  r. *Wheatgrass-Needlegrass*
  s. *Grama-Wheatgrass-Needlegrass*
  t. *Grama-Buffalograss*

Functional Relationships

53. MAJOR DISPERSAL AGENT—the primary agent of seed dispersal (descriptor states self-explanatory).

  a. *birds*
  b. *mammals*
  c. *insects*
  d. *gravity*
  e. *water*
  f. *wind*
  g. *other*

54. MYCORRHIZAL RELATIONSHIP—the nature of the relationship of a plant to a mycorrhizal association.

  a. *endomycorrhizal*—mycorrhizal association having a loose network of fungal hyphae enclosing the root and intracellular hyphae penetrating the cortical cells of the root.
  b. *ectomycorrhizal*—mycorrhizal association having a dense fungal sheath enclosing the root and intercellular hyphae penetrating the root cortex.
  c. *ectendomycorrhizal*—mycorrhizal association having a dense fungal sheath enclosing the root and both inter- and intra-cellular hyphae penetrating the root cortex.
  d. *endo/ecto*—refers to plants reported as being both endomycorrhizal and ectomycorrhizal.
  e. *ecto/ectendo*—refers to plants reported as being both ectomycorrhizal and ectendomycorrhizal.

161

f. *nonmycorrhizal*—refers to either plants that have been examined for mycorrhiza with none found, or plants that occur in families considered to be classically nonmycorrhizal (Aizoaceae, Amaranthaceae, Brassicaceae, Caryophyllaceae, Chenopodiaceae, Commelinaceae, Cyperaceae, Fumariaceae, Juncaceae, Nyctaginaceae, Polygonaceae, and Urticaceae).

55. NODULE FORMING—occurrence of root nodules on a plant's roots.

    a. *reported*—reported as nodule forming by observation or in the literature.
    b. *possible*—reported as possible nodule forming in the literature.
    c. *no*—plant is reported in the literature as not forming nodules.

56. NITROGEN FIXING—a plant that can assimilate and fix the free nitrogen of the air with the aid of microorganisms.

    a. *yes*—plant fixes nitrogen, as reported in the literature.
    b. *maybe*—the plant may fix nitrogen, but has not been reported as such in the literature.
    c. *no*—plant is known not to fix nitrogen.

57. POTENTIAL BIOMASS PRODUCTION—the relative ability of a plant to produce plant material by weight on an annual basis as a major component of an established stand, within a *comparable lifeform*.

    a. *high*—plant possesses ability to produce a yield of dry plant material comparable to moist forests or tall grass prairies.
    b. *medium*—plant possesses ability to produce a yield of dry plant material comparable to mountain forests or mid grass prairies.
    c. *low*—plant possesses ability to produce a yield of dry plant material comparable to semi-arid woodland, shrublands, or short grass prairies.
    d. *very low*—plant possesses ability to produce a yield of dry plant material comparable to slow growing plants from arid climates.

Population Dynamics

58. ENDEMIC—confined to a certain area or region, having a comparatively restricted distribution.

    a. *Colorado*—plant population is confined to the state of Colorado.
    b. *Montana*—plant population is confined to the state of Montana.
    c. *Wyoming*—plant population is confined to the state of Wyoming.
    d. *CO-WY*—plant population is confined to Colorado and Wyoming.
    e. *WY-MT*—plant population is confined to Wyoming and Montana.
    f. *CO-WY-MT*—plant population is confined to Colorado, Wyoming, and Montana.

59. RARE-CO—a plant which has a small population in its range in Colorado. It may be found in a restricted geographic region, or it may occur sparsely over a wider area.

    a. *yes*—the plant is rare in Colorado.
    b. *no*—the plant is not rare in Colorado.

60. RARE-MT—a plant which has a small population in its range in Montana. It may be found in a restricted geographic region, or it may occur sparsely over a wider area.

    a. *yes*—the plant is rare in Montana.
    b. *no*—the plant is not rare in Montana.

61. RARE-WY—a plant which has a small population in its range in Wyoming. It may be found in a restricted geographic region, or it may occur sparsely over a wider area.

    a. *yes*—the plant is rare in Wyoming.
    b. *no*—the plant is not rare in Wyoming.

62. SPECIES STABILITY-CO—the relative vulnerability of a plant to extinction in Colorado. Descriptors 62-64 use the following states.

    a. *endangered*—any plant which is in danger of extinction throughout all or a significant portion of its range, due to change in habitat, disease, predation, exploitation, etc.
    b. *threatened*—any plant which is likely to become an endangered species within the foreseeable future throughout all or a significant portion of its range. This includes species categorized as rare, very rare, or depleted.
    c. *vulnerable*—a plant which should be monitored for possible decreases in range and/or number and is not yet considered endangered or threatened.
    d. *good*—a plant whose population is stable or increasing.

63. SPECIES STABILITY-MT—the relative vulnerability of a plant to extinction in Montana.

64. SPECIES STABILITY-WY—the relative vulnerability of a plant to extinction in Wyoming.

Economic

The attributes of a plant which have either positive or negative monetary value.

Human Health and Nutrition

65. ALLERGENIC—inducing an allergic response in humans.

    a. *yes*—the plant is reported in the literature as allergenic.
    b. *maybe*—the plant is reported to possibly cause allergic response; thought to be allergenic but not yet proven so.
    c. *no*—the plant is known to be definitely not allergenic in any circumstances and reported so in the literature.

66. EDIBLE—a plant which can be eaten as food by humans.

    a. *yes*—the plant is edible.
    b. *yes-qualified*—it is edible only after a specific preparation or in certain seasons. User should consult an expert.
    c. *no*—the plant is not edible but not poisonous.
    d. *poisonous*—the plant contains toxic substances or potential toxic substances that would prove harmful if ingested.

Revegetation Planting

67. CULTURE—a text descriptor on the planting requirements of a species.

68. EROSION CONTROL POTENTIAL—a plant which commonly exhibits growth habit, plant structure, and/or biomass that potentially reduces soil erosion.

    a. *high*—plant has aggressive growth habits, persistent plant structure, high potential biomass, and good soil-binding root or root-rhizome-runner systems in established stands.
    b. *medium*—plant has moderately aggressive growth, moderately persistent plant structure, good potential biomass, and good soil-binding characteristics in established stands.
    c. *low*—plant has growth, persistence, biomass, or soil-binding characteristics which make it generally inadequate for erosion control.

69. ESTABLISHMENT REQUIREMENTS—the relative extent of cultural practices which must be employed to insure a successful planting of the species on sites to which it is adapted.

    a. *high*—requires elaborate, energy intensive cultural practices.

b. *medium*—requires special cultural practices of short duration and/or moderate resource inputs.

c. *low*—only minimal cultural practices are required.

70. VEGETATION POTENTIAL—the ability of a plant to become established and persist on sites to which it is adapted.

   a. *high*—plant demonstrates good growth, cover, reproduction, and stand maintenance characteristics.

   b. *medium*—plant demonstrates fair growth, cover, reproduction, and stand maintenance characteristics.

   c. *low*—plant demonstrates poor growth, cover, reproduction, and stand maintenance characteristics.

71. WEEDINESS—a plant considered undesirable, unattractive, or troublesome; especially one that is growing where it is not wanted.

   a. *noxious*—plants which are listed on official noxious weed seed lists of Colorado, Montana, and Wyoming. User should consult authorities for individual state lists. Has priority over economic.

   b. *economic*—plants whose growth and reproduction cause economic loss.

   c. *colonizing*—a plant which has attributes enabling it to become easily established in suitable habitats.

   d. *non-weedy*—not a weed.

Wildlife and Livestock

Cover Value—the degree to which the plant provides environmental protection for an animal. Descriptors 72-77 use the following states:

   a. *good*—readily utilized for cover when available.

   b. *fair*—moderately utilized for cover when available.

   c. *poor*—rarely utilized for cover when available.

72. ELK COVER VALUE

73. MULE DEER COVER VALUE

74. ANTELOPE COVER VALUE

75. GAME BIRD COVER VALUE

76. SMALL NON-GAME BIRD COVER VALUE

77. SMALL MAMMAL COVER VALUE

Food Value—the relish and degree of use that an animal shows for a particular plant, or plant part as well as its availability. Descriptors 78-86 use the following states.

   a. *good*—readily available in the plant's range and consumed to a high degree.

   b. *fair*—readily to moderately available in the plant's range but consumed only to a moderate degree.

   c. *poor*—may or may not be available but the plant is consumed very little by the animal.

78. CATTLE FOOD VALUE

79. SHEEP FOOD VALUE

80. HORSE FOOD VALUE

164

81. ELK FOOD VALUE

82. MULE DEER FOOD VALUE

83. ANTELOPE FOOD VALUE

84. GAME BIRD FOOD VALUE

85. SMALL NON-GAME BIRD FOOD VALUE

86. SMALL MAMMAL FOOD VALUE

Forage Value—the relish and degree of use (i.e. palatability) that an animal shows for a particular plant or plant part. Descriptors 87-95 use the following states.

    a. *good*—highly relished and consumed to a high degree.
    b. *fair*—moderately relished and consumed to a moderate degree.
    c. *poor*—not relished and normally consumed to only a small degree.

87. CATTLE FORAGE VALUE

88. SHEEP FORAGE VALUE

89. HORSE FORAGE VALUE

90. ELK FORAGE VALUE

91. MULE DEER FORAGE VALUE

92. ANTELOPE FORAGE VALUE

93. GAME BIRD FORAGE VALUE

94. SMALL NON-GAME BIRD FORAGE VALUE

95. SMALL MAMMAL FORAGE VALUE

Toxicity

96. POISONOUS-LIVESTOCK—a plant which contains or produces, under natural conditions, physiologically active or toxic substances in sufficient amounts to cause harmful effects in animals.

    a. *acute*—animal shows immediate symptoms after ingestion of a small amount of the plant. Ingestion of the plant may cause rapid death.
    b. *cumulative*—animals usually require repeated ingestion of poisonous plants and concentrations in their system to show symptoms.
    c. *geographically variable*—the plant is poisonous in certain parts of its range. User should check with expert.
    d. *seasonably variable*—the plant is poisonous during certain phases of its growth cycle or as a response to abnormal weather conditions. User should consult with expert.

Information Handling for Southern Africa's Rare and Endangered Species Survey

A. V. Hall[1]

About 65% of Southern Africa's 1,915 known threatened and critically rare plant species are concentrated in a small region in the south and south-west, forming less than 1% of the total area of the subcontinent. Studies of this phenomenal concentration and threatened plants in adjacent regions are supported by an information system. This consists of descriptive dossiers on each species, cross-indexed by a computer program operating on a short-record data bank. The computer gives guides to the survey team on the best areas to visit and what will be in flower in each local area. A geographical grid-reference system using latitude and longitude allows easy sorting of the records as well as a high level of archival permanence. The value is emphasized of adapting the information system to a dynamic role of following changes in the conservation status of species.

The first survey of Southern Africa's rare and endangered plant species was set in motion in February 1974. It owed its origin to Dr. R. Melville and his co-workers of the Threatened Plants Committee of the International Union for the Conservation of Nature and Natural Resources (IUCN). Dr. Melville rightly sensed that numbers of endangered species would be found in South Africa, especially in the south-western Cape Province. Through the Science Co-ordinator of South Africa's National Programme for Environmental Sciences, Mrs. Enid P. du Plessis, Cape botanists were asked what species might be threatened or endangered in their area. The task proved larger than expected, so proposals were made to the National Programme for Environmental Sciences for funds, guidance, and assistance that would set the survey on a firm footing.

The proposals were accepted and a survey was started in the rich floras of the south-western part of the country, in an area of the Cape Province bounded in the north by the banks of the Orange River and in the east by the 26°E meridian, close to Port Elizabeth. More recently, other parts of the country have come under scru-

[1] Bolus Herbarium, University of Cape Town, Rondebosch 7700, South Africa.

Based on the presentation "Information Handling for South Africa's Rare and Endangered Species Survey" by Dr. Hall at the Symposium on Geographical Data Organization for Rare Plant Conservation, held 15-17 November 1977 at The New York Botanical Garden.

Pages 167-184 *in:* Larry E. Morse and Mary Sue Henifin, eds., *Rare Plant Conservation : Geographical Data Organization*, The New York Botanical Garden, Bronx, N.Y.

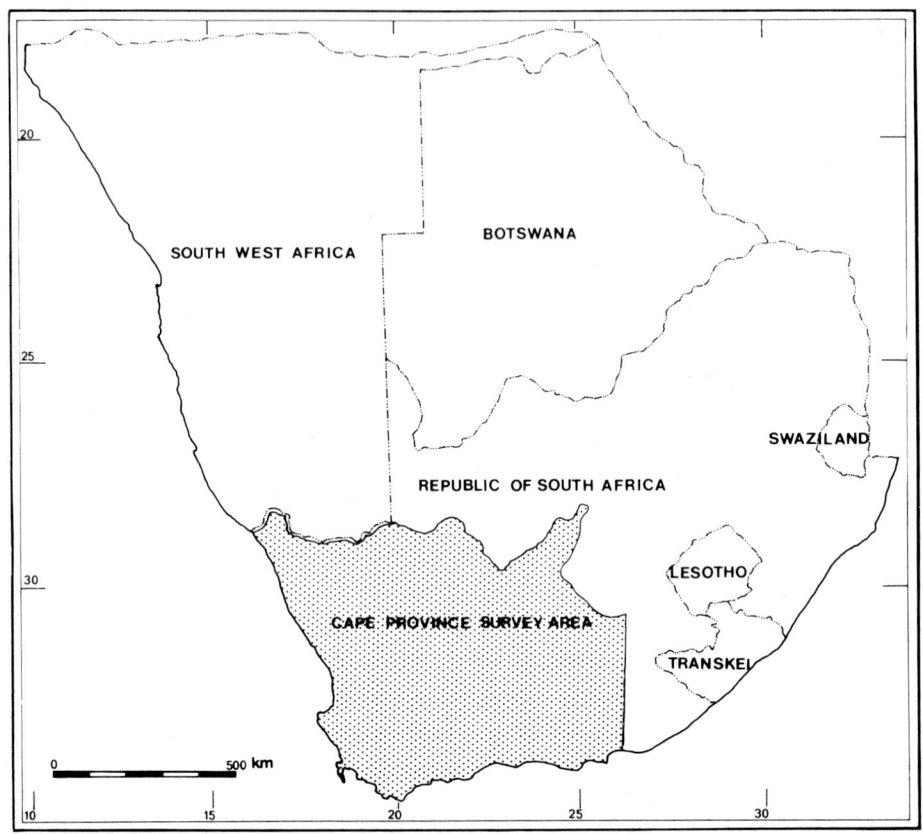

Figure 1. Countries in Southern Africa for which a preliminary list of rare and endangered plant species is in preparation. The information system for the Cape Province Survey area is the main subject of this paper.

tiny. A preliminary list of rare and endangered plant species in South Africa and neighbouring territories has been prepared. It contains 1,915 taxa (Hall *et al.,* in press). Of these, 1,244 (or 65% of the total) are concentrated in the coastal plains and mountains of the south and south-west (see Figures 1 and 2).

It is in this region that the Cape Macchias are found, which form one of the richest concentrations of flowering plant species in the world (Good, 1964). About 6,000 species are confined to an area formerly covering 46,000 square kilometers (Acocks, 1953). Satellite photographs now show the areas of wild vegetation to be reduced to remnants mostly along narrow mountain corridors. These areas total only 18,000 square kilometers, a reduction of about 61% (Hall, 1978). Such a severe reduction in area gives concern for the conservation of individual species, especially as many have rather limited geographical ranges. This concern is accentuated by the fact that plant geographers recognise the Cape Macchias, an assemblage of hard-leaved shrubs and herbs, as forming on their own one of the six Floristic Kingdoms of the world (Good, 1964; Walter, 1968; Takhtajan, 1969). Most of the former area has been taken up by farmland. Important local ecosystems have been replaced by wheatfields,

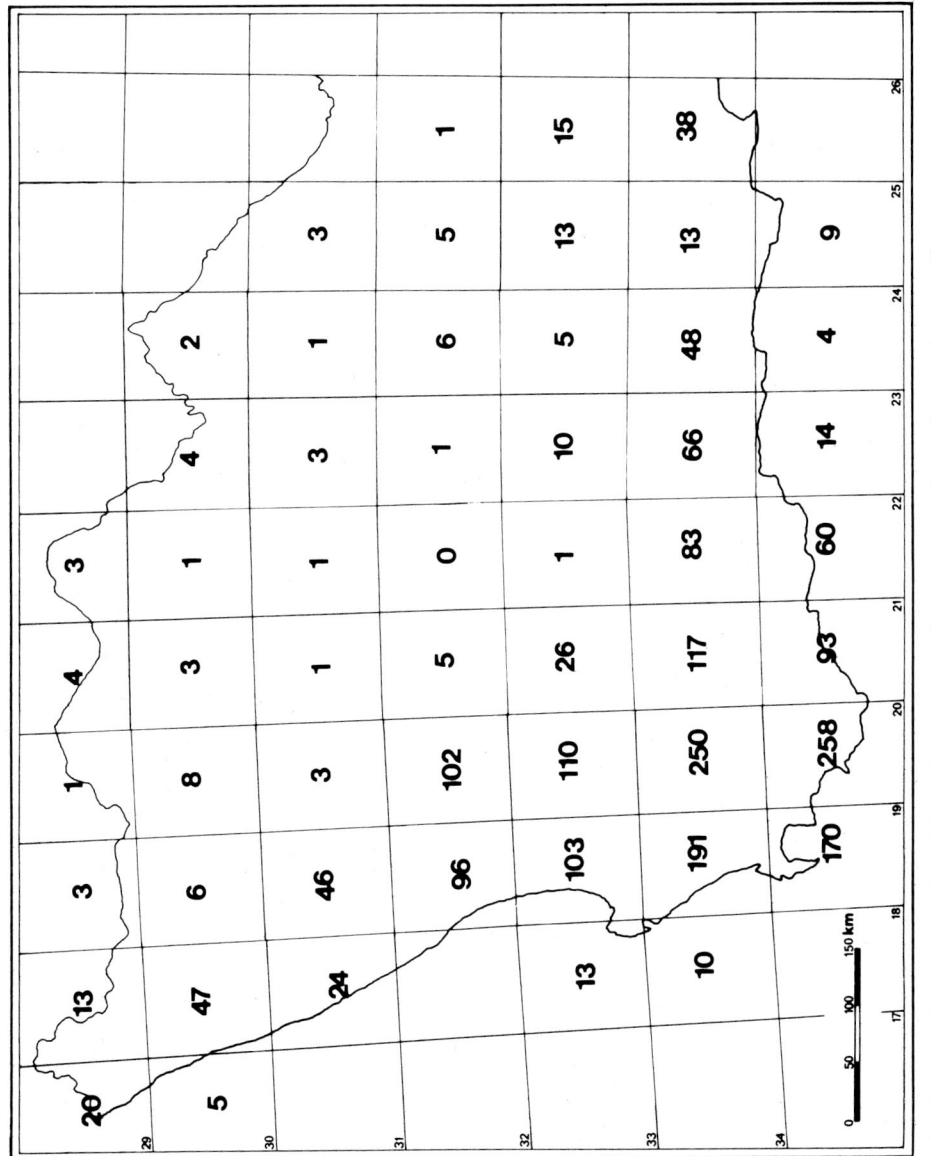

Figure 2. Occurrences of candidates for rare and endangered status in one-by-one degree grid areas in the Cape Province Survey area.

169

timber plantations, towns, dams and the other reasonable needs for a human population that in South Africa as a whole is increasing by 60,000 persons per month (Hall, 1978). The vegetation has long boundaries of contact with human pressures. These include damage from accidental fires, the destruction of pollinators by farmers' insecticides, and a widespread invasion by escaped exotic trees and shrubs that have been introduced from Australia and southern Europe. Some 24% of the remaining area for the Cape Floristic Kingdom is lightly to heavily infested with these exotics. Thickets of the exotics *Pinus pinaster, Hakea, Acacia,* and *Leptospermum* can be seen spreading widely over the mountains and flats that form the remaining home of this superb flora.

This situation looked ominous before intensive surveys started in this region. However, it was not expected that so many species would be found to be threatened in the study area. Special information-handling measures were needed, and the experience gained in designing them is the chief subject of this paper. The measures have followed the three broad phases of the work: herbarium studies, field surveys, and proposals for conservation.

Herbarium Studies

The Cape Province survey had to be based initially on information gleaned from specialists and collections in herbaria. The first task was to examine the herbaria for the region, which together hold about half a million specimens. With the aid of specialists in various groups, taxa were listed as potential candidates for endangered status if they were restricted to one or a few very localized areas not more than several square kilometers in extent, or had been found more widely but at very low frequencies. Taxa were included that had habitats that were under human pressures such as agriculture, forestry, town development, or flower-picking. In view of the incompleteness of some of the taxonomic revisions in South Africa, difficult groups such as the succulent Mesembryanthemaceae presented special problems. Recent revision work in this family had shown that many species represented by one or two specimens should be reduced to synonymy (R. P. Glen, pers. comm.). Taxa in such families could only be admitted to the list if endangerment was firmly suspected by present-day specialists. While some endangered taxa may have been overlooked by this compromise, there was no reasonable alternative. It was much easier to check that the suspect taxon was not redundant in better classified groups where its close relatives were clearly shown. A few infra-specific taxa had to be included in the list, in cases where it was likely that future taxonomic revisions might elevate them to species status. However, no systematic consideration has yet been made of threatened infraspecific taxa generally.

This first listing process gave a total of nearly 2,000 candidates for some status of rarity and endangerment. A second process, the compilation of a dossier on each candidate, brought with it a more critical scrutiny that eliminated many taxa, so that the total at the time of writing stands at 1,450.

The dossier consists of a manila folder containing a descriptive 210 × 298 mm card and copies of field notes, maps, and any other information that comes to hand on the taxon. The descriptive card is designed for recognizing the taxa in the field and carries photographs of herbarium specimens, copies of drawings and descriptions, and tables of distribution, ecology and flowering times. Carried in transparent envelopes in a ring-back hard-cover field file, these cards have been of great help in

Table 1.
List of categories and candidates for rare and endangered status in the area of the survey in the Cape Province, South Africa.

Extinct (38 species): species no longer known to exist in the wild after repeated searches of all former and other possible localities. This category is also used for the species that are extinct in the wild but survive in at least some form in cultivation.

Endangered (68 species): species in immediate danger of extinction if the causal factors continue operating. Included are taxa with populations so critically reduced that a breeding collapse due to lack of genetic diversity becomes possible, whether or not they are threatened by human activity.

Vulnerable and Declining (84 species): species that had been more widespread but are on the decline, and are likely to become endangered if the causal factors continue operating.

Rare (278 species): a species with a relatively small world population that is not declining and is under no known threat. Because of its rarity, such a species should be checked regularly for a decline due to some unexpected pressure.

Indeterminate (218 species): a species known to be in one of the categories Endangered, Vulnerable, or Rare, but not studied fully enough to be placed convincingly in one category in preference to another.

Uncertain Whether Safe or Not (773 species): a species for which there is so little information, due to lack of field evidence, that there is a real possibility that it could prove to be safe. Species seldom collected due to the remoteness of their habitat may appear to be rare or even endangered from available records and must be placed temporarily in this category to await the arrival of more information.

Neither Rare nor Threatened Outside the Survey Region (no species): species which belong to one of the above categories but are neither rare nor threatened where they naturally exist elsewhere. (No figure is given for this for the Cape Province survey as it is being integrated with studies in neighbouring territories, and none of the taxa listed for the Cape are found elsewhere).

---

seeking rare taxa in the Cape's rich floras. In difficult groups such as the Restionaceae, where minute characters are needed to confirm identifications and match separate male and female plants, close-up photographs on the cards proved helpful. In the Mesembryanthemaceae, the cards carry statements that distinguish the candidate taxon from look-alikes in the region concerned. This has proved helpful in other groups.

The dossier-card also carries the best current estimate of the taxon's conservation status according to the criteria of the Threatened Plants Committee of the IUCN (Lucas & Walters, 1976). In view of the wide range of interpretations given in the scientific and legal literature to the words expressing endangerment, the IUCN criteria are of special interest as they effectively offer a world standard. The numbers of taxa are given in Table 1 for each category for the total area of the survey, about 468,000 square kilometers, in the Cape Province, South Africa.

Locality Information

A precise way of describing localities is an essential adjunct to rare plant studies. The method should be usable with a wide range of maps and should have enough permanence for archival use. Names of areas are often used for broad regional location, followed by a more accurate designation. If the area-names refer to districts or counties, the boundaries, and even the names themselves, may change with time. Using them in studies of old records can be frustrating, especially when almost unobtainable rare maps need to be repeatedly consulted. It is especially misleading when a currently well-known district-name, such as Uitenhage in South Africa, has undergone major previous reductions in area. With a long history of these imperfections, South African botanists recently agreed to use the grid-areas bounded by the lines of latitude and longitude. These are given on all better-quality maps and are likely to be as permanent as the needs of future archival studies require. Edwards and Leistner (1971) refined this method by proposing the following standards, which are now widely used:
1. The main grid units are areas bounded by lines of one degree of latitude and longitude.
2. A one-degree grid-unit is referred to by the latitude and longitude of its corner that lies closest to the equator and the Greenwich Meridian, respectively. The name of the chief town or feature of importance may be added to give clarity to what would otherwise be a row of numbers in a list. The one-degree grid-unit for the location of this Symposium would therefore be 40 73 New York.
3. The one-degree grid can be divided into four quarters, designated by letters A to D. Quarter A is set in the corner giving the reference coordinates of the one-degree grid—the corner nearest the equator and the Greenwich Meridian, Quarter B lies next to A, C is above or below A, and D above or below B, depending on whether the locality is in the Northern or Southern hemisphere, respectively. The region of this Symposium could then be given more accurately as 40 73 D New York. A further quartering gives still greater accuracy: 40 73 DD New York.

No standard has been set for the many ways used to describe the exact location of a rare plant. Very local place-names may have a limited life in the changing patterns of land-use and may only appear for a few decades on large-scale maps. Once again, latitude and longitude coordinates offer better accuracy and precision for long-term reference, and are being used as much as possible by the Southern African Rare and Endangered Plant Species Survey. Of great assistance has been the publication of a gazetteer of the names appearing on all important maps of South Africa, including a few from those of the early collectors (Leistner & Morris, 1976). With 42,324 entries, this document has shown clearly the fallibility of the place-names method: on page after page there are lists of homonyms referring to quite different places, sometimes even in the same district.

Computer Aids

In spite of the effort in preparing the dossiers, they would have effectively remained

Table 2
Commands for the computer program for producing expedition guides for the Cape Province rare and endangered plant species survey.

COMMAND	DESCRIPTION OF ACTION
NEW:	Start a new bank in a named disc file, placing in it the records that follow.
ADD:	Place the records that follow this card in the named disc file started by the command 'NEW'. In the case of a larger bank, a string of up to eight files may be used, giving in total space for 14,400 records, adequate for the present purposes.
CHANGE:	This command is followed by the record-number where alterations must start in the bank, then the altered records.
SEARCH:	This command calls for inspection of the data-bank and the retrieval and sequencing of records according to the following criteria.
MONTHS:	This criterion is followed by the range of months (flowering-periods) for which the records must be retrieved.
SQUARE:	This criterion is followed by the geographical one-by-one, half-by-half or quarter-by-quarter degree unit for which extraction of records is required. Up to sixty of these grid criteria can be given for an extraction run, set in any sequence, such as along a planned expedition-route.
CLOSE:	This command terminates a run.

Table 3
Computer-card data fields for records in the data-bank for the South African rare and endangered plant species survey.

CARD COLUMNS	CONTENTS OF DATA-FIELD
1-30	Name of the species
31-32, 37-38	Maximum range of flowering months
33-34, 35-36	Most likely full-flowering months
39	Symbol for IUCN endangerment category
40-42	Year last seen at the locality
43-74	Abbreviated locality description
75-80	Reference to geographical grid area

an informational "data-crypt" (Hall, 1974) without effective cross-indexing. In particular, the localities, as in an herbarium, would have remained classified by species, making the planning of field studies a frustrating task. The computer offered a good way of providing this cross-indexing. Previous experience had shown the computer's value in providing up-to-date, sequenced listings from a growing and changing data-bank; it had also demonstrated the variety of ways in which cross-indexing might be achieved (Hall, 1972a, b). In deciding to use the computer for the survey, it was recognized that an elaborate program would not be needed. The aims were in fact fairly narrow. It was hoped to provide a purpose-orientated output that would list the records in sequences of geographical grid-areas, and within these, a sub-classification by the ranges of months in which the species is in flower. The output would be restricted to any chosen period of months or sequence of geographical grid-areas in a region of study or along an expedition-route. The best months for a visit to a grid-area would be proven, shown by a tally of the number of records of plants of interest in flower for each month of the year. The aim was to provide an efficient expedition-guide, something all collectors should have had years ago to provide a rational coverage of their regions of study. If any other cross-indexing needs arose, they were to be met by the writing of additional programs.

These aims were achieved with a compact, 420-line FORTRAN IV program written for a UNIVAC 1106 computer having a 131 K-word core and extensive back-up storage from drums, discs and magnetic tape. The strategy was for the program to be a small module answering the seven simple commands given in Table 2.

Experience has shown that the program, which should be usable almost directly on other computers of similar or larger core-size, gives reasonably short run-times. The longest runs, for sorting and sequencing nearly 4,000 records over 54 grid-areas, takes less than four minutes of central processor-unit time. This speed is the outcome of keeping records to a limited size of 20 computer-words, and using most of the core for inspecting complete bank subfiles brought in from back-up storage, and a sort-file for the records retrieved for a grid-area.

The small record is in fact a principal feature of the system. With the large storage capacity of modern computers it is tempting to load the electronic data-bank with all available information. While there may be long-term benefits in this, the disadvantages are very significant: large records imply long hours of keying-in information and extensive proofreading which are together the Achilles' heel of nearly all computer-based data banking; further, the longer programs that are needed to extract search-and-sort thesauri may drain efforts in programming and upkeep that could be better directed elsewhere; and lastly, computer charges generally rise steeply with longer records. The electronic data-bank can be a cost-effective servant to the scientist up to the point where only strictly necessary information for cross-indexing is admitted to it. In the present case, most of the information can be handled satisfactorily in the paper-based data-bank of the dossiers, filed automatically by family, genus, and species. The short computer record designed for the survey fits onto an 80-column punched card, as shown in Table 3.

The program is run in batch mode, as the output is voluminous, running to seventy pages for a complete retrieval of all the present 4,000 records. In general, a run for the entire region is made once a month for use by the field-survey team. The data-cards are stored alphabetically by family and by species. This allows separate banks to be created and geographical listings of records to be made for specialists in certain groups. These listings help stimulate professional interest and activity in the

Table 4.
Extract from a print-out of an expedition guide for the Cape Province rare and endangered plant species survey. For details of the records see Table 3.

RARES EXPEDITION GUIDE, PREPARED ON 7/7/1977.

GRID-SQUARE — 34 18

Collecting-months:	1	2	3	4	5	6	7	8	9	10	11	12
Record-frequencies:	9	2	7	14	7	0	1	2	9	27	25	26

Species				Year	Status	Locality	Grid-square
Herschelia barbata	—	9-10	—	1883	Uncertain	Cape Peninsula sandy heath	34 18 A
Amphigena tenuis	3	3-4	5	1968	Rare	Table Mt. slopes above Llandudno	34 18 AB
Amphigena tenuis	3	3-4	5	1943	Rare	Constantiaberg	34 18 AB
Satyrium guthriei	—	10-11	—	1890	Extinct	Btw. Retreat Station & Tokai	34 18 AB
Acrolophia ustulata	—	11-12	—	1882	Extinct	Muizenberg Mt. 1300 ft.	34 18 AB
Acrolophia ustulata	—	11-12	—	1921	Extinct	W slopes of Paulsberg	34 18 AD
Disa stokoei	—	10-11	—	1923	Endangered	Hottentots Holland Mts.	34 18 BB

175

endangered species problem, and give lines of communication for keeping the stored information up-to-date. An extract from a print-out from the orchid data-bank is given in Table 4. The grid-area shown, 34 18, one of several in the run, had many more records of rare and endangered orchids than the sample of seven given in the Table. The row of record-frequencies shows that the austral winter months of June and July would be poor times for hunting orchids in this area; also February, which is in the warm and windy, dry Mediterranean-climate summer. The primary sequencing is by the letters in the last column, that gives the half-by-half and quarter-by-quarter degree grid-area designations. Within these, for example 34 18 AB, the sequencing is by the first of the main flowering-months, here from March (3) to November (11). For more general use, the species can be listed in grid-areas as in Table 5. The same data-bank is used for this, acted upon by two small special-purpose programs.

Table 5
Example of Listing of Species in Grid-areas from the Data-bank.

```
SURVEY REGION :    SOUTHERN CAPE PROVINCE
```

```
AREA NAME:  MOSSEL BAY
BOUNDARIES: 34-35S.: 22-23E.
NUMBER OF TAXA LISTED:    14.
```

ENDANGERED:	DIOSMA ARISTATA
	SATYRIUM MUTICUM
VULNERABLE:	THAMNOCHORTUS MUIRII
UNCERTAIN:	ASPALATHUS OBTUSIFOLIA
	CYPHIA TORTILIS
	ERICA PEARSONIANA
	GEISSORHIZA BURCHELLII
	HERSCHELIA MULTIFIDA
	MASSONIA BOLUSIAE
	MASSONIA CANDIDA
	ORNITHOGALUM TORTUOSUM
	PENTASCHISTIS BURCHELLII
	STAPELIA BIJLIAE
	WAHLENBERGIA CILIOLATA

```
AREA NAME:  KNYSNA
BOUNDARIES: 34-35S.: 23-24E.
NUMBER OF TAXA LISTED:    4.
```

ENDANGERED:	SATYRIUM MUTICUM
INDETERMINATE:	AGATHOSMA ALARIS
UNCERTAIN:	ACMADENIA ALTERNIFOLIA
	ERICA KEETII

Field-Work: Verification and Proposals for Conservation

With the relatively rapid depletions of wild vegetation in many important parts of the Cape Province survey area, only a considerable increase in staffing could allow many of the species to be studied in time for adequate conservation measures to be proposed. The present research budgets to not allow such staffing so that compromise priorities have had to be made. A leading priority is the confirmation of endangerment in high-impact areas. The field team has concentrated its first efforts on depleted habitats near the fast-expanding urban areas around the City of Cape Town and in the farmlands on the adjacent coastal plains. The team has been helped here and in the drier areas of Namaqualand by members of the Provincial Department of Nature and Environmental Conservation. The results have so far largely verified the findings of the herbarium studies: a serious endangered-species problem indeed seems to exist in the south-western Cape.

In one area north of Cape Town, 28 endangered species exist in an 8 square kilometer patch of vegetation, a few extending outside it as minor populations in small remnants of wild vegetation in the wheatlands. Some species are suffering a variety of impacts. Fig. 3 shows the flowers of *Gladiolus aureus*, a species that had been reduced to 18 adults and 17 seedlings in 1977 from over 60 flowering plants in 1975. Its habitat had been invaded by weedy *Acacia saligna* from Australia and *Pinus pinea* from Europe. It had been trampled and picked at its only site, which was next to a path, a children's play-area, and two picnic-places. Finally, its drainage patterns and soil conditions had been changed by bulldozing for gravel in strips on the slope above.

Some species have a very small distribution range and can be easily destroyed by a local impact. This happened to *Moraea loubseri* which grew on a granite hill that was quarried away for a harbour development near Saldanha. The spectacular flower of this species, which now survives only in cultivation, is shown in Fig. 4. *Erica fairii* (Fig. 5) is confined naturally to a one-hectare site on a plateau on the Cape Peninsula, and was until recently menaced by a spreading thicket of exotic *Pinus pinaster*. The pine trees were cut down by a party of volunteers from the Mountain Club of South Africa, and subsequently the entire area was accidentally burned by a damaging fire at the height of a summer draught. This would have destroyed seedlings and surface seed of the pines. With the remarkable fire tolerance of the Cape heaths, *Erica fairii* may well reappear in at least local abundance, regenerating from its seed-store in the soil and perhaps from rootstocks.

An information system is needed for the wide range of problems emerging from the field-work. At present the first aim is to find the geographical extent of populations of candidate species, and to seek as much information as possible about their habitat requirements and conflicts with pressures that have originated with human activity. This information is written onto a generalized form at the time of the site visit. Recommendations for urgent action to conserve the species may also be written on a form, copies of which are passed, after consultation, to landowners and public authorities. The form carries a recommended date on which a return visit should be made by the conserving authority or the survey team to ensure that the action taken has been effective, and that no new pressures have appeared.

A program of future visits is extremely important. Especially in South Africa, which has one of the highest human population growth-rates in the world, impacts that are at first quite minor emerge after a few years as major local destructive

Figure 3. *Gladiolus aureus*, a species brought almost to extinction by a variety of pressures.

Figure 4. *Moraea loubseri,* probably extinct in the wild due to stone-quarrying at its only locality.

Figure 5. *Erica fairii,* a very local endemic known from only a one-hectare area. Such endemics are easily made extinct by human interference.

forces. Lacking extensive nature reserves in most of its seventy kinds of vegetation (Edwards, 1974), South Africa's floras have become seriously endangered in this way. The country's information system for endangered plant species should be closely in touch with these changes, and should help promote the regular checking of the conservation status of rare plants. It is becoming clear that the present small teams of scientists, with their limited resources, will have to be aided by volunteers. Keeping track of all such activities could be an important task for an information system.

Besides monitoring, other activities may need to be recorded. A seed-bank is currently being planned as an off-shoot of the survey program to help avoid the extinction of important rare taxa should all other methods fail. Seeds will first be desiccated over silica gel for about three weeks and then stored at -20 to -25 degrees Celsius. A laboratory for finding the optimal germination conditions for testing each species' seed is under consideration. Horticultural practices may also play a role. Where a species has had its habitat completely destroyed by human activity, the only way of conserving it may be through a temporary phase of cultivation. The immediate aim would be to provide a 'bridge,' for transferring as much as possible of the former gene pool to another wild area where the species is likely to flourish as much as it used to before the advent of destructive human pressures. Such work is no doubt full of hazards: part of the gene-pool may be lost through the processes of artificial germination and cultivation methods, pollinators may not be present in the new environment, or, perhaps worst of all, unsuspected hybridizations may cause the ultimate loss of the species as a separate genetic entity. Despite all these problems, temporary cultivation as a route to a new environment seems well worth trying (Simmons *et al.*, 1976).

In South Africa, much of the territory where endangered plants grow lies on privately-owned land. Land-owners hold far-reaching rights over the wild plants and animals on their properties and their good will is becoming an important factor in the conservation of rare and endangered species. Much has already been done in urgent action to conserve dwindling herds of wild animals on private land, and many farmers now derive pleasure and profit from controlled hunting, leasing areas to huntsmen and selling meat from wild game. For plants, an industry for marketing cut-flowers in South Africa and overseas has brought landowners in the Cape Macchia region the equivalent of over three million dollars in revenue in the past year. This has come mostly from spectacular and common members of families such as the Proteaceae and Ericaceae, generally grown for cutting in specially cultivated semi-wild areas. A few rare species, notably the Blushing Bride, *Serruria florida* (Fig. 6), are being cultivated as valuable commodities for this new industry. However, it will often be emotional concern that will be the most important factor in conserving rare and endangered plants on private land. Some system of awarding certificates of guardianship may have to be developed to promote this, perhaps along the lines of the U.S. National Natural Landmarks Program (Waggoner, this volume). Independent and regular monitoring will be needed in a supporting role and an information system may be required for recording the schedules for visiting each species. As with other features of the field survey and conservation program, views are still being considered on how much of the data-base should be held in the computer and how much can be kept on paper. In the dynamic situation faced by many of South Africa's endangered species, the flexibility of the computer-based approach will certainly make it an important ally.

Figure 6. *Serruria florida,* a local endemic that has become widely cultivated for its attractive flowers.

Summary and Conclusions

After about four years of study, it has been shown that a large number of Southern African plant species are threatened or critically rare, most of these being concentrated in the Macchias of the south-west Cape Province. The data supporting this conclusion have been obtained from specialists in various plant groups and from collectors' records in herbaria. A total of 1,459 taxa are known or likely to be threatened or critically rare in the area of intensive survey in the Cape Province south of the Orange River and west of the 26° E line of longitude. A total of 1,915 threatened or critically rare taxa have so far been recorded for the rest of Southern Africa, defined as the area south of the northern and eastern boundaries of South-West Africa, Botswana, the Transvaal, Natal, and Swaziland (Hall *et al.,* in press).

The survey of South Africa's rare and endangered plant species has been coordinated by the country's National Programme for Environmental Sciences. This has a working group that meets at intervals to discuss problems as they arise. The National Programme keeps in close touch with the Threatened Plants Committee of the IUCN at the Royal Botanic Gardens at Kew in England. Two assistants are employed full-time on the Cape Province survey, and six other persons are making part-time contributions in surveys in the Cape and in other parts of the country. The rate of depletion of the flora suggests that the field studies are urgent and should be completed more rapidly than at present. Conservation work in existing reserves and in wild lands owned by the Department of Forestry is progressing well. However, there are far too few nature reserves and the prospects of creating new ones, especially in high-impact areas where they are most needed, are worsening steadily due to population growth and rising land-prices.

Studies of the biology of rare species would be valuable for designing long-term conservation measures. The problem here is to find funds and staff for such work in the light of other pressing priorities. Could we afford spending three person-years on the study of the conservation status of a single group of pollinators, or is such a project so fundamental that it should be pressed for without question? There seem to be good grounds for believing that such projects would make a profitable field for post-graduate study at South African universities.

A centralized information system has given the essential focus for accumulating data in the early stages of the survey. This has consisted chiefly of manual files indexed using the computer. The cross-indexing was made in a way that gave field-guides for the survey, on the species that should be sought in each area at certain times of the year. This system should be easily adaptable to scheduling the times of future visits to monitor the effects of changing environmental pressures and the counter-moves for species conservation. The work is becoming critically urgent. At stake is the possible extinction of many members of one of the world's richest floras.

Literature Cited

Acocks, J. P. H. 1953. Veld types of South Africa. *Bot. Surv. S. Africa* 28: 1-192, with map.

Edwards, D. 1974. Survey to determine the adequacy of existing conserved areas in relation to vegetation types: A preliminary report. *Koedoe 17:* 3-38.

Edwards, D. and O. A. Leistner. 1971. A degree reference system for citing biological records in Southern Africa. *Mitt. Bot. Staatssamml. München* 10: 501-509.

Good, R. 1974. *The Geography of the Flowering Plants.* 3rd Ed. Longmans, London.

Hall, A. V. 1972*a*. Computer-based data banking for taxonomic collections. *Taxon* 21: 13-25.

_____. 1972*b*. The use of a data-banking system for taxonomic collections. *Contr. Bolus Herb.* 5: 1-78.

_____. 1974. Museum specimen record data storage and retrieval. *Taxon* 23: 23-28.

_____. 1978. Endangered species in a rising tide of human population growth. *Trans. Roy. Soc. South Afr.* 43: 37-49.

Hall, A. V., M. de Winter, B. de Winter, and S. A. M. van Oosterhout. In Press. *Threatened and Critically Rare Plants of Southern Africa.* South Africa National Scientific Programmes Report, C.S.I.R., Pretoria, South Africa.

Leistner, O. A., and J. W. Morris. 1976. Southern African place names. *Ann. Cape Prov. Mus.* 12: 1-565.

Lucas, G. Ll., and S. M. Walters. 1976. List of rare, threatened, and endemic plants for the countries of Europe. I.U.C.N., Morges, Switzerland.

Simmons, J. B., R. I. Beyer, P. E. Brandham, G. Ll. Lucas, and V. T. H. Perry. 1976. *Conservation of Threatened Plants.* Plenum, New York.

Takhtajan, A. L. 1969. *Flowering plants: Origin and Dispersal.* Oliver and Boyd, London.

Walter, H. 1968. *Die Vegetation der Erde.* Band II. Gustav Fischer Verlag, Jena.

Specimen Locality Indexing by the New England Botanical Club

Larry E. Morse,[1] Michael Lamson,[2] Alice F. Tryon,[3] and Russell R. Walton[4]

The work of the New England Botanical Club described here involves the development and implementation of a simplified, economical means for indexing museum specimens from a limited area as documentation for distribution maps and related products. One computer card is used per specimen, and minimal information is recorded in a strictly coded format. Among the information indexed is the state, county, town, year of collection, species name, and herbarium. More detailed locality or habitat data must be obtained from the specimen labels when needed.

The detailed documentation of plant species distributions is essential to comprehensive plant conservation programs as well as to informed biogeography. The ability to correlate records on distribution maps with particular voucher specimens adds considerably to the long-term usefulness of such reports. The work of the New England Botanical Club described here involves the development and implementation of a simplified, economical means for indexing museum specimens from a limited area as documentation for distribution maps and related products.

In 1899 the New England Botanical Club (NEBC) began publication of checklists of New England plants in which distributions by states were documented, if possible, by personal examination of specimens, with lists of such vouchers being deposited in the Club archives (Anonymous, 1899). In the next few decades, the various Committees on Plant Distribution of the NEBC produced detailed hand-prepared distribution maps for about a third of the vascular plant species in the six New England states, largely under the leadership of M. L. Fernald, C. A. Weatherby, C. E. Knowlton, and R. C. Bean.

[1] Cooperative Parks Study Unit, The New York Botanical Garden, Bronx, N.Y. 10458, and Gray Herbarium, Harvard University, Cambridge, Mass. 02318; present address: The Nature Conservancy, 1800 N. Kent St., Arlington, Va. 22209.
[2] Management Systems, Univ. Massachusetts, Amherst, Mass. 01003; present address: 33 Carriage Lane, Amherst, Mass. 01002.
[3] Gray Herbarium, Harvard University, Cambridge, Mass. 02138.
[4] 22 Divinity Ave., Cambridge, Mass. 02138.

Based on the presentation "Specimen Locality Indexing by the New England Botanical Club" by Mr. Lamson at the Symposium on Geographical Data Organization for Rare Plant Conservation, held 15-17 November 1977 at the New York Botanical Garden.
Pages 185-191 in: Larry E. Morse and Mary Sue Henifin, eds., *Rare Plant Conservation: Geographical Data Organization*, The New York Botanical Garden, Bronx, N.Y.

The present committee was organized by Alice F. Tryon in 1974 with the primary purpose of revising these maps for publication. This committee decided to concentrate first on the ferns (Filicales), a small, discrete group of general interest. Committee membership during this project has included Judith Vickers Burlingame, Clifford David, Martha Fisher, Walter Judd, Aminta Kittfield, Michael Lamson, Larry Morse, Mary Perry, Alice Tryon, and Russell Walton.

In New England, the town (or township or city) is the primary political division of the county. Each locality belongs to a "town," and this information is frequently recorded on specimen labels and often used in local floristic studies. On the other hand, very few New England specimens bear latitude/longitude or Universal Transverse Mercator (UTM) coordinates on their labels. Appreciation of this historical perspective was fundamental to the committee's decision to continue the traditional use of the town as the basic data-collection unit, rather than a theoretically more preferable system such as a latitude/longitude grid. The original intent of the Filicales project was to map the species by county only, with subdivision of a few of the largest counties, such as Aroostook in northern Maine. However, preliminary results indicated much of the pattern of the distribution reports was lost at the county level, compared with mapping directly by towns.

Scope, Content, and Strategy

There was considerable discussion of the kinds of data to be recorded or indexed. At one extreme, all information on the specimen labels and annotations might be copied, but that would require a considerable effort per specimen to record, edit, and verify the information, much of it unnecessary to the purpose at hand, namely the summarization of distribution information to the town level of precision. Furthermore, the labels themselves are readily available for consultation in the herbarium, so detailed locality and habitat information can be obtained from them directly whenever needed.

Since there were about 6,000 fern specimens to be processed in the NEBC herbarium, not to mention additional specimens in other collections, a simple system was needed to put the available volunteer effort into indexing as many specimens as possible with the least amount of effort consistent with accurate work. It was decided that some kind of computer-assisted indexing procedure would be more flexible than indexing with manual file cards. Listings and reports could then be produced easily with the information sorted in different ways, such as by species or by locality. The use of a computer-based approach would also allow for future revisions and corrections to be made readily, and provide the ability to incorporate the New England information into a larger data base at some time in the future. A pilot specimen-indexing project developed by David S. Conant at the University of New Hampshire provided a general strategy from which the committee members developed more refined specifications and procedures in 1974 for the specific needs of the NEBC project.

The principal data fields to be indexed would be scientific name, locality (state, county, and town), date, and an abbreviation for the herbarium in which the pertinent voucher specimen is deposited. A short abbreviation of the name of the plant family was also included in each data record, to allow for searching and sorting at the family level as well as at the genus and species levels. To handle the various cases of three-word scientific names such as names of subspecies and varieties as well as interspecific hybrids, the committee developed the convention of adding a single-letter rank indicator to each such name to identify the particular cases. Later, a unique identifying number was added to each record to facilitate editing and revision.

There was considerable discussion regarding the degree of precision appropriate for recording dates of collection, and even whether these dates were necessary to the goals of the project. One view held that the dates should be recorded precisely, to the exact day, because this information would show times a particular species was conspicuous in the New England flora. However, this information would be relatively useless without knowledge of the phenological state of the specimen (*cf.* Crovello, 1972), and was in any case considered information peripheral to the primary purpose of the committee's work. Furthermore, many specimens lack exact dates on their labels. The conclusion was reached that recording the month and date would be of little use in the project. However, it was readily recognized that the year a specimen was collected did provide significant information concerning distributions and historical patterns of collecting, and it was decided to record the year whenever known, taking the earliest vouchered report available for each town or equivalent locality. (Recent reports of taxa considered rare or endangered in one or more New England states are being summarized under the Club's Committee on Rare and Endangered Species, as discussed by Countryman, Dowhan, and Morse in this volume; other groups undertaking comparable projects might consider recording both earliest and most recent dates in one file.)

Finally, the group had to agree on a particular taxonomic classification to follow in organizing and reviewing the data. Several floristic works in current use provide classifications for the New England ferns (*cf.* Lawyer *et al.,* in press). However, these do not agree in all details, and are all somewhat out of date. Alice Tryon therefore developed a checklist of the ferns of New England specifically for this project; this was published recently in the Club's journal, *Rhodora* (Tryon, 1978).

Data-Coding Conventions

The formal specifications of our data-coding format are presented in Table 1. This format provides for concise recording and storage of the pertinent data fields on 80-column data-processing cards, using a fixed-field format in which each data element is consistently placed in a particular part of the card.

State, county, and town or locality names are recorded separately on the cards. States are abbreviated following the U.S. Postal Service's two-letter system, and counties are identified by the first seven letters of their names. (Block Island is treated by name as if it were separate from the rest of Washington Co., Rhode Island.) Town names are generally spelled out in full, except in certain specified cases. Locality names unidentifiable to the town level are edited and included parenthetically in the town field, such as Mt. Mansfield, Vt., which lies on a town line.

Taxon names are encoded as abbreviations to minimize space requirements yet maintain legibility. The family abbreviations are taken primarily from an unpublished list developed several years ago for the Flora North America Program at the Smithsonian Institution. Tryon's (1978) list gives the abbreviations for scientific names, which are codes adapted primarily from an unpublished list developed by

Table 1.
Formal Specifications for Data Cards for NEBC Specimen-Locality Indexing Project.

Columns	Length	Content
1-2	2	State Abbreviation
4-10	7	County Abbreviation
12-15	4	Year Collected *(If no year given, use 0000)*
16	1	Blank, or * if year ambiguous
17-36	20	Town Name, or other locality indication
37-41	5	Family Name Abbreviation
42-61	20	Taxon name*
62-75	14	[*Vacant; reserved space*]
76-80	5	Herbarium abbreviation

* Taxon names are subformatted as follows:

42	X if hybrid, else blank
43-47	Genus
48	X if named hybrid, else blank
49-53	Species epithet, or first hybrid-formula name
54	[Blank]
55	Rank indicator if 3-word name, else blank
56	[Blank]
57-61	Infraspecific epithet, second hybrid-formula name, or comment

RANK INDICATORS for Infraspecific Epithets and Formula Hydbrids:

S = Subspecies	X = Hybrid formula
V = Variety	N = Nothomorph of named hybrid
F = Form	C = Cultivar

David S. Conant at the University of New Hampshire and provided for our use.

Names of herbaria are abbreviated following *Index Herbariorum* (Holmgren and Keuken, 1974), as is customary in plant systematics.

Strategy for Data Collection

This card format was designed to support an innovative data-collection strategy that avoids repeated typing of words frequently used in the data base, such as the county abbreviations and the species names. Editing the inevitable keypunch errors in such repetitive information has proven a major difficulty in many earlier specimen-indexing projects, and handling of variations in spelling add further difficulties, topics discussed by Shetler (1974), Brummitt (1975), and Morse (1977).

Ideally, each unique word in a data base should be typed and proofread just once, then automatically duplicated in the correct form in all the appropriate places in the data file. In practice, few projects can attain this goal completely, but often the most repetitive words (or the most difficult ones) can be handled in this fashion. Martin (1975), Chamberlain (1976), and Date (1977), among others, provide further technical discussion of such "relational data bases."

Our specimen-indexing procedure requires typing of scientific names, herbarium codes, and state and county abbreviations only once, rather than separately for each specimen as would be required in conventional computer-based indexing. The only data that are typed (keypunched) in full for every specimen are the town or locality name and the year of collection. The towns and localities are too numerous to standardize readily beforehand; instead, general guidelines were developed for recording them. The occasional spelling variants and keypunching errors are corrected later. The years, entered as four-digit integers, are simple to type and have a low error rate.

Two factors strongly influenced the development of these specific data-collection procedures. First, the project was limited geographically to a well-defined area, and each pertinent specimen would be from one of the 67 New England counties. Second, processing of specimens could be done in lots, each being a series of collections of a particular herbarium. Thus, the species name and the herbarium code could be added once to the entire batch of records for a particular taxon from a particular herbarium, providing the dates and localities were first recorded individually for the specimens in that lot. Furthermore, the small total number of counties being considered suggested that the state and county codes could be pre-recorded on decks of cards, one deck per county, from which cards could be selected quickly as needed while indexing a batch of specimens.

Procedure for Initial Data Collection

A supply of cards pre-punched with state and county codes was first produced. Fifty or 100 cards were initially prepared for most counties, and more for heavily collected counties such as Middlesex or Barnstable in Massachusetts. These cards were produced either at a keypunch using a program card (drum card), or by a simple computer program. The card decks were then sorted and stored in the herbarium alphabetically by county within state. Additional batches of pre-punched county cards were produced later as needed.

For each taxon to be indexed, a data recorder first examined the specimens, checking for obvious misidentifications or other possible problems. Acceptable specimens were then indexed by selecting a pre-punched card for the appropriate county and writing the year of collection and the town name (or locality indication) directly onto this card. Atlases or topographic maps were consulted if necessary in identifying the town.

After all the NEBC specimens of a taxon had been examined, the cards were bundled together and the top card labelled with the taxon name, then taken to the computer center. The town and year were then keypunched onto them, and the data then printed for proofreading. Finally, the taxon name and herbarium code were added to the card deck for that taxon, using a reproducing card punch programmed for gang-punching.

As discussed earlier, this procedure minimizes the amount of repetitious typing since the state and county codes are pre-punched onto the cards beforehand and the taxon names and herbarium codes are gang-punched afterwards; only the town name and year must be punched individually onto the cards. In this way, the four major kinds of information were recorded for each pertinent specimen: its scientific name, its locality (state, county, and town), its year of collection, and its herbarium depository.

Processing and Revising the Data Files

Various standard program packages were then used to revise the file and produce neatly formatted listings and reports. For these purposes, the card decks were copied to computer tape as a fixed-format sequential file containing the several thousand eighty-column records. Tape was used because it is cheap and portable, and the technology for processing sequential fixed-format files is so mature that special-purpose programs are not needed to handle the file and produce simple listings.

One program used to generate reports from the files was the software product called Data Analyzer, one of the many excellent programs available commercially for inquiry against sequential files. Data Analyzer interprets commands such as *select, sort,* and *print,* then creates a specific program (in FORTRAN) to perform the requested functions. Execution of this program then produces the desired report.

One advantage of a program such as Data Analyzer is that the compact codes in the data file can be decoded or expanded before printing, and columns of data can be rearranged, so more readable reports can be produced than is possible by simple listing of the cards. Such programs can also answer more complex questions, such as listing the counties in which several species are all reported to occur, or tallying the total number of collections reported from each county or each town.

In order to add additional data from other herbaria, a listing of currently recorded localities for each species is sent to the cooperating herbarium, and any additional reports are added there. These new records are then added to the master computer file, either by typing cards for each revision, or (preferably) through use of an on-line editing program.

Some Cautions

We initially encountered a number of minor problems in punching and handling the computer cards, since most of the volunteers had no prior experience with computer-assisted data processing. For example, one problem was the continuation of long locality names into the data field reserved for the scientific name codes, which led to various unexpected consequences in data processing. Another problem was physical damage to a few of the card decks, mainly from misuse of rubber bands in bundling them together.

A major difficulty in this work was obtaining correct identifications of the specimens and applying currently accepted names to the taxa. In some cases, we asked a specialist to review the specimens of a group before processing. For example, Alice Tryon checked many of the identifications in critical groups such as *Dryopteris*.

Preparing the Atlas

We are presently preparing town distribution maps by hand for the various species, and are already seeing many patterns in the reported occurrences. Some of these reflect natural features of the New England landscape, but other patterns show concentrations of collections near colleges and universities, population centers, resort areas, and major roadways and railroads.

These maps are based on the specimen index, and cannot be more accurate than it. A fairly high degree of uniformity has been achieved by having one individual (R.R.W.) map the town localities, although in some areas (such as metropolitan Boston) the towns are so small they cannot be identified accurately on the base map. The printed lists of localities remain the primary index to the collections studied, and the maps are a visual summary of the information compiled. The printed lists, rather than the maps, can be consulted if details such as the exact town and the herbarium and year of collection are needed.

The planned publication of these preliminary maps should stimulate field studies in particular counties and towns, focusing on undercollected areas and searching for unrecorded species. Only after such intense field efforts will the maps more closely reflect the actual distributions of species rather than the distribution of botanical activities. Production of county species lists to complement the maps should also provide an important contribution to New England floristics and an additional incentive to further fieldwork.

Literature Cited

Anonymous. 1899. A projected check-list of New England plants. *Rhodora* 1: 91-92.

Brummitt, R. K. 1975. Some thoughts on the computerization of label data in major herbaria, with particular reference to type collections. *Boissiera* 24: 403-410.

Chamberlain, D. D. 1976. Relational data-base management systems. *Comput. Surv.* 8: 43-66.

Countryman, W. D., J. J. Dowhan, and L. E. Morse. 1981. Regional synthesis of rare plant information by the New England Botanical Club. *(This volume, pages 123-131.)*

Crovello, T. J. 1972. Computerization of the Edward L. Greene Herbarium (ND-G) at Notre Dame. *Brittonia* 24: 131-141.

Date, C. J. 1977. *An Introduction to Database Systems* (2nd ed.) Addison-Wesley, Reading, Mass. *xxiii* + 536 pp.

Holmgren, P. K., and W. Keuken. 1974. Index herbariorum. I. The herbaria of the world (6th ed.) *Regnum Veg.* 92: 1-397.

Lawyer, J. I., A. M. Miller, L. E. Morse, and J. T. Kartesz. *in press.* A guide to selected current literature on vascular plant floristics for the contiguous United States, Alaska, Canada, Greenland, and the U.S. Caribbean and Pacific Islands. *Mem. Torrey Bot. Club.*

Martin, J. 1975. *Computer Data-Base Organization.* Prentice-Hall, Englewood Cliffs, N.J. *xviii* + 558 pp.

Morse, L. E. 1977. Some information-management problems raised by the international nature of systematic biology data. Pp. 323-327 *in* Dreyfus, B., ed. *The Proceedings of the Fifth Biennial International CODATA Conference.* Pergamon Press, Oxford and New York.

Shetler, S. G. 1974. Demythologizing biological data banking. *Taxon* 23: 71-100.

Tryon, A. F. 1978. The New England ferns (Filicales). *Rhodora* 80: 558-569.

The California Native Plant Society Rare Plant Project

W. Robert Powell,[1] Thomas Duncan,[2] and Alice Q. Howard[3]

The California Native Plant Society Rare Plant Project was founded in 1968 by Dr. G. Ledyard Stebbins. Since then extensive information about rare plants of California has been developed. Much of this information has been published as the California Native Plant Society's Special Publication No. 1, "Inventory of Rare and Endangered Vascular Plants of California." Information categories and priorities developed in preparation of this Inventory and ongoing efforts to expand and update the data base are described. Future plans for the Rare Plant Project relating to computerization and the impact of recent California legislation are also discussed.

The California Native Plant Society Rare Plant Project (CNPS-RPP) was founded in 1968 by Dr. G. Ledyard Stebbins while president of the Society. The initial goal was to compile a list of rare plants for California based on Philip Munz's (1959) *A California Flora.* Included were those species whose ranges in the state appeared to be 100 miles or less. During the next three years, five successively more refined lists were prepared with the help of many botanists, lay as well as professional, and distributed in-house by coordinator Roman Gankin for further evaluation. In 1971 the first rare plant list for external distribution was issued (CNPS, 1971); it included 520 taxa. Also in 1971, a move to computerize part of the available information was begun under the new coordinator, W. Robert Powell. Under his direction, current information priorities were developed and information categories defined. These efforts culminated in 1974 with publication of the California Native Plant Society's Special Publication Number 1, "Inventory of Rare and Endangered Vascular Plants of California" (Powell, 1974). This Inventory includes 704 taxa of very rare and rare and endangered taxa, as well as 556 rare and not endangered taxa and 134 taxa that are not rare but are mostly of limited distribution. Recently proposed additions now being evaluated have brought the total number of very rare and rare and endangered

[1] Dept. Agronomy and Range Science, University of California, Davis, Calif. 95616.
[2] Dept. Botany and University Herbarium, University of California, Berkeley, Calif. 94720.
[3] University Herbarium, University of California, Berkeley, Calif. 94720; and California Native Plant Society, 2380 Ellsworth St., Berkeley, Calif. 94704.

Based on the presentation "Information Priorities of the California Native Plant Society Rare Plant Project" by Dr. Duncan at the Symposium on Geographical Data Organization for Rare Plant Conservation, held 15-17 November 1977 at The New York Botanical Garden.
Pages 193-197 *in:* Larry E. Morse and Mary Sue Henifin, eds., *Rare Plant Conservation: Geographical Data Organization*, The New York Botanical Garden, Bronx, N.Y.

vascular plant taxa in California to about 800, approximately one-quarter of the nation's total and the largest number for any state other than Hawaii.

During the course of almost ten years, the CNPS-RPP has grown extensively. Growth has been not only in the number of dedicated people involved with rare plant conservation in California but also in the amount and kinds of data that have become available through field, herbarium, and library studies carried on by the Society.

Data Base and Information Priorities

For each taxon the known distribution within California can be documented and some assessment made of the current potential for survival. This information is necessary to take action to protect those rare plants that are endangered and to monitor the status of all rare plants. Basic needs are to define what is meant by "rare," to decide how to document distribution, and to develop means to assess potential for survival. These priorities are an expansion and extension of the original concept of Stebbins and require for each taxon detailed information and a critical evaluation of status.

To evaluate rarity and potential for survival, a four-part code system called the Rarity-Endangerment Code, or REVD code, was developed by Powell. First, rareness is assessed in terms of both numbers of individuals and manner and extent of distribution. Four categories of rarity are recognized, as follows:

1. Rare, of limited distribution, but distributed widely enough that potential for extinction or extirpation is apparently low at present;

2. Occurrence confined to several populations or one extended population;

3. Occurrence in such small numbers that it is seldom reported or occurrence in one or very few highly restricted populations;

4. Possibly extinct or extirpated (P.E.).

Second, endangerment or the possibility of a taxon's being threatened with extinction or extirpation is evaluated as *(1) not endangered, (2) endangered in part of the range, or (3) endangered throughout the range.* Third, the vigor or dynamics of the population(s) is evaluated as *(1) stable or increasing, (2) declining, or (3) approaching extinction or extirpation.* Finally, distributions of the taxa are assessed as *(1) not rare outside California, (2) rare outside California, or (3) endemic to California.*

The source of geographical information is specimens housed mainly in California herbaria. In the initial stages of the project, geographical information from labels of collections of those species on early lists was hand-copied by lay volunteers. This was found to be time-consuming and unsatisfactory because of the chance for inclusion of misidentified specimens. Therefore, the procedure was altered to include critical examination of all specimens by professionals and label information was recorded by photography. As needed, annotations for project records were included in the photograph. From these efforts a primary data base of photographs of herbarium labels was accumulated. As locality information from this file was examined and mapped, a more detailed statement of the locality was included on a separate form. In this way the area of collection was pinpointed as closely as possible on 7-1/2' or 15' United States Geological Survey topographic quadrangle maps for California. Each quadrangle was given a code number by means of which mapped

194

presence of a given taxon upon the quadrangle is indicated in the Inventory. Additional geographical information included for each plant is a code number for the county(ies) or coastal island(s) where collections have been made. Currently details from labels, including collectors, detailed locality information, and date of collection are not included in the computerized information or in the published Inventory (though rough temporal information is given by the nature of the symbol used in plotting a given locality on the associated maps). Flowering time is given in the Inventory but is not a part of the computerized information. Finally, a code for each taxon is taken from Reed *et al.* (1963) or assigned as needed, and a vernacular name is given. Two examples that illustrate how these types of data appear in the Inventory is given in Table 1.

The computer-assisted data bank has been designed and developed by Powell. From it any combination of the descriptors described above (taxon, REVD code, map quadrangles, counties) can be retrieved. This data bank is currently housed on a Burroughs 6700 computer at the University of California, Davis. The inventory published in 1974 was largely generated from this data bank. Updates to the informational files have been on-going and particularly intensive during 1977, when many of the state's herbarium collections previously omitted were photographed and those collections already reviewed were gleaned again for information about proposed additions to the Inventory. All new locality information is being mapped and added to the data bank for incorporation into a revision of the Inventory.*

During the last few years additions to the primary label data base and the derived data bank have included a file on each taxon that contains copies of pertinent literature, status reports (to be discussed shortly), current field information as received from members of CNPS and other cooperators, additional locality information obtained from numerous sources, illustrations, and individual distribution maps. A collection of color slides of rare plants is being developed to include both habit and habitat photos. Howard has been directing compilation of these additions to the data base.

Mechanisms for updating and expanding the data base and associated materials include field studies by members of the various chapters of CNPS. Each of the 18 chapters is encouraged to take on a territory to investigate and undertake to rediscover old, pinpoint vague, and discover new localities. Preparation of status reports is a good way to identify questions and needs for further investigations and to add locations from literature sources which are not documented by specimens examined. Currently, status reports for almost 250 of the some 800 rare and endangered plant taxa in California are completed or in final stages of preparation. These reports have been prepared under a cooperative agreement with the U.S. Forest Service's California Region to provide information on a specified subset of the rare and endangered plants of California. The reports contain information on synonymy and history of the taxon, its distribution, a description, habitat notes, assessment of endangerment factors, management suggestions, references, maps, and illustrations as available. An example of a status report is provided as Appendix I. It is hoped that similar efforts will be continued to include all 800 taxa.**

---

* Revision to be available from CNPS in May, 1980.
** As of March, 1980, status reports were completed for 426 taxa, with an additional 50 in preparation.

Table 1.

Two Examples of the Uses of CNPS-RPP Information Categories

Plant Code	PLANT NAME Scientific/Vernacular	R-E-V-D Codes	Flower Period	County(s) Code	Quadrangle(s) Code
ARMC	Arabis mcdonaldiana Eastw./McDonald's Rock-Cress	3-2-2-3	May-Jun	23	600B
SICO-1	Sidalcea covillei Greene/ Owens Valley Sidalcea	3-3-3-3	May-Jun	14	351A, 372C

Future Computerization Plans

The CNPS-RPP is at initial stages of computerizing the information available. Discussions are underway as to how to proceed. Initial indications are that computerized mapping is of most interest. Currently, the computer facility at the University of California at Davis has graphics capabilities that could be used. Also, at the University of California at Berkeley a remote computing facility with interactive graphics capabilities has recently been implemented. This facility might also be used by the CNPS-RPP. Applications currently under consideration include the use of graphics tablets to digitize map coordinates directly for general purpose mapping. This would be based on the quadrangle and locality information currently available. The production of high quality maps using a digital plotter available at either Berkeley or Davis is feasible in the relatively near future.

Impact of Recent California Legislation

During 1977, State Senate Bill Number 308, the Nejedly Endangered Plant Bill, for protection of rare native plants was passed by the California State Legislature. Passage of this bill was facilitated by the work of the California Native Plant Society and other conservation organizations in cooperation with State Senator John Nejedly and his staff. The bill requires that the State Department of Fish and Game identify and report to the legislature and governor a list of rare plants in the state and evaluate their status, and has various other provisions. This legislation should have tremendous impact on the CNPS-RPP. With over nine years experience, the CNPS is in an excellent position to develop and provide the kinds of information needed to comply with this statute. At this time it is unclear what part CNPS will play. With the shadow of California's Proposition 13 looming over all state programs, it is uncertain how implementation of the bill will proceed. The State Department of Fish and Game seems intent on listing taxa as rapidly as possible. Therefore, CNPS's role is likely to be concentrated on completing more status reports for this purpose. In addition it will be aiming toward a revision of its 1974 Inventory and, for this purpose, working to include additional information from herbarium collections, the literature, and field work. Further computerization does not at present appear to be a high priority consideration on the part of either the State or CNPS in comparison to the foregoing needs.

Literature Cited

California Native Plant Society. 1971. *Inventory of Rare, Endangered, and Possibly Extinct Plants of California.* California Native Plant Society, Berkeley, Calif. 46 pp.

Munz, P. A. 1959. *A California Flora.* Univ. Calif. Press, Berkeley. 1681 pp.

Powell, W. R., ed. 1974. *Inventory of Rare and Endangered Vascular Flora of California.* Spec. Publ. 1. California Native Plant Society, Berkeley, Calif.

Reed, M. J., W. R. Powell, and B. S. Bal. 1963. *Electronic Data Processing Codes for California Wildland Plants.* U.S. Forest Serv. Res. Note PSW-N20. x + 314 pp.

Appendix I. Sample Status Report

CALIFORNIA NATIVE PLANT SOCIETY   Rare Plant Status Report
Compiler: Mary DeDecker   *Date:* 1977

Name: SIDALCEA COVILLEI Greene
                           Owens Valley Sidalcea

Family: Malvaceae: Mallow Family
CNPS Taxon Code: SICO-1
Status: CNPS (1974): 3-3-3-3   Federal Register (1976): Endangered

Synonymy and History: *Sidalcea covillei* Green (1914); *S. malvaeflora* (DC.) Gray var. *covillei* (Greene) Roush (1931). Type, 20 Jun 1891, *Coville & Funston 1004;* US. [SICO-1 is treated in Mason (1957) under *S. neo-mexicana* var. *parviflora,* in Abrams (1951) under *S. neo-mexicana* var. *covillei;* these names involve differences in interpretation and are not true synonyms.]

Distribution: Type collection from Haiwee Meadows, Inyo Co., Calif., now under Haiwee Reservoir, part of the Los Angeles aqueduct system. Other collections from near Bishop (1927), near Lone Pine (1897), and near Independence (1974), all in Owens Valley bottomlands. At some of these sites it was exceedingly plentiful, in bloom like "a large meadow of Dodecatheon." However, changes in the hydrological regimen attendant upon development of the water resource for the Los Angeles Department of Water and Power have resulted in SICO-1's extirpation from all known localities except the recently discovered one near Independence. There seepage from the aqueduct still provided the needed moisture, but heavy grazing pressure has, since 1974, caused SICO-1 to disappear. Thorough searches were conducted by DeDecker 1975-1977 without success in locating any SICO-1 at this locality. Historical distribution was at elevation of ca. 4000 ft. Maps: USGS Bishop, Lone Pine, Independence, and Haiwee Reservoir, all 15'.

Description: Perennial herb from *fleshy roots;* stems several, 2-6 dm tall, somewhat *pubescent* with fine or coarse, *branched or simple hairs.* Leaves mostly basal with small stipules, fleshy, with hairs 2-rayed or stellate, shallowly lobed, coarsely crenate to lobed half their width, the cauline leaves divided more deeply; petioles of basal leaves up to 2 dm long. Inflorescence simple to compound, elongate, loosely many-flowered, finely stellate; pedicels 2-8 mm long; calyx 5-lobed, 4-8 mm long, densely stellate; petals 5, *pinkish-lavender, 10-15 mm long;* carpels ca. 2.5 mm long, prominently reticulate but not pitted on sides, 1-seeded and dehiscent. Style branches are needle-like and have long stigmatic surface on inner side. Stamens many, fused into a tube around the styles. Flowering time: May-Jun.
        The only other mallow found on alkaline sites in the area is *Sida leprosa* var. *hederacea,* and the

197

two would not be confused. The *Sida* is a low plant, usually decumbent, with leafy stems and rather inconspicuous yellowish flowers.

Habitat: Moist alkaline, grassy places, with *Distichlis spicata* and *Juncus balticus*.

Endangerment Factors: Man-caused alteration of hydrological regimen upon which SICO-1 depends. Haiwee Meadows were covered by a reservoir; water from the Owens River was diverted miles above the Lone Pine and Independence sites; extensive pumping of ground water in the 1960's lowered the water table until any moist areas on the river bottom dried up. The Independence site receives some seepage from the aqueduct but is also heavily grazed. The Department of Water and Power was requested by DeDecker in 1975 to fence off SICO's habitat and agreed to do so but, preoccupied with preparation of an EIR to increase ground-water pumping in Owens Valley, procrastinated until 1977, when the California Native Plant Society added its voice to DeDecker's.

Management Suggestions: The last known site is now fenced. One can only hope that some remnants of SICO-1 may still be able to regenerate from roots or seed scattered in past years now that grazing pressure has been removed.

References:
Greene, E. L. 1914. Manipulus malvacearum. *Cybele Columbiana* 1: 33-36. (orig. descrip., p. 35)
Hitchcock, C. L., & A. R. Kruckeberg. 1957. A study of the perennial species of *Sidalcea*. *Univ. Wash. Publ. Biol.* 18: 1-96.
Roush, E. M. F. 1931. A monograph of the genus *Sidalcea*. *Ann. Missouri Bot. Gdn.* 18: 117-244.

Introduction to a System for Ecological Diversity Classification

Albert E. Radford[1]

The preservation of rare plants has to be based upon a knowledge of the critical habitat and distribution of each rare species. The use of this holistic, comprehensive diversity classification system enables the scientist to analyze the habitat for critical factors and to thoroughly inventory one or more provinces for geographic and physiographic data. An outline of this classification system is presented here.

Natural diversity is the foundation of our natural heritage. A holistic, comprehensive classification of the diversity features in our environment is essential to the basic inventory of species, communities, and habitats in that heritage. A system of classification of natural diversity is necessary for comparable and consistent descriptions of our natural areas and sites. A classification is basic to the identification of elements of diversity in the landscape. A system is fundamental to habitat analysis and synthesis for populations and communities. A comprehensive natural diversity classification system is necessary for perspective in ecosystem analysis and categorization. This classification is also fundamental to studies in ecology, taxonomy, evolution, and resource management. Inventories based on such a system of natural diversity classification are the first step in determining which species, communities, and habitats are endangered or rare.

Any conservation effort must be based upon a thorough survey of biotic and abiotic features. All types of communities, from the pioneer to the climax, developed during time over different rock and water types should be included in the representative site samples of an area. The successional communities, the topo-edaphic climaxes and the continua should be part of the master ecosystem study. Biogenesis has to be integrated with pedogenesis in explaining the present and past development of species and communities; climatogenesis and phylogenesis have to be coupled with succession and soil formation to explain the present composition and distribu-

[1] University of North Carolina, Chapel Hill, N.C. 27514.

Based on the presentation "A Natural Diversity Classification: Relevance to Rare Plant Conservation and Geographic Data Organization" by Dr. Radford at the Symposium on Geographical Data Organization for Rare Plant Conservation, held 15-17 November 1977 at the New York Botanical Garden. For a more complete description of Dr. Radford's system see "Ecological Diversity Classification and Inventory" in Radford, Otte, and Otte (1978) or *Natural Heritage: Classification, Inventory, & Information* (Radford *et al.,* in press).

Pages 199-205 *in:* Larry E. Morse and Mary Sue Henifin, eds., *Rare Plant Conservation: Geographical Data Organization,* The New York Botanical Garden, Bronx, N.Y.

© 1981 The New York Botanical Garden

tion of biotic assemblages. We must try to conserve the *total diversity of species in as broad a range of habitats as possible within the different climates in each province* in order to understand the origin, migration and evolution of species, floras, and faunas, as well as the productivity and composition of present communities. Sound conservation depends upon a classification of natural diversity.

The natural diversity for any province or area includes: (a) Vegetation (with animal dependents or communities), (b) Climate, (c) Soils, (d) Geology, (e) Hydrology, (f) Topography, and (g) Hydrography. All of these components of the habitat are interacting but conceptually independent systems that compose the Ecosystems, Biomes and Natural Areas. Vegetational composition (with animal dependents), distribution, and development are dependent upon climate, soils, geology and topography acting through time. Climate (microclimates) is dependent upon vegetation, soils, geology and topography. Soil composition, distribution and development are dependent upon vegetation, climate, geology, and topography acting through time. Geological structures, formations, and sedimentary rocks are dependent upon vegetation, climate, soils, topography and time. Topographic land forms and features, structures and development are dependent upon vegetation, climate, soils, geology and time. Since these interrrelated components are related to the various habitat types, they have to be major parts of any classification scheme for natural diversity.

Who knows which component of our habitat is more significant than another in the maintenance of species diversity? In the origin, migration and evolution of species, faunas, floras and communities? Who understands the role of species diversity in maintaining habitat diversity? Who knows which element of the biotic or abiotic habitat will be of great natural resource value to humans? These fundamental questions can be answered only if we preserve total species/habitat diversity in carefully selected Natural Areas. Identification of that total species/habitat diversity depends upon the development of the best possible classification of that diversity.

Many systems of classification have been devised for components and elements of diversity. The reasons for development of those systems, however, have been mapping, productivity, cover, etc. rather than specifically for diversity classification. No commonly used system has as its goal the delineation and delimitation of ecological diversity. Most systems are hybrids, combining various factors inconsistently. These systems, hybrid or otherwise, can be used, however, in part or whole, in the construction of a classification system for natural diversity.

The purpose of classification is to arrange elements, components, objects, or taxa in a way that gives the greatest possible command of knowledge, makes the most efficient and effective use of information, and leads most directly to the acquisition of more data, information, and knowledge. We believe that the hierarchical component approach, as we have developed it (see Appendix I), has enabled us to accomplish most effectively the goals for the classification of natural diversity.

Reasons for the Adoption of the Hierarchical Component Approach for a Natural Diversity Classification

1. The establishment of separate ecological component classes permits the description of natural areas in terms of independently variable factors. This will en-

hance the recognition of varying ecological character combinations that leads to identification of maximum ecological diversity.

2. The hierarchical component approach enables the developers of this system to utilize the professional expertise from each of the respective disciplines and to definitely take advantage of the diversity knowledge in those fields gained through years of experience.

3. Many presently used systems of classification can be included easily in this system, *e.g.*, SAF cover types and the Braun-Blanquet system (slightly modified) at the type level in the biological component; the Köppen climatic classification at the system level in the climate hierarchy; the Soil Conservation Service soil classification at all levels; traditional rock classification at various levels in geology; the wetlands classification at several levels in the hydrology hierarchy; the geological panel classification at several levels in the topographic-landform component, etc.

4. An inventory of ecological diversity within a state or region can be evaluated effectively for thoroughness using the component classification system, *e.g.*, in checking the inventory of pioneer, transient, and climax communities in relation to each soil series, rock type, landform feature, etc., within each climatic regime in a physiographic province.

5. The component approach to diversity classification can be used easily as a basis for comprehensive, complete, comparable, and consistent habitat analysis for populations, communities, and ecosystems.

6. This is an open-ended system that can be done as thoroughly as the time and experience of the investigator will permit. Appropriate data or information can be added efficiently and changed or deleted at the proper hierarchical level for each component whenever available or necessary.

Objectives of a Natural Diversity Classification System

At present no means exist for the identification of the full array of diversity in our natural heritage. Nearly all classification systems dealing with the subject invariably tend to homogenize diversity, usually inconsistently mixing distinctly different factors such as vegetation, climate, or soils. The resulting composite classes inherently negate the recognition of diversity.

Our system (Radford, Otte, and Otte, 1978) has internal comparability and consistency and the potential for effective and efficient compatibility with world, national, regional, and state classifications used in the inventory of a variety of natural elements. Our system provides a classification that is holistic and comprehensive. The primary objectives are:

1. To develop a system of classification which will form the framework for a comprehensive survey of ecological diversity.

2. To design a system which will provide the classification base for an inventory of biotic, geologic, topographic, and hydrologic features as well as natural areas and sites.

3. To devise a system which can be the basis for inventory and identification of all types of successional communities; all types of mono- and topo-edaphic climaxes; and all typical as well as rare, endangered, threatened, relict, or disjunct species, communities, and ecosystems.

Guiding Principles for the Development of the Natural Diversity Classification System

The principles used in the development of our classification system are included to assist the user in understanding the conceptual basis for the system, to aid in understanding the application of the classification in inventory, and to provide a device for evaluating the effectiveness of the system. These basic assumptions are:

1. That the fundamentals of classification be used as rigorously as possible.

2. That the existing classification systems of various scientific disciplines be utilized as much as possible.

3. That a hierarchical component classification system is most useful for efficient handling of information.

4. That the classification system have precisely-defined classes and elements, so that information will be treated consistently and comparably.

5. That the classification system be compatible with other pertinent systems, as much as practical.

6. That the system be comprehensive enough to encompass all elements of ecological diversity.

7. That the system be open-ended and capable of being expanded and changed.

8. That each species is selectively and uniquely adapted to a habitat.

9. That species diversity is related to habitat diversity within an area.

10. That habitat diversity is related to the diversity of climate, soils, geology, hydrology, topography, and biology within an area.

11. That species assemblages are recurring combinations under similar habitat conditions within an area at a given time.

12. That species assemblages are the result of the interaction of species and habitat diversity in an area through time.

Inventory Use of a Natural Diversity Classification System

An inventory—using a standardized classification system—of the species, communities, and habitat diversity (Natural Diversity) in natural areas and sites will provide the baseline data and fundamental information for the following efforts:

1. Conservation of species, community, and habitat diversity.

2. Ecological characterization of species, communities, and habitats.

3. Formulation of a predictive system for species, communities, and habitats (*i.e.,* what's where under what conditions).

4. Production of perspective in species biology studies (*e.g.,* endangered and threatened), community analyses, and habitat significance.

5. Interpretation of the origin, migration, and evolution of species, floras, faunas, and communities.

6. Foundation for research in applied and advanced problems in many fields of endeavor (*e.g.,* hydrology, pedology, habitat cover, food productivity, etc.).

7. Stimulation of production of integrative classifications of many types (*e.g.,* map systems, habitat productivity, trout stream catch, etc.).

8. Decisions on land use, impact evaluation, and management problems of many types.

9. Establishment of priorities through environmental analysis for formulation of land use policies, land classification, and management programs.

10. Establishment of priorities through natural area basic inventory analyses for natural area acquisition and resource protection.

General Comments on Development of the System

The development of this system of diversity classification is based on inventory experience gained during the past six years in studying some 200 natural areas from Florida to New Jersey. These sites represent most of the plant biotic systems; the estuarine, lacustrine, riverine, and palustrine wetland systems; the alfisols; entisols, histosols, inceptisols, spodosols, and ultisols among the soil orders; the basic, acid, calcareous, ferruginous, carbonaceous, salinaceous and siliceous rock classes; and a wide variety of climatic, topographic, and hydrographic features in at least eight distinct physiographic regions. Most of the data for inventory reports have been collected by members of plant ecosystematics classes at the University of North Carolina at Chapel Hill and by natural area research associates sponsored by the Highlands Biological Station.

We are indebted for many ideas and suggestions to a panel which was convened in Washington, D.C., March 20-24, 1978, under the auspices of the Heritage Conservation and Recreation Service of the Department of Interior, for the development of a natural classification system for ecological diversity. Many parts of the draft report (Radford, *et al.*) made by the panel have been, directly or indirectly, incorporated into this presentation. A large number of publications were also used as background resources in the development of this classification system.

Much work remains to be done on the system from a classification standpoint and particularly in the development of a glossary for the diversity elements. There is an absolute necessity for professional input into the component classifications. A great need exists for the incorporation of field experience at the state and local levels for the lower orders of the system, particularly the biological.

Organization of the System

The following are the major components of the diversity classification system with each indicated by a Roman Numeral:

I. Biology	III. Soils	V. Hydrology	VII. Physiography
II. Climate	IV. Geology	VI. Topography	

The following are hierarchical levels within the system, each level being indicated by one or two letters:

A. System	B. Class	C. Generitype
AA. Subsystem	BB. Subclass	CC. Type

The element entries for the levels of each diversity component are indicated by an arabic number; *e.g.,* I.A.*1.* Forb System, III.AA.*17.* Andepts, IV.B.*11.* Ferruginous sandstone, V.BB.*7.* Excessively drained. The *forb system* is the element entry at the System level of the biological component; *andepts* is the element entry at the Subsystem level in soils; *ferruginous sandstone* is the element entry at the class level in the geological component; *excessively drained* is the element entry at the subclass level in the hydrological component.

With this 3-part code (Roman Numeral, Letter, Arabic Number), or any similar code it is possible to encode, store, and retrieve any element of diversity in the classification system. This classification system is outlined in Appendix I.

Literature Cited

Radford, A. E., *et al.* 1978. Ecological classification in the National Heritage Program. (First working draft) U.S. Dep. Interior Heritage Conservation and Recreation Service, Washington, D.C.

Radford, A. E., D. K. S. Otte, L. J. Otte, J. R. Massey, and P. D. Whitson. in press. *Natural Heritage: Classification, Inventory, and Information.* Univ. North Carolina Press, Chapel Hill, N.C.

Radford, A. E., L. J. Otte, and D. K. Otte. 1978. Ecological diversity classification and inventory. Part I *in* Radford, A. E., L. J. Otte, and D. K. Otte, eds., *Natural Heritage Classification and Inventory Systems,* Univ. North Carolina Student Stores, Chapel Hill, N.C.

Appendix I. Natural Diversity Classification System

CODE	HIERARCHY	INCLUSION

I. Biology

I. A.	Biotic System	Plant and Animal System
I. AA.	Biotic Subsystem	Plant and Animal Physiognomy
I. B.	Biotic Class	Community Cover Class
I. BB.	Biotic Subclass	Community Class
I. C.	Biotic Generitype	Community Cover Type
I. CC.	Biotic Type	Community Type

II. Climate

II. A.	Climatic System	Climatic Regime
II. AA.	Climatic Subsystem	Climatic Subregime
II. B.	Climatic Class	Annual Sunshine & Precipitation
II. BB.	Climatic Subclass	Freeze-Free Period
II. C.	Climatic Generitype	Max. & Min. Temperature & Precipitation
II. CC.	Climatic Type	Temp. & Precip. Relationships

III. Soils

III. A.	Soil System	Soil Order
III. AA.	Soil Subsystem	Soil Suborder
III. B.	Soil Class	Soil Great Group
III. BB.	Soil Subclass	Soil Subgroup
III. C.	Soil Generitype	Soil Family
III. CC.	Soil Type	Soil Series & Family

IV. Geology

IV. A.	Geological System	Rock System
IV. AA.	Geological Subsystem	Rock Subsystem
IV. B.	Geological Class	Rock-Sediment Chemistry Class
IV. BB.	Geological Subclass	Rock-Sediment Occurrence
IV. C.	Geological Generitype	Rock-Sediment Type
IV. CC.	Geological Type	Natural Area Geologic Site Characteristics

V. Hydrology

V. A.	Hydrologic System	Hydrologic System
V. AA.	Hydrologic Subsystem	Hydrologic Subsystem
V. B.	Hydrologic Class	Water Regime
V. BB.	Hydrologic Subclass	Drainage
V. C.	Hydrologic Generitype	Water Chemistry Class
V. CC.	Hydrologic Type	Natural Area Water Color and Chemistry

VI. Topography and Hydrography

VI. A.	Topographic-Hydrographic System	Landscape Forming Process
VI. AA.	Topographic-Hydrographic Subsystem	Major Topographic-Hydrographic Feature
VI. B.	Topographic-Hydrographic Class	Local Landform-Water Body
VI. BB.	Topographic-Hydrographic Subclass	Natural Area Landform-Water Body
VI. C.	Topographic-Hydrographic Generitype	Natural Area Landform-Water Body Relief Feature
VI. CC.	Topographic-Hydrographic Type	Natural Area Topographic Site Characteristics

VII. Physiography

VII. A.	Physiographic System	Physiographic Region
VII. AA.	Physiographic Subsystem	Physiographic Province
VII. B.	Physiographic Class	Physiographic Section
VII. BB.	Physiographic Subclass	Natural Area Topographic Landform
VII. C.	Physiographic Generitype	Age and Name of Underlying Geologic Formation
VII. CC.	Physiographic Type	Natural Area Landform-Water Body

Forest Service Programs and Activities for the Conservation of Rare and Sensitive Plant Species

Andrew F. Robinson,[1] L. E. Horton,[2] and Edward Schlatterer[3]

This paper describes Forest Service activities and programs of three different geographical areas pertinent to rare and sensitive plant species. The section on the Southeastern Area (SA) describes the types of information which are being developed by State and Private Forestry for use by the 13 State Foresters' staffs. In California (Region 5), the recently issued Region 5 Supplement to the Forest Service Manual is discussed, including policy, standards, information needs and management directions for implementing the Regional plant program. In the Intermountain Region (Region 4), the program is in its infancy. The emphasis there is on creating a data base and developing future programs for the implementation of the Endangered Species Act of 1973.

Introduction

Section 12 of the Endangered Species Act of 1973 (P.L. 93-205) directed the Smithsonian Institution to prepare a list of endangered and threatened plant species to be considered for formal classification. This list was submitted to Congress and published in the Federal Register (Vol. 40, No. 127, pages 27824-27924) on July 1, 1975. A second revised listing of plant species proposed for endangered classification was published in the Federal Register (Vol. 41, No. 117, pages 24524-24572) by the Fish and Wildlife Service on June 16, 1976. However, by late 1977 only sixteen plant species have been formally classified as endangered or threatened.

Section 7 of the Endangered Species Act of 1973 requires that all Federal agencies:

---

[1] U.S. Department of Agricultural Forest Service, 1720 Peachtree Road, N.W., Atlanta, Ga. 30309. Present address: U.S. Fish and Wildlife Service, P.O. Box 95067, Atlanta, Ga. 30347.
[2] U.S. Department of Agriculture Forest Service, 630 Sansome St., San Francisco, Calif. 94111.
[3] U.S. Department of Agriculture Forest Service, 324 25th Street, Ogden, Utah 84401.

Based on the presentation "Forest Service Programs and Activities for the Conservation of Rare and Sensitive Plant Species" by Messrs. Horton, Robinson, and Schlatterer at the Symposium on Geographical Data Organization for Rare Plant Conservation, held 15-17 November 1977 at The New York Botanical Garden.

Pages 207-216 in: Larry E. Morse and Mary Sue Henifin, eds., *Rare Plant Conservation : Geographical Data Organization*, The New York Botanical Garden, Bronx, N.Y.

1. ". . . utilize their authorities in furtherance of the purposes of this Act by carrying out programs for the conservation of endangered species and threatened species . . . ." and

2. Take " . . . such action necessary to ensure that actions authorized, funded, or carried out by them do not jeopardize the continued existence of such endangered species or threatened species or result in the destruction or modification of habitat of such species which is determined . . . to be critical."

Overall responsibilities for program direction and implementation of the Endangered Species Act of 1973 are assigned to the Department of Interior, Fish and Wildlife Service, Office of Endangered Species. Regulations regarding Section 7 consultations were published recently (Federal Register, Vol. 43, No. 2, pages 869-876).

Responsibility for Forest Service program direction is assigned to the Director, Wildlife Management, Washington, D.C. Office.*

In the Eastern United States most of the forest lands are in private ownership and forestry has been practiced extensively since the late 1800's. Conversely, in the west much of the forest and range lands are administered by federal agencies and private forestry is less significant.

The South, in 1970, accounted for 45 percent of the nation's output of roundwood products and net growth of timber. Of this, 2/3 was supplied by small private landowners, 24 percent by industrial forest lands and 5% from publicly owned lands. Recent trends in land use patterns point to a smaller area of forest land available for timber production. With a decreasing forest land base in the South, timber production per acre must be increased. Most of the intensified forest management practices are currently being done on industrial lands. Current emphasis is being placed on increasing production on privately owned lands. In 1970, 15,000 acres were in pine plantation and managed intensively using various forest management practices.

Because of these differences in type of land ownership, and the nature of disturbance, the Forest Service approach to the implementation of the Endangered Species Act differs from one geographical area to another.

The role of the Forest Service in the Southeast Area is primarily one of providing technical assistance to State Foresters, whereas in the western regions it is more concerned with direct program activities on National Forest lands.

This paper describes Forest Service activities and programs in the Southeastern Area, California (Region 5) and the Intermountain Region (Region 4).

Southeastern Area

The 13 states of the Southeastern Area extend from Texas and Oklahoma eastward to Virginia and south to Florida. The Commonwealth of Puerto Rico and the U.S. Virgin Islands are also included. In the South, separate or group species description-management guides have been compiled for 160 of the 373 forest-related south-

---

* Note: In January, 1980, the U.S. Forest Service issued its policy and management directives on threatened and endangered plants and animals, as Chapter 2670 of the *Forest Service Manual*. Selections from that chapter are reprinted as an appendix to this volume. In April, 1980, at a meeting of Forest Service botanists in Milwaukee, further revisions and clarifications to FSM-2670 were discussed.—The Editors.

Table 1.
Expected effect of selected management practices on *Balduina atropurpurea* Harper

Expected* Effect on the Species		Management Practices						
	Prescribe Burn	Bulldoze or Root Rake	Bed	Chop	Thin over-story	Cut over-story	Establish Plantation	Graze
Destroy		X		X			X	
Damage			X					
No Lasting Effect								X
Beneficial if Done Properly	X				X	X		

Other Comments:
Expected effect on the species is an *estimate* made by Dr. Robert Kral based on his knowledge of the habitat and on knowledge gained from personal field observations. Estimates are "rough" in many instances. Results of practices may be modified depending upon the degree of application, intensity of treatment, nearness to plant communities, etc. A management practice for which no entry is made indicates a lack of sufficient information from which to predict expected results. As observations are made in the field by users of the data, the expected effect will be refined.

ern plant species listed on the Smithsonian Institution Report. The basic data for each species description-management guide included the family name, scientific name, common names, synonyms, botanical description, distribution, flowering season, special identifying features, habitat, management implications, reference, and county distributional maps.

Table 1 is an example of a management implication chart which is supplementary to the "Habitats and Management Implications" section of the species description-management guides.

In addition to the individual species description-management guides, tables depicting the possible impacts of silvicultural activities on proposed threatened and endangered species which grow in a particular ecosystem will be provided. The anticipated effect of each forestry management practice on a species was estimated using the following scale:

> 1 = Destroy
> 2 = Damage
> 3 = No lasting effect
> 4 = Beneficial if done properly
> NA = Not applicable
> — = "blank" insufficient information

Table 2 is an example of anticipated effects of forestry management practices on selected proposed threatened and endangered plant species known to grow in pine flatwood ecosystems. The species considered are all geophytes, plants which have underground stems. Acceptable management practices for the conservation of these threatened and endangered species are quite obvious—burning, thinning or cutting of the overstory affected the proposed endangered or threatened plants least, and in most cases were beneficial. Drainage, root-raking, bedding, chopping and establishing pine plantations had the greatest negative effect on these species. Grazing had a mixed effect on these proposed threatened and endangered species.

The final package of material in the description management guides is a list of species collected from or observed within a county. The lists are being prepared in tabular form with the names of those U. S. Geological Survey (U.S.G.S.) quadrangles across the top in which endangered or threatened species have been observed or collected. The proposed threatened and endangered species of the county are listed alphabetically by scientific name down the left margin of the table. The habitats which the species were collected from are coded into the table under the appropriate (7-1/2 minute) U.S.G.S. quadrangle. For example, habitats occupied by proposed endangered and threatened species in Gulf County, Florida (Table 3) included *Hypericum* bogs, cypress ponds, shallow flatwood ponds, ditches, pine flatwoods, sandy oak-pine woods, pinelands, pine savannahs, prairies, and roadsides. Pine flatwoods (habitat code 7 and 8) had the greatest number of threatened and endangered species (six). *Justica crassifolia* (Chapm.) Small, *Macbridea alba* Chapm., and *Scutellaria floridana* Chapm. have been collected only from this ecosystem. Within Gulf County, Florida, the White City quadrangle has the greatest number of species (six). *Oxypolis greenmanii* Mathias and Const. had the widest distribution, occurring in five quadrangles.

The above materials are being developed for the Service Forester of the 13 Southeastern Area States. This material will be used by the Service Forester in pre-

Table 2.
Some forest management practices rated for their effects on selected plant species of pine flatwood ecosystems.

	Drainage	Burn	Root Rake	Bed	Chop	Thin Over-story	Cut Over-story	Establish Plantation	Graze
GEOPHYTES—Rhizomes									
Harperocallis flava McDaniel	1	4	1	1	1	4	4	1	—
Hartwrightia floridana A. Gray ex. S. Watson	1	4	1	1	1	4	4	1	2
Justica crassifolia (Chapm.) Small	1	4	1	2	1	4	4	1	—
Lachnoncaulon beyrichianum Sporleder ex Korn	1	4	1	2	1	4	4	1	2
Macbridea alba Chapm.	1	4	1	2	1	4	4	1	—
Physostegia veroniciformis Small	1	4	1	2	1	4	4	1	3
Scutellaria floridana Chapm.	1	4	1	2	1	4	4	1	1
GEOPHYTES—Bulbs									
Nemastylis floridana Small	1	4	1	1	1	4	4	2	1
Nolina atopocarpa Bartlett	1	4	1	2	1	4	4	1	—
Sphenostigma coelestina (Bartr. ex Wild) R. C. Foster	1	4	1	2	1	4	4	2	1
Zephyranthes simpsonii Chapm.	2	4	1	1	1	4	4	1	3
Zephyranthes treatiae S. Wats.	2	4	1	1	1	4	4	1	3
GEOPHYTES—Tubers									
Cuphea aspera Chapm.	1	4	1	2	1	4	4	1	1

211

Table 3.
Selected plant species collected in Gulf County, Florida, habitat affinity and distribution

Plant species of Gulf County, Florida	U.S.G.S. Quadrangle Map							
	Indian Pass	Jackson River	North Allentown	Overstreet	Port St. Joe	Ten Mile Swamp	Wewahitchka	White City
Cuphea aspera Chapm.					3, 7, 19			
Gentiana pennelliana Fern.							20	
Justica crassifolia (Chapm.) Small	7	7						7
Liatris provincialis Godfrey					17			21
Macbridea alba Chapm.	7		7		7			7
Oxypolis greenmanii Mathias & Const.			21	22		8	4, 21	3, 8, 21
Rhododendron chapmanii A. Gray					10			
Sarracenia psittacina Michx.								7, 20
Scutellaria floridana Chapm.								7

*Habitat codes:
3 = *Hypericum* bog
4 = Cypress ponds
7 = Pine flatwoods
8 = Shallow flatwood ponds
10 = Pinelands
17 = Sandy oak pinewoods
19 = Roadsides
20 = Pine savannahs
21 = Ditches
22 = Prairies

paring management plans for private landowners. As observations are made in the field by users of these guides, additional data on management effects, distribution etc., will be forwarded to the Southeast Area office through their respective State Forester. This data will be incorporated and revised guides issued periodically.

California (Region 5)

In March, 1975, the Forest Service's California Region issued a policy statement concerning threatened and endangered plants which read as follows:

"All plant species on national forest lands in California that are being considered for threatened or endangered classification will be regarded as sensitive and will be treated in all planning and land management activities as though they were already classified. All necessary steps will be taken to assure that such actions do not jeopardize the continued existence of these species or their habitats until such time as their official status is determined."

Interim guidelines to implement this policy were issued a few months later and have now been in effect for over two years. From the experience thus gained, the guidelines were revised and recently re-issued as a Regional Supplement to the Forest Service Manual. The Supplement outlines formal direction for a Regional plant program to implement the requirements of Section 7 of the Endangered Species Act of 1973.

Activities of the Sensitive Plant Species program include completion of status reports for all sensitive species, field verification of known and reported locations, preparation of individual population records, identifying and field surveying potentially suitable habitats to locate new populations, field reconnaissance of projects and input to Environmental Assessments and Environmental Statements, implementing a records maintenance and reporting system, preparing unit management plans, identifying basic research needs, monitoring of selected populations, and preparing Individual Species Management Guides.

In addition, the State and Private Forestry function includes current exchanges of all new information with the State of California, furnishing program support to the State of Hawaii in their inventory and survey program, and making available a training package in program management to interested parties.

A 10-year plan has been prepared which envisions a three-phase program, namely, (1) inventory to expand the initial knowledge base, (2) interim management to maintain present populations, and (3) recovery management to remove the causes of endangerment. Tasks and activities by program phases have been identified and budget and personnel needs to carry out the program are estimated.

Of particular interest to this Symposium are those aspects of the program that concern information needs and the development of an adequate information base. Three kinds of records are being developed: (1) status reports, (2) population records, and (3) map displays. The remainder of this paper highlights the principal elements in each of these records.

The status report provides a starting point and is intended to assemble, in brief summary, the state of present knowledge concerning a species and to serve as a working tool for non-botanist field personnel in day-to-day land management activities. Status reports have been completed for over 300 taxa. These reports consist of a

3-5 page synopsis and contain the following information elements derived mainly from the literature, herbarium collections, and firsthand knowledge of botanists:

1. Plant name, including synonymy, history, and references.
2. Status, including classification, vigor of populations, and endangerment factors.
3. Descriptions—technical descriptions, guide to distinguishing characteristics, similarly appearing species in the area, flowering dates, and illustrations such as color photos and line drawings.
4. General distribution and range as presently known.
5. Habitat characteristics and ecological requirements.
6. History of area as it has affected the species.
7. Management suggestions.

In contrast to the general overview nature of status reports, population records are concerned with specific and detailed information developed from field investigation and verification of each individual population of each species. Each of the populations is assigned a unique alpha-numeric code identifier for record control. Population records contain the following sorts of information:

1. Location information—land ownership, legal description, narrative description of where the population is and how to find it.
2. History—who previously located the population and when? Who did the current verification and when?
3. Site and habitat information—elevation, slope, exposure, topographic setting, surrounding plant community, associated species.
4. Population information—area covered (size), detailed map, number of plants, density, vigor, age class distribution, activities in area, and hazards observed.

Negative population records are also maintained to document unsuccessful searches.

A standard format for map records has not been required and individual field units have developed a variety of ways to handle their map information. However, the typical basic map record consists of 1/2-inch per mile scale maps of each National Forest on which all known, reported, or suspected locations of sensitive plants are plotted and identified by plant code. As these locations are field verified, they are color-coded to so indicate and are assigned a population code number to cross-reference with the written population record.

To date, most of the effort has been devoted to preparation of status reports, field verification of reported locations, and field reconnaissance of proposed project activities. Some effort is now shifting to other potentially suitable habitats in search of previously unknown populations. Records are being manually maintained as physical files for some 325 taxa of sensitive plants occurring on National Forest land in California and for more than 500 additional taxa found on other land throughout the State.

Intermountain Region (Region 4)

In contrast to the Southeastern Area and California, the Forest Service program in the Intermountain Region is in its infancy. For the most part, the Intermountain Region has not been botanized intensively and new species are still being described annually. The range and distribution of all species is poorly known. The program emphasis, therefore, is on filling these data gaps and developing positive future programs.

The program for threatened and endangered plant species in the Intermountain Region includes the following major phases:

1. *Literature and herbarium searches for existing information on distribution and ecological requirements of the plant species listed or proposed for listing by the Office of Endangered Species.*

Intensive searches of all major university and college herbaria in Idaho, Nevada, and Utah have been completed. Most of the large California herbaria and some of the herbaria in Oregon, Washington, Colorado, and Wyoming have also been partially searched. In addition, an intensive literature search for basic descriptive, distribution, and ecological information for all species proposed for listing as threatened or endangered has been completed.

The summary information of these herbarium and literature searches is stored on permanent IBM mag cards. A one-page summary sheet for each species has been prepared and distributed. The base data on these summary sheets includes distribution by state, counties, and National Forest, generalized locality information, cover type and associated species, special habitat requirements, elevational range, and manual references. The summary sheets are in loose-leaf form so that additions, deletions, and corrections can easily be made. An annual update of this summary information is planned.

2. *Provide for extensive field searches for each of the plant species listed or proposed for listing to define the population distribution, increase the number of known site locations, and build a data base on the ecological requirements of the species.*

National Forest System lands in the Intermountain Region encompass more than 33 million acres. The Region's action program proposes extensive plant surveys on approximately two-thirds of this acreage by the mid-1980's. Priority is being given to localities in imminent danger of habitat disturbance by exploration and mining, oil and gas development, geothermal development, vegetation manipulation, and other activities potentially damaging to the habitat of plants proposed for listing as threatened or endangered. Extensive field searches are currently being undertaken through contracts with universities and other organizations with competence in plant systematics and taxonomy. The Region anticipates contracts will be awarded for plant surveys on approximately 2 million acres per year, beginning in 1978.

3. *Provide for intensive site-specific plant surveys where habitat alteration is to take place.*

Intensive site-specific plant surveys are to be conducted on locations where actual disturbance is to take place, such as oil and gas drill sites, strip mine locations, road and powerline rights-of-way, vegetation manipulation projects, timber harvest, etc. An estimated 50,000 acres or more of National Forest lands in the Intermountain Region are so disturbed annually. To comply with Section 7 of the Endangered Species Act, intensive field searches for threatened or endangered plant species will be required on all acres of critical habitat where site disturbing activities are proposed.

4. *Develop positive programs for the conservation of plant species once critical*

habitat is defined and implement recovery efforts to achieve restoration of the species and its habitat.

Until such time as critical habitats are designated by final rule-making action, Forest Service policy is to consider all such species and their habitats as though they were already listed. In the interim, the Intermountain Region will attempt to manage lands and habitats occupied by such species to avoid destruction of critical habitats by accident or oversight.

Five major interim management alternatives will be implemented, depending on the needs of the species and its habitat: *First,* essentially doing nothing except recognize that the species is present. On many acres of National Forest System lands, little or no activities take place and the habitat of a species does not appear to be jeopardized by the activities of man. *Second,* minor modification of management activities that could potentially adversely modify critical habitats. Examples of modifications include utilization of alternative sites by minor relocations of roads, harvesting timber from alternative areas and modifying grazing use patterns to accommodate a species. *Third,* major modification, cessation, or prohibition of activities. Where the land use or activity is so drastic a modification of habitat, and no alternatives are possible, complete prohibition must be implemented to meet the provisions of the Act. *Fourth,* policing for commercial exploitation. The Forest Service may prohibit or regulate the removal of plants from National Forest lands. The Region currently sells timber, issues permits for grazing or wood cutting, and can issue, or not issue, plant collecting permits. Regulations are expected shortly on private and commercial collecting of exploited species. *Fifth,* establishment of special management areas during the Region's land use planning activities. Potentially damaging activities and uses will be restricted. In addition, two special designations may be used: Botanical Areas and Research Natural Areas. Each designation would provide absolute protection to the habitat of a species by withdrawing all nonconforming uses.

Information Systems for Use in Studying the Population Status of Threatened and Endangered Plants

Paul D. Whitson[1] **and J. R. Massey**[2]

The designation and preservation of threatened and endangered plants should be founded upon biologically sound population information. The urgent need to efficiently acquire and evaluate comparable population data necessitates the development of a comprehensive information and procedural program. A model system is described which provides organization and direction for the acquisition of specific information essential to the evaluation of the biological status of identified populations. The population status is determined by the identification, substantiation and analysis of the characteristics for the four basic life cycle stages—reproduction, dispersion, establishment and maintenance. The model consists of a series of priority questions to direct investigations, a comprehensive information system of related answers, and suggested documentation and evaluation elements. An overview of the relationships of this model to other natural diversity preservation efforts is discussed.

". . . use . . . of all methods and procedures which are necessary to bring any endangered species or threatened species to the point at which the measures provided pursuant to this Act are no longer necessary."

Section 3(2)
Endangered Species Act of 1973

We have witnessed, in the spirit of the Endangered Species Act, the development of many state and national threatened and endangered plant species lists and the acquisition of related general information by many individuals and organizations. This required activity and basic information have not only stimulated our interest, awareness, and appreciation of many plant species, but have also prompted us to evaluate the status of our knowledge, understanding, and preservation-conservation efforts.

---

[1]   University of Northern Iowa, Cedar Falls, Iowa 50613.
[2]   University of North Carolina, Chapel Hill, N.C. 27514.

Based on the presentation "Information Systems for Use in Studying the Biological Status of Threatened and Endangered Plant Populations" by Dr. Whitson at the Symposium on Geographical Data Organization for Rare Plant Conservation, held 15-17 November 1977 at The New York Botanical Garden.
Pages 217-236 *in:* Larry E. Morse and Mary Sue Henifin, eds., *Rare Plant Conservation: Geographical Data Organization*, The New York Botanical Garden, Bronx, N.Y.
© 1981 The New York Botanical Garden

Although the many efforts are to be lauded, some concern still persists over the lack of an acceptable system of minimal information elements supported by adequate and proper element documentation. Such an information-documentation base would seem essential to assist (1) management of information, (2) designation of threatened or endangered species, (3) evaluation of present status designations, and (4) provision of information necessary to initiate detailed species studies. Perhaps the time has come for a group such as the participants of this symposium to provide leadership by drafting and recommending a unified system toward these objectives.*

Following these initial efforts, it is now apparent that we must focus some of our attention and energies toward a new plateau—the identification of the biological status of recognized populations and their supportive factors, both biotic and abiotic. The incorporation of this information into the selection and evaluation of essential or critical habitats and its use in remedial resource management activities should in some cases result in the eventual removal of species from immediate concern. It is to this new plateau that our recent efforts have been directed and the following model is presented for your review and comments. At this time we wish to acknowledge the helpful comments and criticisms of our colleagues, especially A. E. Radford and J. F. Matthews. The invaluable contributions of several students who have endeavored to apply and relate the model to their own personal research projects are sincerely acknowledged.

Guiding Philosophy

During the development of this model we have held a rather consistent philosophy in our approach to the problems of threatened and endangered species and to the identification of alternative solutions. To assist in understanding and evaluating the model, we feel that it is important that at least a synopsis of this philosophy be presented.

Preservation of natural diversity—Protection and consequent preservation of natural diversity is a societal issue which must be viewed in a biological, political, sociological and economic context. The threatened and endangered species problem is an integral part of this larger issue, and effective solutions must take the often disparate components into account. For this reason, the best information which the biological community can provide should be available for evaluation in the complex arena of judicious decision-making.

Species biology—Threatened and endangered species preservation must be founded upon sound biological information which the biological community can best secure from a holistic perspective. We regard this broad perspective as species biology—the study of individuals, populations, and population systems of a species by utilizing the structures, processes, and habitat relationships of each major life cycle phase within a particular time reference.

Multidisciplinary effort—The acquisition of the broad spectrum of essential

---

* The "Guidelines for the Preparation of Status Reports on Threatened and Endangered Plant Species" (Henifin *et al.*, this volume) were initiated at the New York Botanical Garden Symposium and present an information system from the perspective of land management agency needs. Development of the proposed guidelines included consultation with Drs. Whitson and Massey and consideration of ideas presented in their paper.—The Editors.

218

species information will require a multidisciplinary effort. To facilitate this effort we feel that a systematic and uniform mechanism would be the most expedient.

Integrated solution—The preservation of threatened and endangered species and their habitats is a complex problem of such magnitude that neither the biologist nor the resource manager can independently provide a solution. The dedication and expertise of both basic and applied scientists must be integrated to achieve a successful solution to this social issue.

Objectives of the Model

Presently, information pertaining to the population status of threatened and endangered species is variable with respect to availability, reliability and detail. To alleviate some of these variables, we have sought to develop a program to:

- Provide an information and method system which is familiar to the plant scientist conducting the field studies and assembling the information;
- Assist acquisition, organization, and documentation of field-derived information;
- Provide for a comprehensive understanding of the species' status, including all major phases of its biology, derived from population data; and
- Assist resource managers in planning and evaluating habitat management alternatives for species preservation.

Guiding Principles of the Model

The principles used to guide the development of our model are included to be of assistance in understanding our concept of the model as well as to be of use in evaluating it. The principles are:

- The model should consist of information, method, and documentation systems for accommodating both general species and specific population information.
- The model should be structured to permit simultaneous multiple investigations without undue overlap of effort.
- Information, method, and documentation systems should be standardized to facilitate inclusion, retrieval, evaluation, and verification of information.
- Information systems should be sufficiently flexible to accommodate additional information categories.
- Information processing should be amenable to both manual and electronic data management.
- Frequent or mid-course evaluation and synthesis of assembled information should be possible.
- Evaluation and synthesis of program information should provide a basis for preservation alternatives.

Elements of the Model

The basic elements of the model which fulfill the objectives and which are consistent

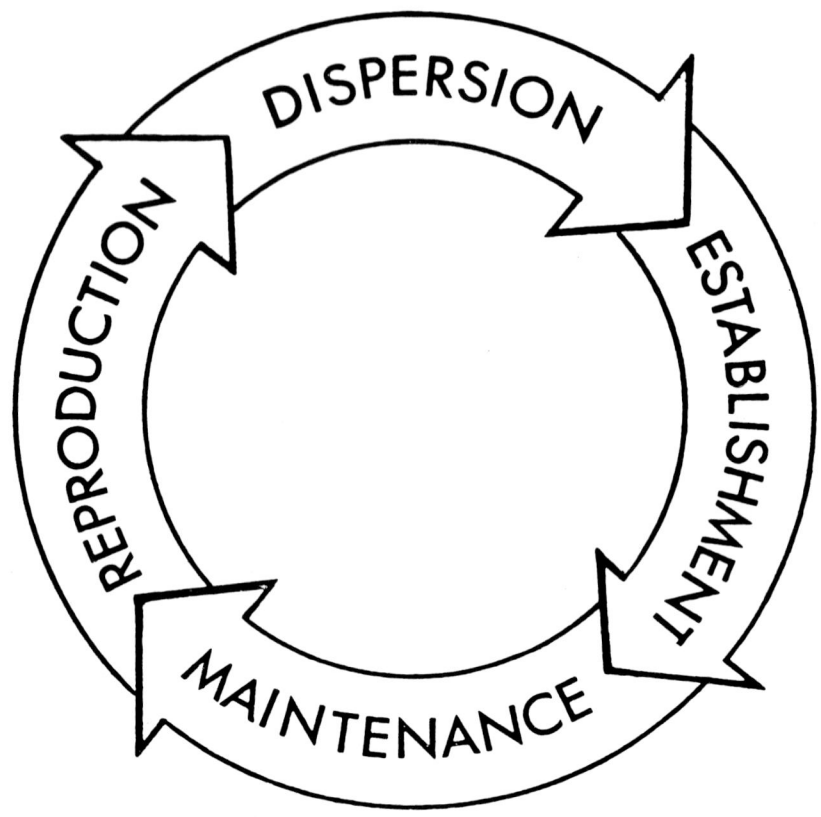

Figure 1. Generalized life cycle. A brief definition or process-product concept of each phase is: (1) reproduction—production of propagules; (2) dispersion—dissemination and distribution of propagules; (3) establishment—origination and settlement of a new, non-selfsufficient individual from a propagule; and (4) maintenance—sustention and interactions of selfsufficient individual.

with the guiding principles are: (1) Species Generalized Life Cycle, (2) Question-Procession System, (3) Answer-Information System, (4) Documentation System, and (5) Evaluation System.

Generalized Life Cycle—To acquire the necessary species biology information, all phases of the biology must be understood or considered during an investigation and evaluation of a species (Massey & Whitson, 1977). The phases used in this model are presented in Figure 1. Because the phases form a continuous cycle, a phase may be individually addressed to determine its status and the specific support it provides to the species. There can be difficulties, as in any organization or classification, in delineating precise phase boundaries or in alleviating overlaps. We feel that boundary definitions should emphasize information acquisition relevant to the organism rather than adherence to a rigid definition.

The acquisition of phase information for specific populations can permit an investigator to: (1) evaluate the biological status of the species, (2) identify the weak phase or phases in the life cycle, (3) compare the status of similarly studied and documented populations, and (4) design specific studies to further evaluate a particular phase or set of sustaining factors. Through the systematic accumulation of population data, plant scientists and resource managers can jointly evaluate research-management activities or priorities.

Question-Procession System—The second major element of the model is the Question-Procession System which is a hierarchical matrix of questions (Table 1) and a series of potential methods for use in answer acquisition (not included). The questions should assist an investigator to address each phase in the biology of the population. By construction, the questions (1) proceed from simple to complex, (2) interrelate the phases, (3) dictate point-in-time information acquisition, and (4) promote multidisciplinary contributions.

A cursory explanation of the rationale for the choice and sequence of the questions seems appropriate. We feel that many species may need only limited remedial attention if the "weak" phase in their biology can be identified. Very specific or directed questions can be utilized to identify this phase. Our intent has been to develop a dynamic view of the population from a single point-in-time analysis. By observing different phase products (seeds, new individuals, reproductive individuals) and their abundance in conjunction with basic processes (reproduction, dispersion, establishment, etc.) within the habitat, a dynamic picture of the population can emerge. A systematic appraisal of these relationships, within the population or across the habitat, can provide significant insight into the heterogeneity of products, processes and sustaining factors.

Our experience with the questions suggests that (1) the questions are occasionally deceiving in simplicity, (2) additional effort is usually required to proceed horizontally because of discipline biases, (3) ancillary questions are frequently stimulated which prove equally as valuable, and (4) every question does not always require an answer to proceed successfully. Even so, rapid and direct progress can be achieved by adhering to the intent of the questions.

Answer-Information System—The third element of the model is the Answer-Information System which provides a series of selected, possible answers to the questions. Four subsystem outlines provide information states of possible population attributes for each of the generalized life cycle phases. These information outlines or answers serve to: (1) facilitate and direct answer selection of population attributes, (2) expedite incorporation of other population attributes, as needed or desired, and (3) promote interpopulation comparisons and species summaries for a spectrum of population attributes.

The major information categories are summarized in Table 2, and the detailed information subsystems (Reproduction, Dispersion, Establishment and Maintenance) are presented in Appendix I. The information system is designed to assist an investigator in identifying the following basic population attributes: (1) population products (kinds, origins, abundance, distribution, etc.), (2) population processes (kinds, relative importance, distribution, etc.), (3) distribution patterns of the products and processes, and (4) the relationship of products and processes with both biotic and abiotic factors. We cannot overemphasize the importance of the quantity and quality of this information since it must form the foundation of our preservation programs.

Documentation System—To emphasize the importance of appropriately documenting each attribute selected from the Information System, we have included a Documentation System procedure in our model. Proper documentation is necessary if the biological community is to expedite the acquisition, evaluation, and

Table 1.
Question-Procession System: Questions

REPRODUCTION		DISPERSION		ESTABLISHMENT		MAINTENANCE
Is reproduction occurring?	+	Are propagules present?	+	Are new individuals present?	+	Is there a range of classes?
What types of reproduction are occurring?	+	What types of viable propagules are present?	+	What are the origins of the new individuals?	+	What are the origins of the classes?
What breeding systems are operative?	+	What dispersal systems are operative?	+	What establishment processes are operative?	+	What are the %'s of each class in the population?
What pollination systems are operative?	+	What are the dispersal units and/or agents?	+	What are the spatial relations of establishment processes?	+	What are the spatial relations of the classes?
What is the reproductive capacity or status of the population?	⊥	What is the dispersal effectiveness of the population?	⊥	What are the establishment effectivenesses based on origin?	⊥	What is the survivorship of each class progressing to the next class?

222

Table 2.

Summary of the Species Biology Information System.

Reproduction

Reproductive system types
- Sexual
- Asexual

Breeding system types
- Selfing
- Outcrossing

Pollination system types
- Pollination types
- Vectors

Reproductive capacity
- Origin[1] status
- Population status

Reproductive effectiveness
- Actual
- Potential

Dispersion

Diaspore system type
- Diaspore origin
- Diaspore type

Dispersal system type
- Release mechanism
- Transport-vector

Dispersal status
- Dispersal unit type
- Dispersal status

Dispersal effectiveness
- Actual
- Expected

Establishment

New individual origin
- Sexual
- Asexual

Pre-establishment processes
- Kind by origin
- Distribution[2]

Establishment processes
- Kind by origin

Establishment distribution
- Origin status
- Population status

Establishment effectiveness
- Population percentages
- Origin percentages

Maintenance

Population classes[3]
- Specific
- Relative

Class origins
- Sexual
- Asexual

Population composition-distribution
- Sociability of individuals
- Distribution status

Maintenance effectiveness
- Vitality/vigor of individuals
- Class survivorship

---

[1] Origin—sexual or asexual
[2] Distribution—spatial location in the population.
[3] Class—a definable subgroup of the population using various characteristics; redefinition may be necessary from population to population; see Maintenance Subsystem in Appendix 1 for exemplary classes.

verification of information and make the effort time- and cost-efficient. For each element of information the appropriate documentation will vary, but should include such items as literature citations, plot maps and photographs, herbarium and other museum specimens, garden records, personal records and observations, and tabulated data sets from natural field populations, experimental gardens, and controlled laboratory experiments.

Presently, we recognize the need for repositories to maintain the information and current species and population summaries. We feel that only selected portions of the information base may need computer banking, while the remainder might best be maintained in manual files.

Perhaps, the most immediate and constructive means of communication would be a publication series, designed for summarized information, similar to the Biological Flora of the British Isles *(Journal of Ecology)*. The leadership and service of an agency, such as the U. S. Fish and Wildlife Service or the National Park Service, with experience in this area should be of invaluable assistance.

Evaluation—Information evaluation should be an integral part of the model. Although we have not developed a system, the objectives of such a mechanism should be to: (1) evaluate the quality of available information, and (2) assist the interpretation of the information with respect to species dynamics through time. As a result of this evaluative process, the biological community should be able to contribute the best available information to management decision-making. Should the decision be made to proceed with management activities, the influence of these activities upon population dynamics can be similarly analyzed and evaluated by the model elements.

Although a broad spectrum of species information is not available for evaluation, several graduate students and a university group are currently applying the model in specific projects. Their comments and results indicate that the model elements can adequately assist in the acquisition of pertinent information essential to the evaluation of population dynamic properties. For example, one population of a threatened fern, *Pellaea wrightiana,* being investigated by James Matthews and several students from the University of North Carolina at Charlotte, is now understood to possess a broad range of individual size classes in its isolated locality. This information, based upon mapped sampling, implies: (1) dispersion capabilities, (2) establishment sites, (3) class maintenance success, and (4) future rate calculation potentials when coupled with resampling results. A second population of this species also under study by the UNC-Charlotte group, possesses no such spectrum of size classes, thus signifying a less dynamic population with quite different potentials.

Application of the model to several common perennial woody species provided remarkable analytical insight for establishment and maintenance phase weaknesses. In each instance documented management options could be cited as specific recommendations for remedial action.

Relationship of Model to Other Efforts

To place the population status model in proper perspective, we feel it is necessary to indicate its relationship to other efforts or programs. Because a compilation of species information is desirable before initiating population studies, we have developed a Species General Information Unit (Appendix II). This unit, like the population status model, includes question, information, documentation, and evaluation elements. The information provided by the unit is similar to that acquired by the efforts referred to in our introductory remarks. Presently we are assembling a synthesized version of the information for a number of Southern Appalachian threatened and endangered species; Appendix III provides an example of these reports.

Utilizing a hierarchical or priority rating concept to evaluate and solve many

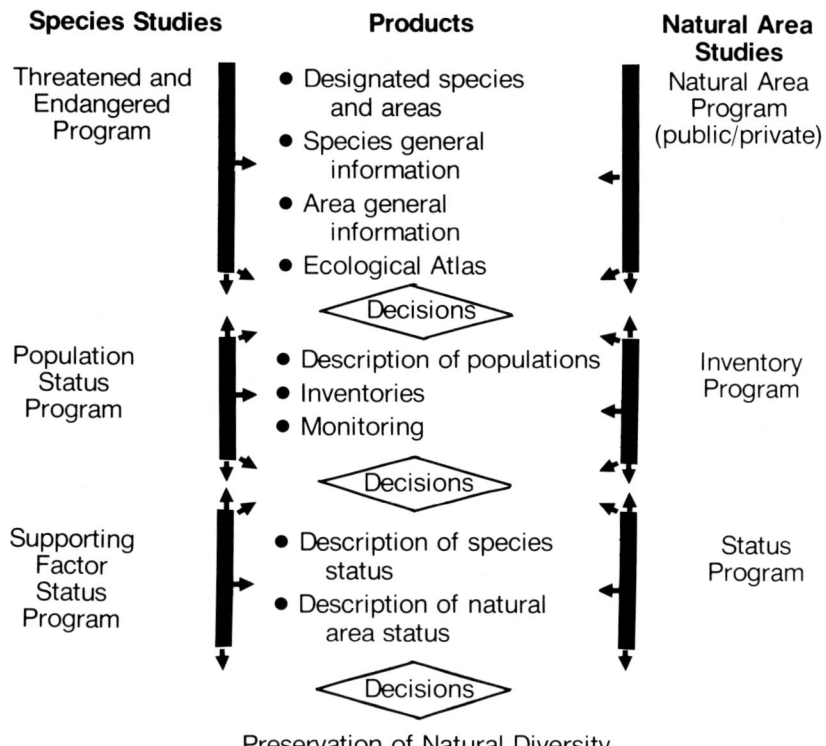

Species Studies	Products	Natural Area Studies

Species Studies

Threatened and
Endangered
Program

Population
Status
Program

Supporting
Factor
Status
Program

Products

- Designated species
  and areas
- Species general
  information
- Area general
  information
- Ecological Atlas

Decisions

- Description of populations
- Inventories
- Monitoring

Decisions

- Description of species
  status
- Description of natural
  area status

Decisions

Natural Area Studies

Natural Area
Program
(public/private)

Inventory
Program

Status
Program

Preservation of Natural Diversity

Figure 2. Natural diversity information flow.

threatened and endangered species problems, we envision the need for a third level unit which can build upon the foundation provisioned by the former two units. This third level will provide identification, analysis, and evaluation of the factors and their relationships which support, sustain, or contribute to the continued existence of a population. The unit will be patterned after the population status model and should significantly assist preservation-management efforts with detailed information.

Because the preservation of species depends on habitat preservation and maintenance, both endangered species and natural area efforts must be integrated. This can best be accomplished by maximizing the exchange of information between the efforts. A flow diagram of this exchange and several selected products is presented in Figure 2. The use of our population status model requires a locality and habitat description for each population studied. The detailed Natural Diversity Classification System developed by A. E. Radford (Radford, 1977; Radford, this volume) for natural area inventory is a superb scheme for this purpose.

Preservation efforts at all levels will be greatly enhanced by developing, testing and evaluating goal-specific models and the information they acquire. Through broader dissemination of the derived information, sound decisions should be assured to make the preservation of natural diversity become a reality.

Literature Cited

Endangered Species Act of 1973. Public Law 93-205, 87 Stat. 884-903.

Massey, J. R. and P. D. Whitson. 1977. Species biology: Definition, direction, data and decisions, pp. 88-94 in *Conference on Endangered Plants in the Southeast Proceedings.* US For. Serv. Gen. Techn. Rep. SE-11. Southeast. For. Exp. Stn., Asheville, N.C.

Radford, A. E. 1977. *A Natural Area and Diversity Classification System.* University of North Carolina Student Stores, Chapel Hill, N. C. 70 pp.

Radford, A. E. 1981. Introduction to a system for ecological diversity classification. (This volume, pages 199-205)

Appendix I. Answer-Information System

Reproductive Subsystem.

1. *Reproductive System*
   1. Amphimixis
   2. Apomixis
   3. Combination

2. *Breeding System*
   1. Autogamy
   2. Allogamy
      1. Xenogamy
      2. Geitonogamy
   3. Combination

3. *Pollination System*
   1. Type of pollination
      1. Anemophily
      2. Entomophily
      3. Hydrophily
      4. Anthropophily
      5. Ornithophily
      6. Malacophily
      7. Other
   2. Pathway
      1. Chasmantheric
      2. Cleistantheric.
   3. Visitor—Plant relationship
      1. Polytropic—polyphilic
      2. Oligotropic—oligophilic
      3. Monotropic—monophilic
      4. Other
   4. Vector(s)
      1. Abiotic
      2. Biotic
         1. Family
         2. Scientific name
         3. Vector sex
            1. Female
            2. Male

4. *Reproductive Capacity*
   1. Population status
      1. Reproductive condition
         1. Asexual
         2. Sexual
         3. Combination
         4. None
      2. Population sector[1]
         1. Entire population
         2. Central sector
            1. Total[2]
            2. Sample[3]
         3. Peripheral sectors
            1. Sector 1, etc.
               1. Total
               2. Sample
      3. Area size
         1. Entire population
         2. Central sector
            1. Total
            2. Sample
         3. Peripheral sectors
            1. Sector 1, etc.
               1. Total
               2. Sample
      4. Asexual census
         1. Entire population
         2. Central sector
            1. Total
            2. Sample
         3. Peripheral sectors
            1. Sector 1, etc.
               1. Total
               2. Sample
      5. Sexual census
         1. Entire population
         2. Central sector
            1. Total
            2. Sample

---

[1] Sector—a definable subdivision of the population; central or peripheral.

[2] Total—a complete or entire enumeration for the defined area.

[3] Sample—a representative enumeration for the defined area.

Dispersion Subsystem.

1. *Diaspore System*
   1. Diaspore type
      1. Sexual
      2. Asexual
      3. Combination
   2. Diaspore structural type
      1. Spore
      2. Seed
      3. Fruit
      4. Vegetative structures
         1. Root
         2. Stem
         3. Leaf
         4. Bulb
         5. Plantlets
         6. Other
      5. Combination
      6. Other

2. *Dispersal System*
   1. Break-off mechanisms
      1. Gravity
      2. Maturation
      3. Ejection
      4. Ablation
      5. Other
   2. Transport (vectors)
      1. Abiotic
         1. Gravity
         2. Wind
         3. Water
         4. Projection
         5. Deposition
         6. Other
      2. Biotic
         1. Scientific name
         2. Position of diaspore
         3. Diaspore-vector relationships
         4. Other

3. *Dispersal Status*
   1. Dispersal unit
      1. Spore
      2. Seed
      3. One-seeded fruit
      4. Multi-seeded fruit
      5. Bulb
      6. Plantlets
      7. Other
   2. Census
      1. No. propagules per dispersal unit
      2. No. dispersal units per unit area
         1. Entire population
            1. No.
            2. Area size

    3.2.2.2. Central sector
        1. Total
           1. No.
           2. Area size
        2. Sample
           1. No.
           2. Area size
      3. Peripheral sectors
        1. Sector 1, etc.
           1. Total
               1. No.
               2. Area
           2. Sample
               1. No.
               2. Area

3. No. propagules per unit area
   1. Asexual
      1. Entire population
      2. Central sector
         1. Total
         2. Sample
      3. Peripheral sectors
         1. Sector 1, etc.
            1. Total
            2. Sample
   2. Sexual
      1. Entire population
      2. Central sector
         1. Total
         2. Sample
      3. Peripheral sectors
         1. Sector 1, etc.
            1. Total
            2. Sample

4. *Dispersal Effectiveness*
   1. No. propagules per population
      1. Sample size
      2. Sample location
      3. Values
         1. Asexual
         2. Sexual
         3. Combination
   2. No. of expected propagules per population[4]
      1. Asexual
      2. Sexual
      3. Combination
   3. Relative effectiveness (observed/expected)
      1. Asexual
      2. Sexual
      3. Combination

Establishment Subsystem.

1. *Origins of new individuals*
   1. Sexual

---

[4]Derived from Reproductive Effectiveness

1. 2. Asexual
   3. Combination

2. *Pre-establishment processes*
   1. Sexual
      1. Stratification
      2. Scarification
      3. Temperature treatment
      4. Leaching
         1. Abiotic (ppt)
         2. Biotic (digestive tract)
      5. Other
   2. Asexual
      1. Cold hardening
      2. Heat hardening
      3. Acclimation
      4. Other
   3. Uncertain
   4. None

3. *Establishment processes*
   1. Sexual
      1. Imbibition
      2. Germination
         1. Hypogeous
         2. Epigeous
         3. Uncertain
      3. Emergence
         1. Radicle
         2. Hypocotyl
         3. Cotyledon
         4. Epicotyl
      4. Other
   2. Asexual
      1. Parental separation
         1. Complete
         2. Incomplete
         3. Uncertain
      2. Root initiation
         1. Primary
         2. Secondary
         3. Other
      3. Shoot initiation
         1. Stem initiation
            1. Primary
            2. Lateral
         2. Leaf initiation
            1. First (pair)
            2. Second (pair)
            3. Other
         3. Uncertain

4. *Establishment distribution*
   1. Establishment states
      1. Sexual
         1. Imbibition
         2. Germination
         3. Emergence
         4. Other
      2. Asexual

4.1.2. 1. Parent separation
       2. Root initiation
       3. Shoot initiation
       4. Other
   3. Uncertain
2. Distribution status
   1. Population sectors
      1. Entire population
      2. Central sector[1]
         1. Total
         2. Sample
      3. Peripheral sectors
         1. Sector 1, etc.
            1. Total
            2. Sample
   2. Area size
      1. Entire population
      2. Central sector
         1. Total
         2. Sample
      3. Peripheral sectors
         1. Sector 1, etc.
            1. Total
            2. Sample
   3. Establishment state census
      1. Sexual states
         1. Imbibition
            1. Entire population
            2. Central sector
               1. Total
               2. Sample
            3. Peripheral sectors
               1. Sector 1, etc.
                  1. Total
                  2. Sample
         2. Germination
            1. Entire population
            2. Central sector
               1. Total
               2. Sample
            3. Peripheral sectors
               1. Sector 1, etc.
                  1. Total
                  2. Sample
         3. Emergence
            1. Entire population
            2. Central sector
               1. Total
               2. Sample
            3. Peripheral sectors
               1. Sector 1, etc.
                  1. Total
                  2. Sample
      2. Asexual states
         1. Parental separation
            1. Entire population
            2. Central sector
               1. Total
               2. Sample
            3. Peripheral sectors

4.2.3.2.1.3. 1. Sector 1, etc.
     1. Total
     2. Sample
  2. Root initiation
   1. Entire populations
   2. Central sector
    1. Total
    2. Sample
   3. Peripheral sectors
    1. Sector 1, etc.
     1. Total
     2. Sample
  3. Shoot initiation
   1. Entire population
   2. Central sector
    1. Total
    2. Sample
   3. Peripheral sectors
    1. Sector 1, etc.
     1. Total
     2. Sample
  3. Uncertain states

5. *Establishment effectiveness*
 1. Population analysis (percentages)
  1. Established individuals
  2. Initial state individuals
  3. Establishment survivorship %
 2. Origin summaries
  1. Sexual
   1. Established individuals
   2. Initial state individuals
   3. Sexual survivorship %
  2. Asexual
   1. Established individuals
   2. Initial state individuals
   3. Asexual survivorship %
  3. Uncertain
 2. Sector analysis
  1. Central sector
   1. Established individuals
   2. Initial state individuals
   3. Central sector survivorship %
  2. Peripheral sectors
   1. Sector 1, etc.
    1. Established individuals
    2. Initial state individuals
    3. Sector 1, survivorship %

Maintenance Subsystem.

1. *Population classes*[5]
 1. Specific age classes
  1. Temporal designation
   1. Hours
   2. Days

1.1.1. 3. Months
   4. Years
  2. Temporal range classes
   1. Class 1
   2. Class 2, etc.
 2. General reproductive classes
  1. Prereproductive
  2. Reproductive
  3. Postreproductive
 3. Relative age classes
  1. Morphological character
   1. Development state of individual
    1. Plant character state
     1. Kind or number of leaves
     2. Kind or number of stems
     3. Kind or number of nodes
     4. Kind or number of roots, etc.
    2. Development range classes
     1. Class 1
     2. Class 2, etc.
   2. Morphometry of individual or portion
    1. Plant character state
     1. Individual
     2. Leaf or portion
     3. Stem or portion
     4. Root or portion
    2. Morphometric state
     1. Height (length)
     2. Girth (dbh, cover)
     3. Thickness
     4. Area
     5. Volume
     6. Weight (destructive sampling not generally recommended)
    3. Morphometric range classes
     1. Class 1
     2. Class 2, etc.
  2. Physiological character of individual or portion
   1. Plant character state
    1. Individual
    2. Leaf or portion
    3. Stem or portion
    4. Root or portion, etc.
   2. Process rate
    1. Qualitative state
     1. Photosynethesis
     2. Respiration
     3. Transportation, etc.
    2. Quantitative state
     1. mg/area unit/time unit
     2. mg/weight unit/time unit, etc.
   3. Process range classes
    1. Class 1
    2. Class 2, etc.
  3. Chemical content

---

[5] see footnote, Table 2.

1.3.2.3.1. Qualitative state
    1. Protein or portion
    2. Carbohydrates or portion
    3. Lipid or portion
    4. Secondary compounds, etc.
  2. Quantitative state
    1. mg/plant character unit
    2. %/plant character unit
    3. Absorption unit/plant character unit
    4. Activity unit/plant character unit
  3. Chemical content range classes
    1. Class 1
    2. Class 2, etc.
4. Other

2. *Class origins*
  1. Class 1
    1. Sexual origin
    2. Asexual origin
    3. Unknown origin
  2. Class 2, etc.

3. *Population composition*
  1. Individuals in each class
    1. Class 1
    2. Class 2, etc.
  2. Individuals in population (sum of classes)
  3. Class percentages
    1. Class 1
    2. Class 2, etc.

4. *Population distribution*
  1. Sociability of individuals
    1. Solitary
    2. Clumps or dense groups
    3. Small patches or cushions
    4. Small colonies or large carpets
    5. Large, almost pure population stands
  2. Distribution status
    1. Population sectors[1]
      1. Entire population
      2. Central sector
        1. Total
        2. Sample
      3. Peripheral sectors
        1. Sector 1, etc.
          1. Total
          2. Sample
    2. Area size
      1. Entire population
      2. Central sector
        1. Total
        2. Sample
      3. Peripheral sectors
        1. Sector 1, etc.
          1. Total
          2. Sample
    3. Class distributions

4.2.3.1. Class 1 census
  1. Entire population
  2. Central sector
    1. Total
    2. Sample
  3. Peripheral sectors
    1. Sector 1, etc.
      1. Total
      2. Sample
2. Class 2 census, etc.
4. Origin distributions
  1. Sexual individual census
    1. Entire population
    2. Central sector
      1. Total
      2. Sample
    3. Peripheral sectors
      1. Sector 1, etc.
        1. Total
        2. Sample
  2. Asexual individual census
    1. Entire population
    2. Central sector
      1. Total
      2. Sample
    3. Peripheral sectors
      1. Sector 1, etc.
        1. Total
        2. Sample
  3. Uncertain individual census
    1. Entire population
    2. Central sector
      1. Total
      2. Sample
    3. Peripheral sectors
      1. Sector 1, etc.
        1. Total
        2. Sample

5. *Maintenance effectiveness*
  1. Vitality or vigor of individuals
    1. Very feeble, never fruiting
    2. Feeble
    3. Normal
    4. Exceptionally vigorous
  2. Class survivorship
    1. Total population analysis
      1. Class summaries
        1. Class 1 survivorship %
        2. Class 2 survivorship %, etc.
      2. Origin summaries
        1. Sexual survivorship %
          1. Class 1, etc.
        2. Asexual survivorship %
          1. Class 1, etc.
        3. Uncertain survivorship %
          1. Class 1, etc.
    2. Sector analysis
      1. Central sector
        1. Class summaries

5.2.2.1.1.1. Class 1 survivorship %, etc.
      2. Origin summaries
        1. Sexual survivorship %
          1. Class 1, etc.
        2. Asexual survivorship %
          1. Class 1, etc.
        3. Uncertain survivorship
          1. Class 1, etc.
    2. Peripheral sectors
      1. Sector 1, etc.

5.2.2.2.1.1. Class summaries
      1. Class 1 survivorship %, etc.
      2. Origin summaries
        1. Sexual survivorship %
          1. Class 1, etc.
        2. Asexual survivorship %
          1. Class 1, etc.
        3. Uncertain survivorship %
          1. Class 1, etc.

Appendix II. Elements of Species General Information System

The elements of the Species General Information System (see section, Relationship of Model to Other Efforts) are presented to illustrate our concept of documentation. The unit is the initial step in the compilation of available data for the development of species status reports and the initiation of species biology studies. Unit information provides a minimal set of organized data essential to the basic understanding of the taxonomic status, general characteristics, legal status, ownership and use status of a species and its habitats. The documentation system specifically identifies each item of information with an Item Code Number and each applicable item response is documented with an appropriate Citation Code and Number. Both codes provide for the convenient addition of information categories and citation sources as well as an up-to-the-moment summary for inspection or duplication.

1. Species taxonomic status
  1. Species name
  2. Description
    1. Original
    2. Recent or revised
  3. Synonomy
  4. Type specimen(s) location
  5. Family
  6. Common names

2. Species general characteristics
  1. Plant habit
    1. Herb
    2. Shrub
    3. Tree
    4. Vine
  2. Inhabitant type
    1. Aquatic
    2. Semi-aquatic
    3. Terrestrial
    4. Epiphytic
      1. host
  3. Nutrition type
    1. Autophytic
    2. Parasitic
      1. host
    3. Saprophytic
    4. Hemiparasitic
      1. host
  4. Life span
    1. Annual
    2. Biennial
    3. Perennial
  5. Vegetative phenophase

    2.5.1. January
       2. February
       3. March
       4. April
       5. May
       6. June
       7. July
       8. August
       9. September
      10. October
      11. November
      12. December
  6. Flowering phenophase
    1. January
    2. February
    3. March
    4. April
    5. May
    6. June
    7. July
    8. August
    9. September
    10. October
    11. November
    12. December
  7. Fruiting or sporulating phenophase
    1. January
    2. February
    3. March
    4. April
    5. May
    6. June
    7. July
    8. August

2.7.9. September
   10. October
   11. November
   12. December

3. Species legal status
  1. Threatened
    1. International
    2. National
    3. State
  2. Endangered
    1. International
    2. National
    3. State
  3. Other
    1. International
    2. National
    3. State
    4. Other

4. Distribution
  1. Biogeographic region
    1. Nearctic
    2. Palearctic
    3. Neotropical
    4. Ethiopian
    5. Oriental
    6. Australian
    7. Antarctic
  2. Physiographic regions
    1. Appalachian highlands
    2. Coastal plains
      1. Atlantic
      2. Gulf
      3. Floridian
    3. Interior highlands
    4. Interior plains
    5. Laurentian upland
    6. Other
  3. Countr(y)ies
  4. State(s) & counties[1]
  5. Localit(y)ies
    1. Location name or number
    2. Latitude
    3. Longitude

5. Ownership status of land
  1. National
  2. State
  3. County
  4. Municipal
  5. Private
  6. Uncertain

6. Land status

6.1. Designation
  1. Research natural area
  2. National forest
  3. State forest
  4. Experimental forest
  5. National natural landmark
  6. State natural area
  7. State park
  8. Commercial forest
  9. Farm
  10. Uncertain
  11. Other
 2. Protection
  1. Protected
  2. Unprotected
  3. Uncertain
 3. Apparent use
  1. Commercial
  2. Recreation
  3. Other

7. Habitat preference
  1. Habitat types
    1. Bog
    2. Swamp
    3. Marsh
    4. Grassland
    5. Savannah
    6. Chaparral—shrub
    7. Desert
    8. Tundra
    9. Ice
    10. Forest
      1. Gymnosperm
      2. Angiosperm
      3. Mixed
    11. Other
  2. Topographic types
    1. Mountains, hills and ridges
    2. Scarps, bluffs, cliffs and escarpments
    3. Benches and terraces
    4. Basins
    5. Lake, bay, ponds and pools
    6. Valley, gorges and channels
    7. Plains and flats
    8. Beaches
    9. Other
  3. Substrate Type
    1. Water
      1. Fresh
      2. Saline
    2. Soil
      1. General type
      2. Texture
    3. Rock

---

[1] State and Counties numbered according to Federal Information Processing Standards Publication (FIPS Pub) 6-2.

7.3.3.1. Origin type
  1. Igneous
  2. Sedimentary
  3. Metamorphic
  4. Uncertain
  2. Specific type
    1. Limestone
    2. Sandstone
    3. Shale
    4. Granite
    5. Other
  4. Humus or organic layers
4. Disturbance types
  1. Biotic
    1. Roads, paths or right-of-ways
    2. Cultivated lands
    3. Waste & spoil areas
    4. Grazed or browsed areas
    5. Mowed
    6. Forest management
      1. Clear cutting
      2. Selective cutting

7.4.1.7. Wildlife areas subject to burrowing,
         rooting and mounding
      8. Abandoned
      9. Chemically treated (weed or
         pest control)
     10. Disease and pests
         1. Types
            1. Pine bark beetle
            2. Dutch elm disease
            3. Other
     11. Uncertain
     12. Other
  2. Abiotic
    1. Fire
    2. Flood
    3. Erosion or slides
    4. Other

8. Habitat development state
  1. Early successional or pioneer state
  2. Middle successional or transient state
  3. Late successional or climax state
  4. Uncertain

Appendix III. Example of Species General Information Documentation

To specifically exemplify the documentation procedure, information is presented on a Southern Appalachian endemic, *Liatris helleri* Porter. This information was compiled under a cooperative agreement (Contract No. 18-606) between the Southern Forest Experiment Station, U. S. Forest Service, and The Highlands Biological Station. This information and that for 48 other southern Appalachian vascular plants in North Carolina represent a portion of a continuing effort on the endangered species and natural areas programs sponsored by the Highlands Biological Station. Specific item responses for *Liatris* are presented below, followed by citation source documentation from literature, statutes, and maps, specimen citations, and observations.

General Information Documentation for *Liatris helleri* Porter[1]

ITEM CODE	DOCUMENTATION CODE	CITATION CODE & NUMBER
1.1	*Liatris helleri* Porter	L1
1.2.1.	"*Liatris Helleri.*—Glabrous, with faint traces of pubescence on the pedicels and along the bases of the leaves; stem from a rootstock irregular in shape, leafy, fifteen or sixteen inches high; leaves, linear, acuminate, diminishing in size and breadth upward, the lowest three lines wide in the middle, not punctate; raceme three to four inches long, loose, inclined to droop; heads six lines high, on short, slender ascending pedicels, seven- to ten-flowered; bracts of the involucre lax, not appressed, light green, with narrow scarious rarely purplish margins, not glandular-punctate, oblong-linear, the tips obtuse, or often so doubled in as to appear acute, lowest short, ovate, acute or acutish; pappus plumose, scanty, weak, scarce half the length of the corolla-tube; achenes as long as corolla-tube and sparsely hairy."	L1

---

[1] Contract compilers: J. R. Massey (Principal Investigator), David Whetsone (Research Associate), Herbarium, Department of Botany, University of North Carolina, Chapel Hill, North Carolina 27514.

1.2.2.	"This species, of a very limited mountain range, seems distinctive in its very short pappus since others having both plains- and mountain-habitats have not shown this variation."	L2
1.3.	*Lacinaria helleri* (Porter) Porter	L3
1.4.	Herbarium (NY)	
	New York Botanical Garden	
	Bronx Park	
	Bronx, New York 10458	L2
1.5.	Asteraceae (Compositae)	L4
1.6.	Heller's Blazing Star	L6
	Heller's Gayfeather	L6
2.1.1.	Herb	L4
2.2.3.	Terrestrial	L7
2.3.1.	Autophytic	
2.4.3.	Perennial	L4
2.5.7.	July	L4
2.5.8.	August (inferred)	L4
2.5.9.	September (inferred)	L4
2.5.10.	October	L4
2.6.7.	July	L4
2.6.8.	August	L4
2.7.9.	September	L4
2.7.10	October	L4
3.3.3.	North Carolina-Candidate Threatened	L6
4.1.1.	Nearctic	L7
4.2.1.	Appalachian Highlands	L7
4.3.	United States	L8
4.4.01.	Alabama	L8
4.4.37.	North Carolina	L4
	Avery (011)	L2
	Burke (023)	H3, L2
	Caldwell (027)	H1
	Mitchell (121)	L2
	Watauga (189)	H2
4.4.51.	Virginia	L7
4.5.1.	North Carolina	
	Avery-Beech Mountain	L2
	Grandfather Mountain	L2
	Burke-Table Rock Mountain	H3, L2
	Caldwell-Blowing Rock	L2
	Grandfather Mountain	L2
	Mitchell-Roan Mountain	L2
	Watauga-Blowing Rock	H2
4.5.2.	35°54′N - 36°07′N	L9
4.5.3.	81°40′W - 82°10′W	L9
5.1.6.	Uncertain	
6.1.10.	Uncertain	
6.2.3.	Uncertain	
6.3.4.	Uncertain	
7.1.6.	Shrub bald	H3
7.2.1.	Mountains, ridges	L2
7.2.2.	Cliffs	L2
7.3.3.1.3.	Metamorphic	L5
7.3.3.2.5.	Gneiss (generalized)	L5
7.4.2.3.	Erosion	H3
8.1.	Early successional state	H3

Citation Sources for *Liatris helleri* Porter

LITERATURE, STATUTES, AND MAPS (L)

L1. Porter, T. C. 1891. A new *Liatris* from North Carolina. Rhodora 18: 147-148.

L2. Gaiser, L. O. 1946. The Genus *Liatris*. Rhodora 48: 165-183, 216-263, 273-326, 331-382, 393-412.

L3. Porter, T. C. 1900. Muhlenbergia 1:6 (Gray Cards).

L4. Ahles, H. E. 1968. Asteraceae. *In* A. E. Radford *et al.* Manual of the Vascular Flora of the Carolinas. University of North Carolina Press, Chapel Hill.

L5. Department of Conservation and Development. 1958. Geologic Map of North Carolina. Williams & Heintz Lithograph Corporation.

L6. Committee on Vascular Plants. 1977. Vascular Plants. Reprinted from J. E. Cooper *et al.* Endangered and Threatened Plants and Animals of North Carolina. Bookstore, University of North Carolina, Charlotte.

L7. Alexander, E. J. 1972. *Lacinaria* Hill. *In* J. K. Small. Manual of the Southeastern Flora. Facsimile reprint of the 1933 edition. Hafner Publishing Company, New York.

L8. Fish and Wildlife Service, U.S.D.I. 1975. Family Lists of Candidate Endangered and Threatened Plant Species in the Continental United States. Excerpt from House Document No. 94-51, the Smithsonian Report. US Department of the Interior, Washington, D.C.

L9. North Carolina State Highway Commission. 1970. Municipal, State, Primary, and Interstate Highway Systems. (Map.) Raleigh?

SPECIMEN CITATIONS (H)

H1. DUKE 35982. Small & Heller, *s.n.* 6 August 1891. North Carolina, Caldwell Co.

H2. DUKE 115290 Seymour 91-8-17. 17 August 1891. North Carolina, Watauga Co.

H3. NCU 55875. Radford 6515. 24 August 1952. North Carolina, Burke Co.

236

Limiting Factors and Pitfalls of Environmental Data Management:

Some Considerations in Developing the Information System for the State Natural Heritage Programs

Helmut P. Moyseenko[1]

Data organization and information systems exhibit complex "system behavior" because they are composed of many coordinated sets of procedures and chains of activities that are linked together to produce an end product—information for analysis. Every link may have its own weaknesses brought on either by technical or perceptual problems. These problems may be sufficient in themselves to lead to unanticipated failure of the entire operational structure. I propose that we need to put more intellectual energy into the overall operational behavior of the whole system. We also need to view the total system structure at its lowest common denominator—the individual data elements and their relationship to future uses and applications. I also attempt to uncover the fundamental flaws and limiting factors that may exist in day-to-day processing operations. To assist future ecological data base organizers, several decision variables The Nature Conservancy was faced with in designing its own information management system will be discussed. A great many objectives and design criteria were formulated to create our system; many decisions made along the way were correctly considered, while others were not, because we did not realize their complete implications. I will present a survey of some important variables that make up an ecological data organization and information system and will also discuss some of the limiting factors and pitfalls extracted from our experiences.

Advances in the computer world over the past several years have prompted the creation of numerous data bases while still many more are being proposed and implemented. Millions of dollars have been spent on data collection, computerization and modeling efforts. Despite all the impressive technical advances, some critics feel that the attempts to implement a large-scale data base management system have been largely wasted because the systems remain inaccurate, unproven, unused, and expensive; it seems that such systems operate in a failure mode most of the time, are ultimately dismantled, or simply collapse.

---

[1]   The Nature Conservancy, 1800 N. Kent St., Arlington, Va. 22209.

  Based on the presentation "Limiting Factors and Pitfalls of Environmental Data Management" by Mr. Moyseenko at the Symposium on Geographical Data Organization for Rare Plant Conservation, held 15-17 November 1977 at The New York Botanical Garden.
  Pages 237-253 *in:* Larry E. Morse and Mary Sue Henifin, eds., *Rare Plant Conservation: Geographical Data Organization,* The New York Botanical Garden, Bronx, N.Y.

There is a popular illusion that the only practical way to overcome past inadequacies is to introduce still more computer speed and storage capacity. Computer specialists are always quick to assert the high-speed capabilities that machines possess, the powerful new storage and output capabilities brought on by new and sophisticated hardware, and the money that can be saved by these advanced data processing techniques. Are these assertions real or imaginary? If the savings are real and the advanced data processing hardware so powerful, then why are so many environmental data bases for complex analyses so weak? Have the advances in computer technology enhanced our foresight, can we get more information for less money? Even if some users have failed to keep a realistic view of automated systems limitations, many others have legitimately found computerization to be an indispensable and fundamental tool for analysis. So how are computer systems best applied?

We need more intellectual energy put into the overall operational behavior of the entire data base and information system structure. We need to analyze the total system structure at its lowest common denominator—the individual data elements, their future uses and applications, and also the decision-making process from a data needs and adequacy standpoint (Young and Tierney, 1976). We need to uncover the fundamental flaws and limiting factors that may exist in day-to-day processing operations. Data management systems exhibit complex system behavior because they are composed of so many coordinated sets of procedures and chains of activities that are linked together to produce an end product—information for analysis. Every link in the chain may have its own weaknesses brought on by either technical or perceptual problems and may lead to an unanticipated failure of the entire operational structure.

The aim of this paper is to show how computerized information systems can be advantageously utilized, rather than utilized in a costly and inefficient way. A survey of the many variables, limiting factors, and pitfalls that The Nature Conservancy has considered in designing its own system will be presented. The basis for this survey will be the data management approach developed mainly under the auspices of The Nature Conservancy's state-oriented Natural Heritage Programs and other ecological inventories in which the Conservancy compiled and systematized data (Jenkins and Moyseenko et al., 1977; Federal Committee on Ecological Reserves, 1977).

Over the last three years, the Conservancy has implemented twelve* such State Natural Heritage Programs. Heritage Programs are conducted in cooperation with state governments or other agencies, usually under a one or two year contract. During the contract period, a systematic process for the collection, management, analysis, and protection planning use of certain ecological data is established for the preservation of ecological diversity in the state. After the initial contract period is over, the program is transferred to the appropriate government agency for operation.

As discussed by Jenkins (this volume), we feel that The Nature Conservancy's State Natural Heritage Programs represent an effective "process" for identifying and increasing our knowledge about significant ecological elements of natural diversity. The information theory facilitates the collection, management and use of ecological data and is the crucial hub of the Natural Heritage Program "process."

The paper is divided into three sections: the first discusses information retrieval

---

* 23 by April, 1980.

choices that are available to us, the second provides the reader with design considerations for an information system, and the third discusses the data organizational structure of The Nature Conservancy's Heritage Program information system.

Information Retrieval

Before addressing questions on system limitations, we need to know something about the information retrieval options available to us. Each option is controlled by specific decision-making needs, and each is strongly associated with the actual content of the data. The three options are fact retrieval, basic statistical inference, and deductive inference (Wiederhold, 1977).

Fact retrieval. Fact retrieval is the traditional aim for information systems and is the primary use of bibliographic data banks. To illustrate fact retrieval, we will use the State Natural Heritage Program concepts as an example. First we establish the objectives of the data base and anticipate what kinds of data must be collected in order to meet them. Then we design a framework for housing the "facts" we wish to record, maintain, and analyze. Figure 1 illustrates the data content of a format used by the Heritage Programs for recording standardized data. In designing such a format, The Nature Conservancy considered the following questions:

Q) *What types of data will have to be collected to achieve the objectives of the data base?*

A) The objectives of a typical Heritage Program data base are to centralize, systematize, and store facts about the existence, numbers, condition, status, and distribution of the elements of ecological diversity. (We define an "element" as meaning a natural feature of particular interest, either because it is unique, endangered, or represents an important type. Examples are plant community types, aquatic types, and endangered, threatened, or rare flora and fauna. An "element occurrence" is a particular place on the landscape where an element is known to exist. For example, "Red Spruce Forest" could be an element, whereas the Red Spruce Forest on Gaudineer Knob, West Virginia, would then be an element occurrence.) The existence of an element is defined by the Heritage Program classification for each state. The numbers pertain to the number of occurrences of an individual element which are discovered during data collection. The condition refers to the degree and types of protection afforded to the element occurrence through ownership and management, while distribution provides the facts on how an element occurrence is related to other occurrences and to other types of geographically distributed data. The information system stores and retrieves data about the existence on the land of element occurrences such as rare and endangered plants, their habitats, and other information that will enable us to protect them and to avoid confrontations with forces of development. This means to avoid such a confrontation by providing information at the onset of the land use process, as discussed by Joe Pardue elsewhere in this volume. The existence and use of a centralized organized body of data on a state's natural diversity makes it possible to avoid some environment-vs.-development controversies by clarifying other alternatives before resources are committed to development.

Q) *How should the data categories be structured?*

A) After we establish what kinds of data are needed for decision-making, we

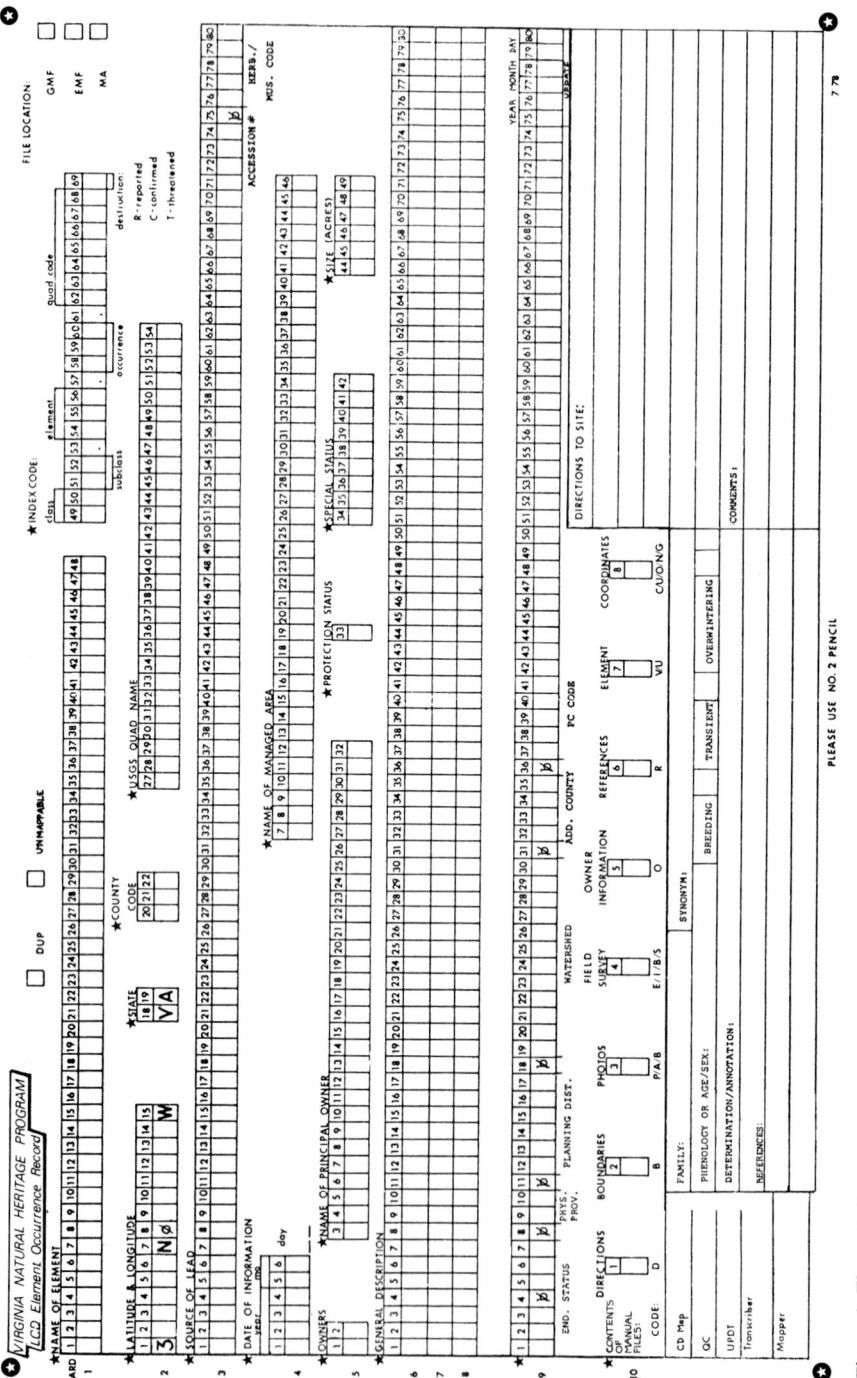

Figure 1. Element Occurrence Record Format for the Heritage Programs.

240

must decide how the data should be structured in order to retrieve information in usable form. The format becomes the structure on which to hang a subset of data from the original body of source material. Note that the format does not achieve an in-depth analysis of the source material, nor does it provide any information on measurements or statistics that may have been part of the original body of literature. Anyone interested in finding out about the life history of a particular species must consult the actual source of data or another file structure. (The concept of multiple, cross-referenceable file structures will be elaborated in a later section of this paper.) If you have multiple needs beyond a single format's capability, you must create multiple formats, each one designed for a specific set of objectives. By limiting the length of each category and of the format as a whole, we divert the additional facts to other file structures. If they do not provide the answers to the decision-making needs outlined for a particular format, they are not made a part of this format. Having established informational requirements for each format, we collect only those data necessary to satisfy those particular objectives; we do not want to collect data not directly relating to the objectives.

A standardized format helps us to achieve the following five objectives:
1. standardize terminology;
2. acquire comparable data coverage for each data category;
3. obtain feedback on significant data gaps;
4. supply sufficient factual data to get a handle on the elements of natural diversity, where they are located, and whether they are protected;
5. improve our speed of access and ability to permute and obtain statistics from the data.

The format could be enlarged to incorporate any number of other data on subject content. However, in the long run, assigning more and more components to a single format in order to capture the complete subject material can result in the generation of more "noise" (unwanted information) than wanted information. If we compile data without any consideration to structuring and understanding how the data will ultimately be used, we quickly discover that the system can only recall the subject material *en masse.* An information system is only as good as its recall capacity. If it returns too much irrelevant information along with the relevant, the system will be ineffective and expensive; having massive amounts of mostly unwanted data that cannot be easily applied to decision-making is just as bad as having no data at all.

Basic Statistical Inference.   Basic statistical retrieval differs from fact retrieval in that the user does not wish to see the detailed contents of any material directly, but rather wishes to obtain statistics on certain aggregations of the data base in order to gain insight. Statistics can be used to reduce, sort out, or summarize large or complex sets of data contained within the format. Statistical inference involves the statistical processing and manipulation of the data categories to get at facts and simple relationships contained in the data constructs. For example, we could find out how many endangered species are found on already-protected sites by simply retrieving data from a format subunit and comparing it with another subunit of the same format through all records of sites. The resulting output can be portrayed in tabular, histogram, or map form. We can only receive statistics on categories of data which are contained in the format.

Deductive Inference. Deductive inference goes one step further than statistical inference in that it attempts to uncover the underlying cause-and-effect relationships that may be inherent among the data. An example of a deductive inference might be how temperature plays a role in species growth. Obviously, such a question is beyond the simple fact retrieval and statistical summation stages and the answer may not be found in the data construct. To organize a data base for uncovering cause-and-effect relationships requires a much finer breakdown of the data categories within the individual subunits of the record format. Finding the relevant data to input and understanding the interaction between the data elements is crucial. Finding the correct data to put into the system is never as simple as it sounds. The majority of problems center around the following four factors:

1. awareness of theory concerning data relationships,
2. data availability,
3. cost of acquiring data, and
4. reliability of data.

De novo searches to obtain the necessary data are always costly endeavors. Lack of theory coupled with expensive and inadequate data makes this alternative in information retrieval the most difficult to carry out successfully. Getting the correct new data and learning how to extract the inherent information often requires more effect than the output is worth. There is always a danger of creating a data bank for which the output and information to be derived from it will be of limited value. The option of creating an ecological data base for "deductive inference" may never be achieved effectively except for solving very simple problems, because too many interrelated, complex technical and theoretical factors are involved (Lindblom, 1959; Lee, 1973; Voelker, 1975).

Design Considerations

Before highlighting The Nature Conservancy's current data organization concept, we think it would be helpful to provide the reader with background on our past experiences with ecological inventory and compilation efforts that used information management systems. We considered many combinations of variables in designing a balanced information management system and ecological data base. A great many objectives and design criteria were formulated to create the best workable system for the Natural Heritage programs. The following are among the most outstanding pitfalls and limiting factors of information system design that past experiences have taught us to beware of. The subsequent section will discuss our findings into three broad subject areas:

1. human factors and their effect on the overall operation,
2. the necessity for a flexible combination of manual and computerized file structures, and
3. general findings on ecological data management.

Human factors. A few of the major system design problems which have been too narrowly defined in the past are: how and by whom the information will be used, how much is actually required, and where to find the relevant data for input. If you don't understand the data and their interrelationships as they enter the data bank, you will certainly not understand the output and its interrelationships. The success of an information system depends largely upon whether or not the system is actually being utilized, after all if the users bypass the system for whatever reason, then the system must be branded a failure even if it is technically sound. This might be called the human factor in that the users must realize how to employ the system on their behalf.

Once a system has been implemented, it can fail in a number of other ways because of human fallibility. We can usually understand and overcome technical problems, but experience suggests that data processing can be severely limited by human factors. Lucas (1975) suggests that the major reason most information systems have failed is because they have ignored organizational behavior problems brought on by people and instead have relied solely on solving the technical problems. If steps are not taken to understand and solve these human factors, systems will continue to fail.

Problems arising from the operations staff concern the human ability to perform the following tasks:

1. securing the source data,
2. transcribing and encoding the data,
3. keying the data into computerized form,
4. verifying and checking the data,
5. modifying and editing the data,
6. producing and supplying outputs,
7. programming and debugging,
8. managing the staff, and
9. providing consistent standards in up-dating.

Because data processing staffs rarely see the whole design of data use, they tend to follow a set of simple rules without having the background for understanding what the data relationships are or how the information will be used in decision-making processes. Erroneous or misleading data may be entered into the system. These problems may not be uncovered until the ultimate users of the system report back that the output has limited or virtually no utility because it is pitted with mistakes and omissions.

Every effort must be made to insert the data into the system correctly the first time. It has been our finding that our good intentions to correct the printouts are often cancelled out by our disinclination to proof-read massive amounts of text. Reading and proof-reading dozens of pages of printout can be tedious, and in fact is a limiting factor, since every new set of printouts may be riddled with errors and omissions. Normally the greatest percentage of a data base operation's cost revolves around salaries of the personnel who perform manual data processing tasks such as compiling, transcribing, coding, proofing, and editing data. If a high degree of quality control is not maintained from the very start, the transcribing, coding, and proofing must then be repeated to correct the errors. As the data system grows larger, lack of quality control can cause an incremental reduction of new data acquisition, bringing about data management bottlenecks, rendering the existing data worthless, or even bringing the entire data management operation to a halt.

Quality control for each data processing task is of the utmost importance. This limiting factor can be ameliorated if the duties of file maintenance are controlled by permanent staffers rather than temporary or untrained employees. A well-defined data control scheme and operational procedures must be developed and strictly adhered to. In addition, it is usually necessary to retain contact with the source materials and locations of ecological data as benchmarks for data accuracy, because data generally undergo some mutation during processing.

Once the data have been assimilated into standard files and formats, a cyclical and continuous process must then be initiated to fill data gaps methodically and to act as a check on accuracy.

Combining Manual and Computerized File Structures.   The ever-popular assertion that computer speed and storage capacity can overcome all obstacles in data management is truly misleading. Larger and faster computers simply provide faster and bigger mistakes (Lee, 1973). It has been our experience that computerized data management is the least satisfactory method of sorting out complex ecological relationships, especially if the system designers have not put enough intellectual energy into indexing and structuring the data. It must be noted that the enlarged capability provided by the computer is not based on any sophisticated procedures but rather on the computer's ability to perform the equivalent manual tasks of searching, comparing, merging, sorting, and reducing data with enormous speed.

It is important to recognize the advantages and disadvantages of various forms of data structures. The need to standardize the terminology and structure the data cannot be emphasized enough. A data structure might be appropriate for one set of objectives yet totally inadequate for others. The choice of structures should be made by thinking of the ways we want to access the data in the future, our analytical needs, and application requirements. Even with high speed computers, an inappropriate data structure can make data processing operations extremely costly because of processing time and programming tasks involved. On the other hand, a well-designed data organization can make the tasks of searching and cross-referencing easier and less expensive. By adequately structuring and indexing the data and maintaining a high degree of input quality control we can be better assured of success. We believe that many information systems fail because they have incorporated too much unstructured data and therefore can only output massive amounts of data that cannot be digested by the decision-makers.

Computer time and related services are very expensive and should be used only when there is sufficient justification. After all, information systems existed long before computers did. We must ask what is sufficient justification for computer use.

We assume that computers are used effectively and economically for repetitive and voluminous tasks of information handling and computation. The real question is: do the ecological data banking problems encountered so far qualify for computerization?

The available information on managed areas, the occurrence of ecosystems, and even on a single species, is very large in some cases. Thus ecological data banking may seem to qualify for computerization due to its sheer volume, but intricate, detailed data are not very amenable to computerization. *Numerous* details about a *single species* are not the same as *few* details about each of *many species*. The strength of the computer system is to handle relatively short records which are replicated many times (such as the format in Figure 1). With records of the same format, the computer performs very efficiently the repetitive searching, sorting, permuting, formatting, and printing. When a project entails voluminous and repetitive operations as described in our assumptions, the data are transcribed and entered into computer storage. At the completion of the project, the computerized data files are retained and become part of the automated system if they are still useful and if the cost of ongoing maintenance is sufficiently low.

General Findings on Ecological Data Management. *Lists of occurrences of common species should not be computerized.* Common species are not germane to uniqueness and so can be omitted from computer-based data banks for unique sites. Data on species which are common and widespread tend to be redundant, and should be structured and stored differently. Any large-scale ecological inventory of the kind described in this paper must consider the problems of how much and what kinds of species data should be collected, and whether comprehensive species lists should be computerized. The answers are determined by the processing needs, analytical needs, and application requirements dictated by the project, but the point is that one should not computerize every bit of species data that is available. The problem we face when doing site inventories is that much of the landscape is highly redundant in the sense that each site contains common species. It is hardly justifiable to systematize and keep records of every deer, duck, or dandelion, *ad infinitum.* This information can usually be obtained from secondary indicators and adds very little to our ability to identify unique sites worth preserving. A more general point to be made here is that such data clogs up the processes of data collection and processing, wastes money, and ultimately leads to irrelevant output. In most cases information on common and widespread species should be kept in manual form and indexed by site; this is far more cost-effective as well as rational approach.

Application-specific computer programs are more cost-effective than large general data base management systems. Our approach to computerization has been to create a wide array of multiple file structures which can be accessed by a set of application-specific programs. (Each program module is designed for one specific purpose. For example, in the Heritage system, the SEARCH module processes user queries, the HISTOGRAM module produces statistics, and so forth.) We have found that, for the volume of data currently being processed, a user-tailored set of programs is far more cost-effective than a large, general-purpose system. Although the application-specific programs do not support "flexible" data handling and analysis, they are the least expensive method of accomplishing a specific set of processing tasks.

The additional programming flexibility offered by general systems may be necessary, especially if many new data categories are entering the system. Fully generalized systems do provide a wide range of options, including the ability to handle new and variable types of data. The drawback, however, is that a general data base management system is usually rented out by the firm that owns it, and the monthly rental charges range anywhere from several hundred to thousands of dollars. Due to the relatively large initial and ongoing investments, then, a careful analysis of the volume of data to be processed must be undertaken before a generalized system can be recommended.

High-speed turnaround is not necessary for most ecological data analysis. For most ecological data analysis high-speed turnaround for simple fact retrieval is simply too expensive and cannot be justified on the basis of speed. We have investigated two basic operational modes, batch and interactive, for their comparative advantages and disadvantages.

1. The batch system is the mode familiar to most people. It is the one where the user travels to the computer center and submits punched cards or tape to the system. The computer does not necessarily process the job immediately, but rather groups or batches the computations, and processes the batch only when a volume of jobs sufficient to justify the computer resources is collected. A disadvantage of the batch

processing system is that there can be a considerable time lapse between entry of the job into the system and delivery of the output. Another is that someone must travel to and from the computing center.

2. The interactive system includes a network of computer hardware, communications equipment, and special programs that permit the user to process programs and data in a conversational mode. The user interacts with the computer almost simultaneously through telephone transmission. This interaction allows the user to get almost instant computer response, and thus rapid turnaround.

Interactive systems simply provide us with the illusion that we control the data more effectively, because it seems as though we have our hands on the data and we are thus able to confront any user queries immediately. In the narrow sense, interactive systems can provide a faster ability to deal with data. However, because more complex systems and techniques are involved, we often become entrapped with working on trivial details of simply maintaining the system. Problems stemming from hardware and software maintenance, documentation, training, debugging, updating, and acquiring new programs because of hardware innovations are some of the disadvantages of securing sophisticated hardware. In addition, a wise axiom of computer science is that the cost and difficulty of operating a system increases exponentially with the size and complexity of the programs. The larger and more complex the system is, the more it can fail in larger and more complex ways.

Most ecological data processing applications do not require high-speed turnaround. For cost and operational effectiveness most ecological data management should be handled by simple non-automated methods such as the batch technique, which involves much less input hardware, is the least expensive option. In our experience the batch system and non-automated methods more than justify their slower turnaround.

A data base must be continuously operated and maintained. We always maintained that it is simply not justifiable to collect, organize, and computerize ecological data in a large statewide inventory project for a single *ad hoc* application nor to consider an environmental data base fully complete by publishing a single hardbound or computerized report. Inventories cannot be one-shot affairs without the data rapidly becoming obsolete. Data must be enriched continuously to reflect the most current and comprehensive environmental data available. User evaluation, feedback, and continuous program operation must be integral parts of the overall data organization and information management system concept.

Structure of the Heritage Information System

In order for us to analyze information concerning the diversity of ecological entities, we have designed a hierarchical, balanced information management system. The system is referred to as balanced because it combines manual, map, and computer techniques. One way of looking at the information system is in terms of repeating file organization, which is diagrammed in Table 1. A basic data paradigm is applied to four "units of record." The unit records are:
1. element,
2. element occurrence,
3. managed area, and
4. information source.

Table 1.
Conceptual Structure of the Balanced Information System

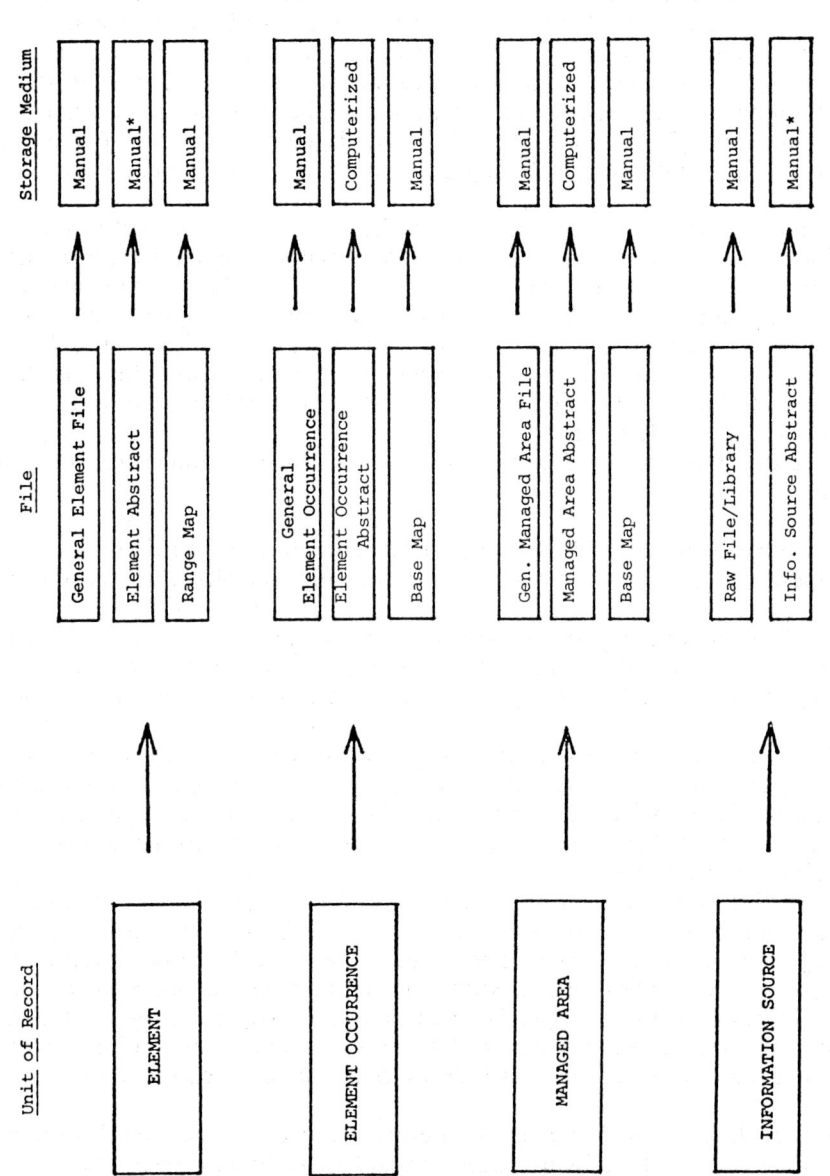

Unit of Record	File	Storage Medium
ELEMENT	General Element File	Manual
	Element Abstract	Manual*
	Range Map	Manual
ELEMENT OCCURRENCE	General Element Occurrence	Manual
	Element Occurrence Abstract	Computerized
	Base Map	Manual
MANAGED AREA	Gen. Managed Area File	Manual
	Managed Area Abstract	Computerized
	Base Map	Manual
INFORMATION SOURCE	Raw File/Library	Manual
	Info. Source Abstract	Manual*

*Computerization of these files is being tested.

247

For each unit of record optimally then there exist three basic file structures: a computerized file, a mapped file, and a manual file.

Element (the Classification).   The Element unit of record consists of three components: the general element file, the element abstracts, and the range maps.

The *General Element File* contains unformatted general information on the specific elements—this includes articles, citations, books, etc., all published and unpublished materials which add to and elaborate on the information contained in the element abstracts. These files contain the actual materials or lists of references which apply to the identification, biology, distribution, status, management, etc. of the elements.

The *Element Abstract* contains specific categories of information including description, nomenclature, range, special data management convention and the like. At present, standardized formats have been created for element abstracts in the classes of plant communities, special animal species, and special plant species. Figure 2 shows a sample element abstract for Plant Communities (PC). The classification abstract document is designed as a handbook or "field guide" on the elements of natural diversity. The document itself is a looseleaf binder and the computerization of the files is being tested.

Associated with these files are *maps* upon which the ranges for the elements are plotted.

Element Occurrence.   The Element Occurrence unit of record also consists of three components: the general element occurrence file, the element occurrence abstract, and the base map.

The *General Element Occurrence File* contains information on each element occurrence, including field survey material, pre-existing records, etc. All general background material applying to a specific element occurrence is filed manually here.

The *Element Occurrence Abstract* incorporates standardized data for analysis of the occurrence, number, status, location, and distribution of the elements defined by the classification. Each record in the file contains information on a separate element occurrence. Figure 1 shows the fact retrieval format for an abstract file used by the Heritage programs.

The 7.5 minute topographic quadrangle *Base Map* is for plotting element occurrences and managed areas. For areas that do not have 7.5 minute topo maps, other topographic or smaller scale maps are used. We have developed coding conventions for plotting and indexing the boundaries of managed areas, element and element occurrences onto these maps. The quad maps are organized by the Universal Map Code location system (Weber and Gregory, 1975). This coding system permits easy map and information cross-indexes to manual and abstract files.

Managed Area.   The Managed Area unit of record also includes three files: the general managed area file, the managed area abstracts, and the base map file.

The *General Managed Area File* contains individual files aggregated by managing agencies and includes protection data, management plans, ecological surveys, etc.

A managed area is an area which merits the protection status "preserved" or "protected" under guidelines based on ownership, special status, legal restrictions,

Name of Element: *Pinus taeda - Myrica cerifera - Persea borbonia* community type.

Description:

The canopy is composed completely of *Pinus taeda* (Loblolly Pine). The subcanopy is dominated by *Myrica cerifera* (Wax Myrtle) and *Persea borbonia* (Redbay). Important associates in this stratum are *Juniperus virginiana* and *Sassafras albidum.* Common herb species are *Ammophila breviligulata, Andropogon scoparius, A. virginicus, Spartina patens,* and *Uniola laxa.* Understory species increase in importance when the pine canopy is not closed. When the pine forms a dense closed overstory a park-like interior develops. This forest develops on barrier islands and coastal areas. Where development is best the Loblolly pine attains the same heights on barrier islands as it does on the mainland.

Ecology:

This forest and variations of it forms the final stage in plant succession on stabilized barrier islands. It develops on mesic areas no longer under the direct influence of seawater flooding or migrating dunes. Salt spray can be of significant importance in limiting the height of the forest canopy; the closer to the fore-dune system the more stunted the vegetation. The western edge of the forest can be virtually identical to some of the mainland communities.

Range:

This community develops on the larger barrier islands from Maryland to North Carolina and as such has a limited habitat and distribution. Its best development in Virginia is on Assateague and Parramore Island.

Status and Management:

The trees can grow rapidly in the conditions but slight environmental changes can eliminate their forests. Where fore-dunes are not stabilized their shifting can bury and kill the forest vegetation. This has occurred many times in the past as evidenced by the many partially covered dead trees frequent along the eastern shores of the islands. This natural phenomenon cannot be prevented but it can be accelerated by human disruption of fore-dune stabilizing vegetation.

Sources of Information:

Dueser et al. 1976. Virginia Coast Reserve Study, Ecosystem Description. The Nature Conservancy.
Godfrey, P. J. and Godfrey, M. M. 1976. Barrier Island Ecology of Cape Lookout National Seashore and Vicinity, North Carolina. National Park Service Scientific Monograph Series, Number Nine.
Harvill, A. M. 1965. The vegetation of Parramore Island, Virginia. Castanea 30: 226-228.
_____. 1967. The vegetation of Assateague Island, Virginia. Castanea 32: 105-108.

Figure 2. Sample Abstract for Plant Community.

etc. Among the managed areas are National Parks, Federal Research Natural Areas, Wilderness Areas, Nature Conservancy preserves, university-owned natural areas, and, in many states, National Wildlife Refuges and State Parks. The scope could be expanded to include less well-protected areas, such as Bureau of Land Management lands, recreational areas, national forests, and wildlife management lands. These areas should be part of the managed area file if reliable information is readily available. Eventually it may be possible to extend our attention to all public lands.

Two additional definitions are provided to clarify what the term implies in a Heritage Program:

1. Usually a managed area will be in public or institutional ownership. It has the additional distinction of being maintained in a manner that will protect the signi-

ficant elements and features occurring within its boundaries.

2. Another definition would be an area with a professional manager who is capable of protecting important element occurrences and other ecological attributes by adopting appropriate strategies for this purpose.

The *Managed Area Abstract* file contains data on managed areas. Each fact retrieval format provides specific data describing a managed area: its location, size, ownership, the level and means of protection afforded the managed area, and the general characteristics of natural diversity contained on the site. This format is similar to the data categories for the element occurrence abstract. A computerized managed area abstract is shown in Figure 3.

The same 7.5 minute *Map* series as for the element occurrence file is used to plot the managed areas.

Information Source.   In order to keep track of all sources of information, a *Raw File and Library* and an *Information Source Abstract* file have been created. In keeping with the data paradigm, the raw file and library maintain undifferentiated, unassimilated data of many types while the information source abstract contains standardized categories of data on individual and other sources of information. The computerization of this file is being tested.

Code Linkages Among Files.   All the files are interrelated by compact codes which permit exceptional flexibility in data storage and retrieval. Search through multiple files is made possible and efficient by the use of index codes. Each site in the system has an index code which serves as a key to all of the related files. The index code is used to organize computer, manual, and map files. Each element occurrence has its own unique code which is used to cross-reference all the data stored in the automated and manual files. The files are thus linked logically even though they are stored in separate media and/or locations. A user can obtain a detailed, composite picture of the data coverage for any given geographic area by simply using the index code to retrieve the data from each file.

Querying the data base.   A general query processing diagram that illustrates how user queries can be handled is shown in Figure 4. A user requests information by telephone or letter. The user query must be specified in such a manner that a decision can be made regarding the appropriate files to search. In addition, the user specifies the type of output desired. A decision is then made as to whether to begin the search in manual and/or computerized files.

If the query does not warrant a computer search, then the information is retrieved directly from the manual files. If the user requires information that can be retrieved more easily by computer, then appropriate computer files and programs are used. For example, a user might request information on all sites in the state known to contain red spruce forest. We would need only to query the computerized files for basic facts on where red spruce forests occur, whether and how they are protected, who owns them, etc. With this information in hand, the user can interrelate data from other file structures simply by applying the index code for each occurrence of red spruce forest as a "key" to related information in the other files. Usually queries will involve a combination of manual and computerized searches. End products include computer printouts, photocopies, maps, and computer plots.

```
SOMERVILLE TABLE RESEARCH NATURAL AREA                          ESTABLISHMENT DATE:
                                                                    1965
COUNTY:                     ACREAGE:              SPECIAL STATUS:
   FREMONT                       680                 SAF NATURAL AREA


LATITUDE:    LONGITUDE:     ELEVATION (FT):
   38.31.  N    105.28.  W      7700 MIN   8750 MAX
CONTACT:
   MGR.,CANON CITY DISTRICT OFFICE,3080 E MAIN ST,CANON CITY,CO 81212   303-275-7494

OWNER/ADMINISTERING AGENCY:
   USA,BUREAU OF LAND MANAGEMENT,CANON CITY DISTRICT

PRIMARY FEATURES:
   BOTANIC: SAF-237 INTERIOR PONDEROSA PINE, SAF-239 PINYON JUNIPER, SAF-210
   INTERIOR DOUGLAS FIR; K-066 WHEATGRASS-NEEDLEGRASS; RELICT - POACEAE-16 SPP.
   ZOOLOGIC: TYPICAL.  AQUATIC: NONE.  LANDFORMS: STREAM-CUT BENCH; FAULT
   ESCARPMENT; FAULT-BLOCK MOUNTAINS; MESA; FAULT VALLEY; STREAM CHANNEL; LAVA
   FLOWS.  GEOLOGIC: IGNEOUS - TERTIARY; BASALT; GRANITE - PRECAMBRIAN; RHYOLITE;
   BRECCIA; ANDESITE; TUFF; SEDIMENTARY; SHALE; SANDSTONE; ARKOSE; METAMORPHIC -
   PRECAMBRIAN; SCHIST; GNEISS.  ALTERATIONS: LAND USE - MINING, HUNTING.
```

Figure 3. Abstract Format for a Managed Area.

Conclusion

Collecting and managing data for informational output is complex because there are so many chains of interrelated procedures to design and implement. Every link in the chain has its own weaknesses that can be brought on by either technical or perceptual problems. Each activity may be sufficient in itself to create a set of problems elsewhere in the system—a chain reaction that may lead to a complete dysfunctioning of the entire operational structure. One must be fully aware of all the complexities and take into account the operational flaws derived from human as well as technical factors that exist in day-to-day information processing and management operations. Once a system is in action there will always be a need to set operational standards. A well defined quality control scheme must be developed and strictly adhered to.

When designing an information management system, keep its purpose in mind. One must know what is most important in order to establish the main organizing principles. Have a clear idea of what individual data elements are needed and how they relate to the informational output, its usage, and applications for decision-making. Be sure to project how much data is actually needed for a given decision-making process. If additional data are collected, examine them in terms of the additional information benefits to be accrued. The data collection process should be tailored systematically for each objective by considering the priorities of the collection of certain data categories. Try not to collect more data than is actually needed for a given decision-making process. Information should be collected in standardized, incremental layers so that it can be compared and analyzed at any stage in the inventory.

Combine the computer capabilities with manual filing structures; remember the computer system limitations. Eliminate heavy reliance on computers if possible, because in some cases they add another level of complexity into an already complicated data management situation. Also, pay attention to structuring the data regardless of whether it will be computerized or not. A data base is useless if we can't get to the

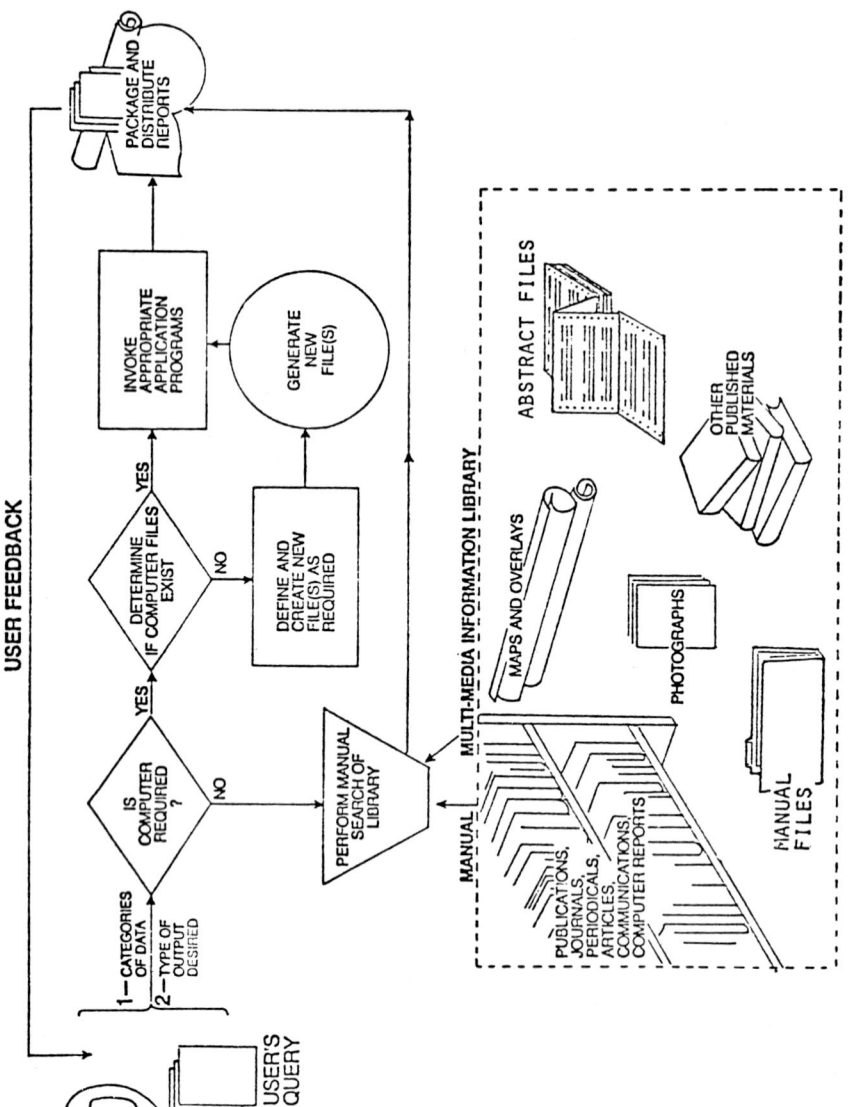

Figure 4. Servicing User Queries.

252

data effectively when we need it. If you simply lump the data into one unstructured file, you are setting yourself up for an informational failure, regardless of whether you have a "powerful" computer. It is crucial when creating an information system to keep financial limitations in mind; exorbitant costs usually follow from any heavy reliance on computer services. For most ecological data analysis, the high-speed turnaround provided by computers cannot be justified on the basis of speed alone. We use the computer only when the cost of using it is less than the manual counterpart required to do the same tasks. Computers are used effectively and economically for handling large numbers of repetitive, short, standardized record formats.

Finally, we must lay aside the notion that ecological inventories can ever be truly complete—they may become refined and mature, but not complete. A continuous program operation therefore must be an integral part of any large data organization and information management concept. The creation of a large data banking system will have little significance beyond the last update unless the operation is maintained. Ecological data banking projects must also be committed from the outset to perpetual operation. We are dealing with dynamic processes and the data base must be cumulatively enriched to reflect the ever-changing environmental mosaic.

Literature Cited

Federal Committee on Ecological Reserves. 1977. *A Dictionary of Research Natural Areas on Federal Lands of the United States of America.* U.S. Forest Service, Washington, D.C.

Jenkins, R. E., H. P. Moyseenko, *et al.* 1977. *Creation of Data Bank on Established Scientific Ecological Reserves,* Final Report to the National Science Foundation. The Nature Conservancy, Arlington, Va.

Lee, D. L., Jr. 1973. Requiem for Large Scale Models. *Amer. Inst. Planning J.,* May 1973.

Lindblom, C. E. 1959. The Science of "Muddling Through." *Pub. Admin. Rev.* Spring 1959.

Lucas, H. C., Jr. 1973. *Why Information Systems Fail.* Columbia University Press, New York and London.

Moyseenko, H. P., J. L. Woodall, and S. A. Woodall. 1978. The Balanced Ecogeographical Information System: A vehicle for data collection, systematization and dissemination. Pages 569-592 *in:* Marmelstein, A., ed. *Classification, Inventory, and Analysis of Fish and Wildlife Habitat.* Report FWS/OBS 78/76. U.S. Fish and Wildlife Service, Washington, D.C.

Moyseenko, H. P., R. E. Jenkins, S. Woodall, and L. A. Miller. 1977. *Lowest Common Denominator Element File: An information management system.* 3rd Edition. The Nature Conservancy, Arlington, Va.

Voelker, A. H. 1975. Some pitfalls of land use model building. ORNL-RUS-1. Oak Ridge National Laboratory, Oak Ridge, Tennessee.

Weber, S. G. and R. P. Gregory. 1975. Universal Map Code (A 7-1/2 minute quadrangle identification system). Techn. Note B-10. Tennessee Valley Authority, Norris, Tenn.

Wiederhold, G. 1977. *Database Design.* McGraw-Hill, New York.

Young, G. K. and G. F. Tierney. 1976. Environmental data management, *Proc. Amer. Soc. Civil Engineers, J. Water Resources Planning Management Div.* 102 (WR2).

Summary and Conclusions

Larry E. Morse,[1,2] Mary Sue Henifin,[1,3] and Jane I. Lawyer[1,4]

The papers presented in this volume indicate the diversity of approaches being taken to organize geographical data for rare plant conservation. Such diverse approaches are the logical outcome of work on different species, in different geographical areas, and through different agencies and organizations. It is now important that methods be developed to improve the coordination, evaluation, synthesis, and communication of these diverse data pertinent to rare plant conservation.

Among the conclusions developed and refined at our symposium and in this volume, the following are of general importance to the biogeographical aspects of rare plant conservation:

1. Site-specific geographical information is essential to the informed conservation of plant populations.

2. Natural history studies of each species are necessary in delimiting areas of essential habitat adequate for the long-term survival of a species' populations. Such areas of essential habitat must include also the habitats necessary to maintain such associates as pollinators, and also provide adequate buffer zones to help protect the sites from edge effects such as invasion by exotics.

3. The conservation status of a species (or other taxon) can best be developed by synthesizing information on each of its known populations, viewed together with any information on changes in historical range and evidence of vulnerability of its characteristic habitats.

4. Planners and land managers have a genuine interest in obtaining site-specific information about possible conflicts between development proposals and conservation priorities; such information is most welcome early in the planning process—when options and alternatives are still being considered.

5. Centralized state repositories for compiling, reviewing, and mapping rare plant information provide an effective planning tool as well as an efficient data-collection procedure.

6. Once a population or a rare or endangered plant species is discovered, local botanists and government agency staff are often best positioned for population

---

[1]   Cooperative Parks Study Unit, The New York Botanical Garden, Bronx, N.Y. 10458.
[2]   Present address: The Nature Conservancy, 1800 N. Kent St., Arlington, Va. 22209.
[3]   Present address: Div. Environmental Sciences, Columbia University School of Public Health, New York, N.Y. 10032.
[4]   Present address: 45 Sturgis Road, Bronxville, N.Y. 10708.

Pages 255-258 *in:* Larry E. Morse and Mary Sue Henifin, eds., *Rare Plant Conservation: Geographical Data Organization*, The New York Botanical Garden, Bronx, N.Y.

studies, monitoring of impacts, and searching nearby areas of similar habitat for additional occurrences.

7. Taxonomic and biogeographic questions concerning rare, peripheral, and disjunct plant populations can best be resolved through continued taxonomic and floristic research; many disjunct populations of common plants were originally thought to be local endemics, while other taxa have remained unrecognized until their genera were studied on a continental scale.

8. Relatively few plant populations are made more vulnerable through public access to information on their localities; among this publicity-sensitive group are plants of horticultural, medicinal, or food value; populations in urban areas or heavily used recreational lands; and populations whose conservation would interfere with financially rewarding developments.

9. Some of the potential adverse effects of publicity can be avoided, when possible, by discussing the site without mentioning the identity of the plants involved.

10. Acquisition of habitat as public conservation land is itself inadequate to assure the continued survival of a rare plant population; further research, monitoring, and management programs are generally necessary to mitigate effects of such factors as visitor impacts, park developments, presence of exotic species, air or water pollution, non-historical fire regimes, or biotic changes due to insular effects.

* * *

The need for improvement in the synthesis of state-oriented plant species status and distribution information was repeatedly stressed at the symposium and in subsequent discussions. Such efforts would improve land management by providing expedited deposition and communication of information, better awareness of current research, and a broader perspective for setting research priorities and making land-use decisions.

Establishment of a formal or informal phytogeographic information network or clearinghouse would help in the difficult task of coordinating the diverse local, state, and regional projects concerned with plant species status and distribution. A national or continental network could also be instrumental in assisting in the development of national or continental guidelines for a refined taxonomic classification and for the documentation and organization of geographical data. Guidelines developed by a nationally recognized network could also assist in organizing and making accessible historical records as well as current status reports on plant species.

A fundamental role of a national clearinghouse would be the promotion and coordination of information transfer at all levels including local, state, regional, national, continental, and global. Two vital areas for information transfer are communication between ongoing state projects and information transfer from researchers to land managers. Failure to communicate between state projects may result in duplication of research efforts and production of noncomparable data which are of little use when regional syntheses are attempted or when data are compared from year to year. Land managers need up-to-date status and distribution information for present and future land planning activities. Often such managers are not aware of the current work of academic or contract researchers, or of work conducted by government agencies.

As well as organizing the national and continental syntheses of plant geographical data and status information, a national network could make recommendations concerning proposed future research. Such recommendations could be based on the perspective gained by comparing information needs in the many states. This perspective would also be invaluable in the designation of state and national plant and habitat conservation priorities. Identifying the most pressing information needs and conservation priorities is particularly important when levels of funding are not adequate to support all worthwhile projects.

Although there is currently considerable ad-hoc and informal information exchange in plant systematics and plant conservation, the lack of a defined and well-organized plant information network of national scope has contributed to the increase of several problem areas in current floristics and plant conservation work. Current and historical data and information are often not known to the people who need it. Information which is incompletely known is often difficult to access or trace. Information which is gathered state-by-state by different researchers or over a period of time by different researchers using different methodologies is often incompatible for use in across-state or across-time comparisons or syntheses. Without centralized review and coordination, it is difficult to set priorities for information needs. Finally, lack of a regional or national perspective makes it difficult to see patterns of rarity and threat; such patterns must be identified if successful recovery plans are to be formulated and implemented.

National coordination could increase continuity and decrease duplication between ongoing state and national programs. For example, detailed locality information gathered by State Natural Heritage Programs should be readily available to and used by groups in adjoining states, national conservation groups, federal agencies, and members of the botanical community. Yet there are many instances of new programs being started which duplicate information-gathering activities of existing State Natural Heritage Programs.

It is logical that state and local projects can best gather specific site information, for no individual or small research team can claim familiarity with all local areas in the nation. Furthermore, it is usually impossible to "work backwards" and deduce specific site information from regional or national projects. It is also prohibitive to consolidate and maintain primary locality information in a centralized data base. On the other hand, it is most efficient for a group with a broad perspective to develop regional or national taxonomic classifications and checklists, generalized distribution maps, and national endangered species listings.

* * *

The following ongoing or proposed projects are offered as examples of the kinds of efforts we believe essential to further progress in improving the synthesis and communication of floristic data pertinent to rare plant conservation:

1. A comprehensive *revision of taxonomic classification* for the vascular plant species of the United States and Canada, as is now being developed by John Kartesz in Pittsburgh (Kartesz and Kartesz, 1980) and also by the Department of Botany of the Smithsonian Institution.

2. An *index to current research in plant systematics,* identifying investigators by taxonomic, geographic, and methodological interests. Such an index is now being developed by the Hunt Institute for Botanical Documentation as an on-going computer-based service.

3. Improvements in *literature awareness* through production or revision of such

works as Solbrig and Gadella's *Biosystematic Literature* (1970), Phillips and Stuckey's *Index to Plant Distribution Maps* (1976), and the guide to floristic literature by Lawyer *et al.* (in press). Authors publishing works pertinent to compilations such as these should be encouraged to send information to particular coordinators for publication of annual supplements and periodic revisions of such bibliographies. Coordination of such work through a computer-based bibliographic service could also provide for selective dissemination of information, a service in which subscribers would receive monthly or quarterly notification of new publications in their indicated areas of interest.

4. An *endangered species information center* could be established to coordinate activities particularly relevent to the conservation of nationally endangered or threatened plant species. The Center could facilitate information transfer on these high-concern species, and synthesize regional and state data on them. The Center could also help prioritize and implement endangered plant species activities in a national context. There are several specific projects such a center could coordinate, such as a national plant conservation newsletter, a periodic summary of state rare plant lists (as initiated by Kartesz and Kartesz, 1977), and an indexed library of status reports on rare or endangered plant species.

5. A clearinghouse could be established for *indexing and reviewing significant plant distribution and locality records,* perhaps using a computer system as an economical means for indexing and summarizing the information. Taxonomic and geographical review of new reports before final acceptance would assure high quality in the data base. Such a project might be closely coordinated with production of a national floristic manual or checklist to assure that comparable taxonomic treatments are used, and that distribution summaries in the floristic work are consistent with those in the phytogeographical data base.

Literature Cited

Kartesz, J. T., and R. Kartesz. 1977. *Biota of North America. Part I. Vascular Plants. Vol. 1. Rare Plants.* Biota of North America Committee, McKeesport, Pa.

——————. 1980. *A Synonymized Checklist of the Vascular Flora of the United States, Canada, and Greenland* (The Biota of North America, Vol. 2). Univ. North Carolina Press, Chapel Hill, N. C.

Lawyer, J. I., A. M. Miller, L. E. Morse, and J. T. Kartesz. in press. Guide to the current literature on vascular plant floristics for the contiguous United States, Alaska, Canada, Greenland, and the U.S. Caribbean and Pacific Islands. *Mem. Torrey Bot. Club.*

Phillips, W. L., and R. L. Stuckey. 1976. *Index to Plant Distribution Maps in North American Periodicals through 1972.* G. K. Hall, Boston.

Solbrig, O. T., and Th. W. J. Gadella. 1970. *Biosystematic Literature: Contributions to a Biosystematic Literature Index (1945-1964). Regnum Veg.* 69. International Bureau for Plant Taxonomy and Nomenclature, Utretcht, Netherlands.

Appendices

Guidelines for the Preparation of Status Reports on Rare or Endangered Plant Species[1]

Mary Sue Henifin,[2] Larry E. Morse,[2] James L. Reveal,[3] Bruce MacBryde,[4] and Jane I. Lawyer[2]

These guidelines for the preparation of status reports on rare and endangered vascular plants are intended to faciliate organization and exchange of information on such species. Rare, threatened, or endangered plants are of urgent management concern to the National Park Service and other federal agencies, and information on the status of populations of these species is vital for synthesis of plant distribution data. This outline integrates population-level distribution data with other information regarding taxonomic classification, biological characteristics, habitat and environment, legal status, and degree of threat or protection for the various populations or portions of range of a species or other taxon. Much of this information is also needed for listing of endangered and threatened species pursuant to the United States Endangered Species Act of 1973 and various state plant conservation laws, for the designation of critical habitats, for the assessment of impacts on populations and habitats, and for the development of management plans and recovery efforts.

The need for a comprehensive status report outline was expressed in August, 1977, at the workshop on endangered plant species convened by the American Society of Plant Taxonomists in East Lansing, Michigan. The first draft of these guidelines was developed in New York in November, 1977, by an ad-hoc group at the symposium on Geographical Data Organization for Rare Plant Conservation, sponsored by the New York Botanical Garden and the U.S. National Park Service. The outline has subsequently been revised following extensive consultation and discussion with numerous people concerned with the ecology and conservation of rare or vulnerable plant species, study of dozens of proposed or currently used status report outlines and other data-collection guides, and circulation of more than 100 copies of a review draft in April, 1978 and many copies of a revised working draft prepared in February, 1979.

[1] A review draft of this chapter was widely distributed in April 1978 as "Proposed Outline for Status Reports on Threatened and Endangered Plant Species."
[2] Cooperative Parks Study Unit, New York Botanical Garden, Bronx, New York 10458. Research supported by the National Park Service, U.S. Department of the Interior.
[3] Department of Botany, University of Maryland, College Park, Md. 20742. Scientific Article No. A2449, Contribution No. 5478, of the Maryland Agricultural Experiment Station, Department of Botany.
[4] Office of Endangered Species, U.S. Fish and Wildlife Service, Washington, D.C. 20240.

Pages 261-282 *in:* Larry E. Morse and Mary Sue Henifin, eds., *Rare Plant Conservation:Geographical Data Organization,* The New York Botanical Garden, Bronx, N.Y.

Particularly useful in our early work on this outline was a review of the experiences of the New England Botanical Club's Committee on Endangered Species, The Nature Conservancy's State Natural Heritage Programs, the Waterways Experiment Station's Sensitive Wildlife Information System (SWIS), the Plant Information Network (PIN) based at Colorado State University, and South Africa's Rare and Endangered Species Survey. Most of these projects are described in detail elsewhere in this volume.[5]

Scope and Organization

These guidelines are intended to demonstrate the diversity of information required for plant population management. The terminology here used is intended for angiosperms (flowering plants); appropriate modification should be made for reports on pteridophytes, gymnosperms, or other kinds of plants. We also phrase this outline for use in the United States; appropriate changes in agency names and other government-related aspects would be needed to implement this guide elsewhere.

The information categories in the status report outline are divided into five sections. The primary presentation of information should be made in Section I, species information. The assessment and recommendations of Section II should then be based on the information presented in the first section. Sections III and IV, information sources and authorship, provide documentation for the data presented in the status report. Section V is designed to organize new information obtained after the preparation of the status report.

Although the scope of the outline may at first appear overwhelming, only those items marked with an asterisk appear to give minimally *necessary* background information for a plant species to be reviewed promptly for listing by the Office of Endangered Species of the U.S. Fish and Wildlife Service.[6] The other, nonasterisked items may be of particular importance in critical habitat studies as well as in developing management plans and recovery efforts. *If this additional information is available we strongly encourage its inclusion in the status report, so that a summary of all pertinent conservation information appears together in one document.*

Status reports should not ordinarily be used as the primary place of publication of significant new scientific findings; articles or notes for the traditional scientific literature should be prepared in such cases.

Suggestions to Authors

Few individuals have the breadth of experience necessary to complete all the information requested in the full outline but it is important to suggest topics that may be

---

[5]   See, respectively, papers by W. D. Countryman *et al.;* P. L. Dittberner *et al.;* and A. V. Hall.

[6]   For a discussion of the Endangered Species Act of 1973 and information needs for listing plant species pursuant to this Act see B. MacBryde's paper in this volume as well as: W. D. Zeedyk, *et al.,* 1978, Endangered plant species and wildland management, *J. Forestry* 76: 31-36; and B. MacBryde, 1979, Plant conservation in North America: Developing structure, *in:* I. Hedberg (ed.). *Systematic Botany, Plant Utilization, and Biosphere Conservation.* Uppsala: Almqvist and Wiksell, pages 105-109; also, E. L. Smith, 1980, Laws and information needs for listing plants, *Rhodora* 82: 193-199.

of particular importance for certain taxa. While this format can be used to organize casually obtained information, development of a status report using this outline is a significant research project. Some of the information requested can be derived from the literature and from specimen labels, but field observations are of greater importance in assessing current range and threats.

We encourage preparation of preliminary reports when more detailed information is not available. Accurate information on a rare or endangered taxon, however brief, is always better than no information. To obtain detailed data, particularly for the population biology and ecology sections, and to document threats to the populations, more than one site visit may be necessary.

Obviously, decisions must be made on the priority of obtaining further information needed for a particular species, since there are very few (if any) species for which all these data items are known and there is rarely time, expertise, or money available to carry out the extensive library, herbarium, and field research needed for development of a complete and fully documented status report. Supplemental information should be supplied later whenever possible.

This report format is designed to synthesize information on the status of a species or other taxon; such information can only come from a population-by-population assessment. Population status reports should be used as documentation for the synthetic information on populations contained in the general status report. Information organized in this manner is also adapted for management decisions made on a population-by-population basis. Because details on populations may differ, and information may be gathered by different individuals, M. S. Henifin *et al.* (this volume) have also developed population status report suggestions in coordination with the general, species-level status report guidelines presented here.

In preparing a status report, information should be organized under the bold faced headings provided in the outline. If no information is available, or if an information category is not applicable to a particular species, this should be indicated in the report under the appropriate heading. No bold-faced headings should be omitted from the status report; their presence will serve as a reminder that additional information might be obtained.

It is of utmost importance that all items of information not based on the personal observation or experience of the author(s) be documented by reference to literature citations, herbarium or museum specimens, field observations, experts consulted, or other sources of information. As a method for footnoting and documentation, we suggest all such items of documentation be listed numerically and referred to by number in documenting information in the report. Alternatively, conventional styles of reference and citation may be employed. However, publicity-sensitive information should be reported only as supplementary material (not intended for general distribution) rather than included in the text or documentation of the report, or, if necessary, referenced indirectly and retained in the author's working files. No publicity-sensitive information should appear in the status report itself unless procedures exist for restricting reproduction and distribution of a particular status report.

Status reports we have prepared using these guidelines have been about 20 to 30 typewritten pages long, and taken several days to organize after the fieldwork, herbarium search, and literature review were completed. Circulation of review drafts of the reports has proven helpful in improving their quality. Among examples of status reports prepared according to these guidelines is the one on *Hudsonia montana* by Morse (this volume).

Acknowledgments

In preparing this outline we reviewed many proposed and currently used status report formats. Through the courtesy of state groups, federal agency personnel, and various other researchers, many published and unpublished status reports and report formats were made available to us. These range from formats developed by units of the National Park Service, U.S. Fish and Wildlife Service, Bureau of Land Management, and Forest Service, to those used by state Native Plant Societies such as those in California, Nevada, and Colorado. Not all of the various specific topics and data-organization techniques were adopted; rather, we attempted to develop a well-coordinated information-organizing system addressing the collective needs of these diverse groups. Solicitation of supplemental information may be appropriate for some specific purposes.

Several published data-organization outlines were major sources of ideas in our work. These include Radford, Otte, and Otte's *Ecological Diversity Classification and Inventory,*[7] La Roi's guidelines for the Biological Flora of Canada,[8] the data sheet format of the Smithsonian Institution's Endangered Flora Project[9] and *The IUCN Plant Red Data Book.*[10]

We especially acknowledge the information and numerous ideas shared by A. V. Hall (University of Cape Town), Paul Whitson (University of Northern Iowa), J. R. Massey (University of North Carolina), A. M. B. Rekas (Waterways Experiment Station, Vicksburg, Mississippi), M. N. Fisher (Oklahoma Natural Heritage Program) and the several members of the New England Botanical Club's Committee on Endangered Species. The presentations and discussions at the November, 1977, symposium on Geographical Data Organization for Rare Plant Conservation stimulated comprehensive consideration of the broad range of topics treated here; to all these symposium participants, our thanks!

For giving freely of their time and experience to share their suggestions, criticisms, priorities, and concerns with us during the preparation of these status report guidelines and related work, we would like to thank T. A. Atkinson, N. D. Atwood, J. Baron, C. J. Boone, S. P. Bratton, N. Brown, G. Bryant, B. Burley, S. C. Buttrick, R. H. Carey, R. Chipley, S. Cochrane, J. Coddington, J. L. Collins, R. E. Cook, G. A. Coovert, C. Corn, W. D. Countryman, G. E. Crow, T. J. Crovello, D. Davy, R. A. DeFilipps, H. DeSelm, J. Dowhan, R. W. Dyer, L. M. Eastman, B. Ertter, M. Evans, J. Fay, C. Feddema, G. Fenwick, K. G. Field, H. Fleet, R. Fletcher, R. Fortney, W. Gallagher, V. C. Gilbert, A. G. Greene, A. M. Greller, S. Griffith, G. A. Guppy, S. Halkin, R. D. Harbison, C. N. Harvey, D. Herbst, L. J. Hickey, W. Hickling, J. E. Hohn, L. N. Hood, A. Q. Howard, H. S. Irwin, J. Jacobi, R. E. Jenkins, M. Johnston, E. Joyal, W. Judd, C. Justice, J. T. Kartesz, R. W. Kiger, M. Koehring, V. LeGarde, R. Loeb, C. R. Long, T. Lucke, M. Mangone, L. Mann, R. Maroncelli, J. Matthews, C. Matti-Natella, G. McKiernan, M. Medley, L. J. Mehrhoff, A. M. Miller, J. Miller, W. L. Milstead, J. K. Moore, M.

---

[7] Part I of *Natural Heritage Classification and Information Systems.* (Chapel Hill: University of North Carolina Student Stores, 1978).
[8] *Canadian Field Naturalist* 91(3): 269-272, 1977.
[9] E. S. Ayensu and R. A. DeFilipps. *Endangered and Threatened Plants of the United States.* (Washington, D.C.: Smithsonian Institution Press and World Wildlife Fund, Inc., 1978), pp. 347-387.
[10] G. Lucas and H. Synge, compilers. Morges, Switzerland: International Union for the Conservation of Nature and Natural Resources, 1978.

Morgan, S. W. Morgan, T. Mulroy, P. J. O'Connor, R. J. Olson, P. Opler, I. Owen, J. Pardue, W. Parker, R. Parmenter, P. Parr, L. Peacock, L. Pendergrass, J. T. Peters, R. Philippe, D. Pittillo, G. T. Prance, R. B. Primack, A. E. Radford, B. Reid, W. Reifer, J. Richards, P. Risser, A. Robinson, C. Rodgers, R. C. Rollins, S. R. Ross, C. Row, R. W. Rutkosky, M. Salk, B. Sanders, J. Scott, C. Shafer, J. Schaller, R. Schuller, S. G. Shetler, C. J. Sheviak, L. M. Shultz, J. Siddall, G. Smathers, C. W. Smith, E. L. Smith, S. D. Smith, A. Solomon, C. Somers, I. M. Storks, C. Stubbs, T. W. Sudia, G. Suter, F. Taylor, M. I. Taylor, R. L. Taylor, E. Thomas, A. F. Tryon, R. M. Tryon, G. Tucker, B. A. Uhl, G. Waggoner, D. C. Wasshausen, N. Weber, S. Werner, R. D. Whetstone, P. S. White, D. Wilson, E. Wofford, L. Woodall, S. Woodall, K. Wright, S. Yost, T. A. Zanoni, W. D. Zeedyk, N. G. Zimmer, and the many others whom we have inevitably omitted here. However, the authors must take responsibility for any errors or omissions in this work.

Guidelines for the Preparation of Status Reports on Rare
or Endangered Plant Species

SUMMARY OF INFORMATION CATEGORIES

Completion of only the items singly asterisked (*) provides an initial summary report on the conservation status of a species and presents the information regarded as minimally necessary for a plant species to be considered for listing by the Office of Endangered Species, U.S. Fish and Wildlife Service. Topics which provide particularly useful supplemental information are indicated by double asterisks (**).

Heading
Table of contents

I. Species Information

* * 1. Classification and nomenclature
* * 2. Present legal or other formal status
* * 3. Description
*   4. Significance
* * 5. Geographical distribution
* * 6. Environment and habitat
* * 7. Population biology
*   8. Population ecology
* * 9. Current land ownership and management responsibility
* **10. Management practices and experience
* *11. Evidence of threats to survival

II. Assessment and Recommendations

*   12. General assessment of vigor, trends, and status
* *13. Priority of listing or status change
* *14. Recommended critical habitat
*   15. Conservation/recovery recommendations
* **16. Interested parties

III. Information Sources

* *17. Sources of information
* **18. Summary of material on file

IV. Authorship

* *19. Initial authorship
*   20. Maintenance of status report
*   21. Record of revisions

V. New Information

---

*Minimally essential information topics
**Topics providing particularly useful supplementary information

Status Report on _____

Heading:

On first page of status report provide a heading indicating significant key words for indexing purposes using following or equivalent style:

Taxon name:
Common name:
Family:
State(s)/Nation(s) where taxon occurs:
Current federal status:
Recommended federal status:
Author(s) of report:
Original date of report:
Date of most recent revision:
Institution, agency or individual to whom further information and comments should be sent:

Table of Contents:

Next, provide a table of contents with page numbers on which each section of text begins.

Text:

Organize and present information under the following boldfaced headings; our italicized instructions, suggestions, and examples are provided under the appropriate headings.

I. SPECIES INFORMATION

1. **Classification and nomenclature.**
    A. **Species or infraspecific taxon.**
        1. **Scientific Name.**
            * a. **Binomial (or trinomial)** with author(s), including rank if trinomial.
            * b. **Full bibliographic citation(s)** of original nomenclatural publication(s). Note that two (or more) references are needed to document later combinations or other names having original author(s) indicated parenthetically.
              c. **Type specimen(s)** and/or other reference specimen(s) having identification recently verified, with indication of herbaria where deposited. Provide reference for any lectotype or neotype designations.
        2. **Pertinent synonym(s).**
            Give frequently or currently used synonyms or alternative names, if any; include author(s) and place(s) of publication for each.
    ** 3. **Common name(s),** if any.
            Give designations used by local residents, field guides, general public, etc.
        4. **Taxon codes** used by government agencies, native plant societies, etc.

5. Size of genus (number of species).
B. Family classification.
    1. Family name.
    2. Pertinent family synonym(s).
        Give frequently used synonyms or alternative names, if any.
    3. Common name(s) for family.
        Give designation(s) used by general public to refer to this family.
C. Major plant group. (Pteridophyte, gymnosperm, moncot, or dicot; designation of orders and subclasses is rarely necessary.)
D. History of knowledge of taxon.
        Include information about date and circumstances of original collection, rediscovery if assumed extinct, etc.
*E. Comments on current alternative taxonomic treatment(s).
        Briefly review alternative taxonomic treatment(s), if any, dealing with the historical application of the name and concept of the taxon. Particularly consider recent monographs, revisions, floras, and checklists; Lawyer *et al.*[1], provide a guide to the state-level floristic literature in the United States and Canada.

2. Present legal or other formal status.
    A. International.
        1. Present designated or proposed legal protection or regulation.
            Indicate whether listed or proposed for listing on appendices to the Convention on International Trade,[2] the Western Hemisphere Convention,[3] or comparable status, and give date(s) of such actions.
        2. Other current formal status recommendations.
            Indicate whether listed or recommended for listing in the IUCN Red Data Book[4] or comparable international status and give year(s) status designated or recommended.
        3. Review of past status.
            Summarize designations and recommendations no longer current.
    B. National. (Repeat by nations if necessary.)
        1. *(2, 3, etc.)* [Name of nation.]
            a. Present designated or proposed legal protection or regulation.
            For example, indicate whether listed, proposed, or under notice of review for listing as endangered or threatened under the U.S. Endangered Species Act of 1973.[5] Indicate also any critical habitat proposed or designated. Give date(s) final designation, proposal, or notice of review published in the *Federal Register.*[6]
            b. Other current formal status recommendations.
            Indicate, for example, whether recommended for national listing as endangered or threatened in the 1975 Smithsonian Institution report,[7] the Smithsonian's 1978 Endangered Flora Project book,[8] or recommended for national status in a state list (for example, the California Native Plant Society's list of

nationally endangered or threatened California species).[9] Give year(s) status designated or recommended.

    c. Review of past status.
Summarize designations and recommendations no longer current such as a prior *Federal Register* proposal.

C. State or equivalent. (Repeat by states if necessary.)
See Lawyer *et al.*[1] and Kartesz and Kartesz[10] for references to many state lists.

    1. *(2, 3, etc.)* [Name of state or equivalent.]

      *a. Present designated or proposed legal protection or regulation.
Indicate whether listed or proposed for state listing as threatened or endangered (or similar designation) under a state endangered species or similar law (such as Michigan's Endangered Species Act of 1974[11]). Give date(s) status designated or proposed.

      *b. Other current formal status recommendations.
Indicate, with reference, whether included in a state list not providing legal protection or regulation, such as McGregor's "Rare Native Vascular Plants of Kansas."[12]

      c. Review of past status.
Summarize designations and recommendations no longer current.

3. Description.

    *A. General nontechnical description.
Give a brief description understandable to the nonspecialist; use nonmetric measurements.

    B. Technical description, perhaps adapted from a recent monographic or floristic publication. Also where appropriate and available include such information as growth rate, longevity, Raunkiaer life form, physiological characteristics, and chromosome number.

    C. Local field characters.
Emphasize differentiation from similar local species with which this taxon might be confused in the field; repeat by population if appropriate. If possible include vegetative characters and characters used to identify seedlings.

    D. Identifying characteristics of material which is in interstate or international commerce or trade, if applicable. (This information is necessary for materials such as wood products or cultivated cacti which may need to be identified by law enforcement personnel.)

    E. Photographs and/or line drawings, if available.
Provide copies if possible, otherwise provide citations to previously published illustrations. Photographs should preferably be $8 \times 10''$ glossy black-and-white prints and/or 35 mm color slides.

      *1. Mature plants in field.

      2. Flowers, fruits, seeds, etc.

      3. Seedlings and immature plants.

      4. Herbarium specimens.

4. Significance of the taxon, as appropriate.

> [Some of these topics appear in section 2(a)(3) of the Endangered Species Act of 1973.]

A. Natural: Ecological, evolutionary, physiographic, etc.

> Give a general statement of the biological significance of the taxon, with particular attention to ecological and evolutionary importance and potential; consider such factors as peculiar adaptations or structures, disjunction or endemism, taxonomic uniqueness, obligate relationships with other species, or roles in stabilizing landforms; indicate if natural significance unknown.

B. Human: Agricultural, aesthetic, cultural, economic, educational, historical, horticultural, medicinal, recreational, scientific, silvicultural, etc.

> Give a general statement of past, present, or potential significance to humans for such uses as food, forage, fiber, chemicals, medicines, ornamental uses, erosion control, wood sources, research on taxonomic relationships, cytological or physiological studies, teaching material, or aesthetic contributions to scenic areas. Possible uses in broadening the genetic base of related economically important species should also be mentioned here.

5. Geographical distribution.

*A. Geographical range.

> Summarize present and past natural range of the taxon by nation(s) and state(s) or equivalents; include county distribution map if useful and not publicity sensitive.

B. Precise occurrence(s).

> Organize geographical data under the following categories:

*1. Populations currently or recently known extant; indicate whether origin from introduction or cultivation is known or suspected.

*2. Populations known or assumed extirpated, with explanation.

*3. Historically known populations where current status not known.

*4. Locations not yet investigated believed likely to support additional natural populations, with explanation.

5. Reports having ambiguous or incomplete locality information.

6. Locations known or suspected to be erroneous reports, with explanation, *e.g.* misidentified specimens and localities.

> Under each of these above six categories, group information geographically and provide the following information for each site, where available and not publicity-sensitive:

a. *(b, c, etc.)* [Name or other designation for site; invent an informal name for publicity-sensitive sites if necessary.]

1. Nation, state, and county (or equivalents).

2. Latitude and longitude, and altitude (optionally add location in UTM coordinates).

3. Range, township, section, etc. (if applicable).

**4. Pertinent U.S. Geological Survey or other map quadrangle(s); indicate name, series (15-minute, 7.5-minute, etc.), and year of map(s).

5. Year of initial discovery, if known.

**6. Year and nature of most recent observation or collection known.

7. Bearing and distance from a well-known landmark, and/or instructions for reaching site from a numbered highway, to extent not publicity-sensitive.

8. Alternative names by which site is known.

C. **Biogeographical and phylogenetic history** of the taxon.

Give a brief account, where pertinent; include such information as fossil records, suspected glacial or interglacial refugia, and migrational patterns.

6. **General environment and habitat description,** as appropriate, repeated by portion of range as necessary. If habitat(s) have been comprehensively described (e.g. by a Natural Area Report following Radford *et al., Ecological Diversity Classification and Inventory*[13]), that information can be referenced here rather than repeated in detail.

A. **Concise statement of general environment and habitat.**

Give summary of the most important aspects of the information presented below, particularly those factors thought crucial to the taxon's survival, distribution, and abundance. Include photograph(s) of characteristic habitat(s) if available.

B. **Physical characteristics.**

A text such as Daubenmire's[14] should be consulted for further information on methods and techniques pertinent here.

1. **Climate.**

a. **Köppen climate classification.**[15]

b. **Regional macroclimate including such information as seasonal and daily ranges in temperature, light, precipitation, humidity, and wind.

c. **Local microclimate** including information on such factors as seasonal and daily ranges in light regime and intensity, exposure to wind, and local moisture regime. See a text such as Rosenberg's[16] for further information.

2. **Air and water qualilty requirements** of plant population(s).

3. **Physiographic province(s)** by Fenneman[17] and/or Hunt[18] system.

4. **Physiographic and topographic characteristics,** such as relief, elevation range, geologic formations, glacial history, slope and aspect, and watershed, stream system, or drainage basin.

5. **Edaphic factors.**

For soils, consider, as appropriate, such characteristics as soil texture, soil moisture and drainage, thickness of litter layer, SCS soil classification,[19] pH, parent material, bedrock type, depth to bedrock or impermeable pan, percentage of rock cover and percentage of rock throughout soil profile, structure and porosity, soil-water potential, chemical composition, nutrient status and availability, and presence of toxic elements.

6. **Dependence of this taxon on natural disturbance** and other dy-

namic aspects of climate, landforms, and other features of physical habitat and its sensitivity to environmental processes and extremes such as erosion, floods, gales or hurricanes, tidal flooding, salt spray, droughts, frosts, or deflation or deposition of sand or soil.

7. **Other unusual physical features** of environment and habitat, such as dependence on groundwater flows.

C. **Biological characteristics.**

   Pertinent texts include those by Mueller-Dombois and Ellenberg[20] and by Daubenmire.[21]

   1. Vegetation physiognomy and community structure.
   2. Regional vegetation type according to systems of Küchler[22], Bailey[23], Society of American Foresters[24], Garrison et al.[25], and/or state or local vegetation map(s) (where available).
   *3. Frequently associated species.
   4. Dominance (cover or basal area) and frequency of this taxon in each community type in which it occurs.
   5. Successional phenomena, including relation to canopy closure. Consider factors such as the colonizing ability, tolerance of disturbance, shade tolerance, and growth on unstable substrates.
   6. Dependence of this taxon on dynamic, periodic, and/or cyclic aspects of biotic associations and ecosystem features, such as tree falls, forest fires, and insect population fluctuations.
   **7. Other endangered, threatened, rare, or vulnerable species occurring in habitat(s) of this taxon.

7. **Population biology of taxon.**

   Summary of information on various populations, referenced to specific status reports on each population, when available. See Whitson and Massey's paper, this volume, for a more complete treatment of species biology. Harper's *Population Biology of Plants*, Grime's *Plant Strategies and Vegetation Processes,* and Solbrig *et al.'s Topics in Plant Population Biology*[26] present additional material that may be helpful.

A. **General summary** of population biology of the taxon.

B. **Demography.**

   *1. Number and geographical spacing of known populations (estimated if necessary), with estimate of currently known number of individuals per population, if available. Define age (or size) classes used and describe census method(s) used to determine area and numbers; see Mueller-Dombois and Ellenberg's text[20] for a description of various censusing techniques. Mention also any pertinent historical observations, such as past estimates of abundance.
   2. General demographic details of each population, including:
      a. *(b, c, etc.)* [Name or other designation of site, as in 5.B.6.]
         1. Area of the population(s).
         2. Number, and age (or size) classes of individuals.
         3. Density (number of individuals per unit area).

    4. Presence of dispersed seeds.

    5. Evidence of reproduction (seedlings, etc.).

    6. Evidence of population expansion or decline such as review of historical accounts or consideration of demographic observations.

C. Phenology.

    1. Patterns and observed times of budding, leafing, flowering (anthesis), fruiting, seed or fruit dispersal, senescence, germination, etc.; also, months identifiable and months conspicuous in the field.

    2. Relation of phenological phenomena to climatic and microclimatic events such as exposure, drought, late frosts, and precipitation patterns.

D. Reproductive biology.

    1. Types of reproduction, relationship to age of plant, and evaluation of relative importance of each type to maintenance of population.

       a. Outbreeding (dioecy, protandry, heteromorphy, self-incompatibility, etc.).

       b. Inbreeding (cleistogamy, autogamy, etc.).

       c. Cloning (nature, rate, and extent).

       d. Other asexual reproduction and/or dispersal (rhizomes, tubers, layering, agamospermy, etc.).

    2. Pollination.

       a. Pollination mechanism(s) (mechanical, insect, wind, water, etc.).

       b. Specific *known* insect (or other) pollinators, or other known agents (e.g. wind or rain), if applicable; state nature of evidence for each.

       c. Additional insect visitors or other possible or suspected pollination agents.

       d. Vulnerability of pollinators (or other pollination mechanisms) to pesticides, pollution, land-use changes, succession, exotic animals or plants, etc.

    3. Seed dispersal.

       a. General mechanisms (wind, water, animal, etc.).

       b. Specific agents.

       c. Vulnerability of dispersal agents or mechanisms to disturbance and/or habitat modification.

       d. Patterns of propagule dispersal (distance, frequency of distribution in a particular area, etc.). Note some propagules may be dispersed to areas in which they cannot grow.

    4. Seed biology.

       (Indicate whether observations based on natural or laboratory conditions.)

       a. Amount and variation in annual seed production.

       b. Seed viability and longevity—percentage of seeds germinating after specific time intervals.

       c. Dormancy requirements, if any known.

      d. Germination requirements, such as scarification, cold temperatures, soil type, moisture, and light.

      e. Percent germination under various conditions.

  5. Seedling ecology.
Include important ecological and microhabitat factors, such as light, moisture, nutrient requirements, and soil disturbance.

  6. Survival and nature of mortality of plants at each life stage, e.g. seedling mortality due to predation or due to intraspecific competition, or mortality from shading of mature plants by canopy growth.

** **7. Overall assessment of taxon's reproductive success.**

8. Population ecology of species.
Summary of information on various populations, referenced to specific population status reports, if available.

A. General summary of population ecology of the taxon.

B. Positive and neutral interactions with mature plants, pollen, seeds, seedlings, and juveniles of this taxon. Include plants or animals with which this species is obligatorily or facultatively associated.

C. Negative interactions with mature plants, pollen, seeds, seedlings, and juveniles of this taxon.

  1. Herbivores, predators, pests, parasites, and diseases (native or exotic) affecting this taxon.

  2. Evidence of competition for light, water, pollinators, or other requirements.

    a. Intraspecific.

    b. Interspecific.

  3. Toxic and allelopathic interaction(s) with other organisms.

D. Hybridization; specify the nature of evidence.

  1. Naturally occurring.

  2. Artificially induced.

  3. Potential for spontaneous occurrence in cultivation.

E. Other factors of population ecology.

9. Current land ownership and management responsibility, repeated by portion of range as necessary.

 *A. General nature of ownership(s) (federal, state, private, etc.).

**B. Specific landowner(s), if known; give name(s) and address(es) if available and not publicity-sensitive.

**C. Management responsibility if different from owner(s).

**D. Easements, conservation restrictions, rights-of-way, special designations, etc., if any.

10. Management practices and experience.
Effects and results, repeated by portion of range as necessary. (Note: Discuss current and potential threats in section 11.)

A. Habitat management.

1. **Review of past management and land-use experiences,** if any, such as prescribed burning, water manipulation, herbicide or insecticide use, exclosures, agricultural tillage, nutrient enrichment, grazing, resort development, road salting, overstory thinning, and regulation of visitor use.
   a. This taxon.
   b. Related taxa, if pertinent.
   c. Other ecologically similar taxa, if pertinent.
2. **Performance under changed conditions,** including observations on plants that appear to survive and reproduce under conditions that are different from those in natural habitat of the taxon.
**3. Current management policies and actions. Current policies, intentions, and actions of landowner(s)—positive or negative—and their effects; indicate extent of awareness of owner(s) of land to presence and requirements of this taxon. Repeat by population if necessary.
4. **Future land use(s);** possible, projected, or anticipated; provide documentation, if available.

B. Cultivation.
1. **Controlled propagation techniques** appropriate, and ease of use for each technique.
2. **Ease of transplanting** cultivated material into natural habitat(s).
3. **Pertinent horticultural knowledge** concerning ecologically similar, related taxa.
4. **Status and location of presently cultivated material** (including commercial stocks, arboreta, botanical gardens, research greenhouses, private gardens, or recent introductions into natural areas, etc.), especially for trade-threatened species.[27]
   a. Specimen plants.
   b. Self-sustaining breeding populations; discuss any possibilities of genetic contamination.
   c. Stored seed (or other stored propagule) banks with history, age, and expected longevity of propagules; also possible sources of seed or other propagule supplies.

*11. Evidence of threats to survival.
Document and discuss threats under one or more of the following five threat factors, basing statements on information presented in earlier sections; state where necessary the portion of taxon's range to which each statement applies. These threat factors are taken directly from section 4(a) of the Endangered Species Act of 1973. Given in brackets under each threat factor are some individual threats we present as examples of each factor, adapted primarily from a list developed by the IUCN.[28]

Provision is made for discussion of past threats here to provide information on trends. Potential threats are those which are likely to become significant in the foreseeable future.
**A. Present or threatened destruction, modification, or curtailment of

habitat or range. [Species-independent threats to habitat or range from such activities as soil disturbance (soil erosion, trail erosion, soil compacting), water manipulation or control (including irrigation, dam construction, flooding, or drainage), fencing, mowing, plowing, burning, logging, nutrient enrichment, herbicide or insecticide use, agricultural tillage, road salting, mining or quarrying, trampling, camping, unnatural vegetation changes, water or air pollution, land clearing, development of habitat for human habitation, and pressures from introduced species.]

    1. Past threat(s).
    2. Existing threat(s).
    3. Potential threat(s).

**B. Overutilization for commercial, sporting, scientific, or educational purposes. [Species-dependent threats from such activities as medicinal uses, wildflower or specimen collecting, horticultural collecting, or traditional rural uses.]

    1. Past threat(s).
    2. Existing threat(s).
    3. Potential threat(s).

**C. Disease or predation. [Such threats as fungal infections, insect pests, or grazing (by native animals, feral exotics, or livestock).]

    1. Past threat(s).
    2. Existing threat(s).
    3. Potential threat(s).

**D. Inadequacy of existing regulatory mechanisms. [Threats such as inadequacy or lack of enforcement of laws restricting detrimental human activities.]

    1. Past threat(s).
    2. Existing threat(s).
    3. Potential threat(s).

**E. Other natural or manmade factors. [Threats resulting from other causes such as lack of pollinators, critically low population numbers, erosion of insular habitat, or climatic catastrophes.]

    1. Past threat(s).
    2. Existing threat(s).
    3. Potential threat(s).

II. ASSESSMENT AND RECOMMENDATIONS.

This assessment by the author(s) of the status of this taxon with recommendations concerning its conservation should be based on the information presented in the preceding sections of this status report. In reaching conclusions here the author(s) may particularly want to consider information presented under topics 5 (Geographic Distribution), 7 (Population Biology), 10 (Management Practices), and 11 (Threats).

    12. General assessment of vigor, trends, and status of habitat(s), population(s), and individual(s). Summarize briefly the general status of this taxon.

*13. Recommendations for listing or status change.
   A. Recommendation to U.S. Fish and Wildlife Service, Department of the Interior; make formal recommendation to Director,[29] unless status report is being prepared for U.S. Fish and Wildlife Service.
      *1. Recommended U.S. federal status: Endangered, Threatened, delist, or not-presently-Threatened.

      "Endangered species" is defined in section 3(6) of the amended Endangered Species Act of 1973 as "any species which is in danger of extinction throughout all or a significant portion of its range. . . . "

      "Threatened species" is defined in section 3(20) of the Act as "any species which is likely to become an endangered species within the foreseeable future throughout all or a significant portion of its range."

      **2. Recommended priority for federal action, on basis of natural vulnerability, present or potential threats, and adequacy of available information supporting recommendation.[30] Indicate any necessity for emergency listing.
   B. Recommendations to other U.S. federal agencies, such as the Heritage Conservation and Recreation Service, Bureau of Land Management, or National Park Service.
   C. Other status recommendations.
      1. Counties and local areas.
      2. State(s).
      3. Other nations (*e.g., Canada, Mexico*).
      4. International Trade Convention[2], Western Hemisphere Convention[3], IUCN[4], etc.). [See cited references for criteria for these status recommendations.]

*14. Recommended critical habitat[31] based on information presented in this status report. Include any necessary buffer zone(s) and/or habitat of pollinator(s), dispersal agents, etc. Consider also amount of suitable habitat presently available for species, what was available in the past, and what is anticipated to be available in the future, where known.[32]

   (For threatened or endangered species, early information on critical habitat is generally required by the U.S. Fish and Wildlife Service so that the final listing and critical habitat determination can be made together.)
      **A. Concise statement of recommended critical habitat including acreage (if known) and formal or informal description of geographical boundaries. (Append detailed map(s) and aerial photos if available and not publicity sensitive.)
      ** B. Legal description of boundaries, if available.
      ** C. Latitude and longitude, range-township-section, and/or UTM designations of area of recommended critical habitat.
      **D. Any comments or data which indicate that an exception should be made to the requirement of section 4(a) of the Amended Act that critical

habitat be proposed concurrent with a proposed listing, because doing so would be imprudent (such as cases where revealing precise location could increase collection or vandalism).

15. **Conservation/recovery recommendations.**
   A. **General conservation recommendations.**
     1. Recommendations suggesting changes in present or anticipated activities that appear to jeopardize the continued existence of the species, such as commercial exploitation, or mining, agricultural, or recreational activities in habitat area.
     2. Specific area(s) recommended for natural area designation, easement, land acquisition, or other protection. Indicate urgency of action for each, based on nature and imminence of threats to the area(s). If land acquisition recommended, review alternatives and the reasons why these are less suitable.[33]
     3. Management and recovery alternatives recommended to reduce or mitigate threats to taxon, such as prescribed burning, removal of competitive exotics, modification of grazing system and fencing, research on propagation of the taxon, and development of cultivated stock as alternative source for commercial use.
     4. Indicate anticipated consequences of publicity associated with the site, including whether site would be placed in serious jeopardy from publicity concerning this taxon.
     5. Other recommendations.
   B. **Monitoring activities** and further studies or research recommended; specify type(s) of research needed with its priority, frequency, and person(s), institution(s), organization(s), or agency(s) suggested to be responsible. Include sites needing field checking.

16. **Interested parties, such as Office of Endangered Species, U.S. Fish and Wildlife Service;[34] other federal and/or state agencies; Smithsonian Institution;[35] IUCN;[36] The Nature Conservancy;[37] botanists with interest in this taxon; and local botanical clubs or native plant societies. Give addresses for each.

III. INFORMATION SOURCES.

 *17. **Sources of information** on this taxon and its habitat and management. Specifically indicate any not yet consulted.
   A. **Publications.**
     1. **References cited in report.**
     2. **Other pertinent publications.**
       1. Technical, such as scientific papers or management reports.
       2. Popular, such as magazines or newspapers.
   B. **Museum collections** (herbaria) consulted; indicate any collections consulted only in part.
   C. **Fieldwork.**
   Summarize fieldwork (if any) conducted in support of this status re-

port, including name(s) and affiliation(s) of researchers, general nature of activities, seasons of site visits, and nature of documentation developed.

D. Knowledgeable individuals.
Persons with pertinent experience or knowledge (including local residents or other individuals familiar with the locality for each population); give name, affiliation (if any), address, and telephone of each, along with a statement of the nature of their pertinent experience or knowledge.

E. Other information sources, such as agency office files.

**18. Summary of materials on file, such as copies of original botanical description and best recent botanical description(s); copies of germane floristic or monographic studies; copies of pertinent correspondence; additional photographs and maps;[38] and copies of other materials of reference value. Indicate location and availability of items not in files of the author(s).

IV. AUTHORSHIP.

*19. Initial authorship of this status report as first formally distributed or submitted. Give name(s), address(es), affiliation(s), and telephone number(s) for author(s).

20. Maintenance of status report.
State person(s), organization(s), institution(s), or agency(s) (if any) taking responsibility for receiving new information, making revisions and corrections to this status report, and distributing revisions to interested parties.

V. NEW INFORMATION.

Append all subsequent findings not in initial status report, with appropriate information category indicated. Record reviews, responses, and revisions of status report, with dates. Indicate to which section new information applies, and (if already incorporated) indicate date revision was made.

21. Record of revisions to this status report; list for each batch of revisions the date of distribution or publication, the name of person making revisions, and the topics affected, for example:
12 Sept. 1979, Topics I.D, 6.B.2, 14.A.4, 17.A, and 20 revised by . . .
This information is needed to maintain a chronological record of changes and to determine whether any information in a particular copy of a status report is no longer current.

---

[1] J. I. Lawyer, A. M. Miller, L. E. Morse, and J. T. Kartesz. In press. A Guide to Selected Current Literature on Vascular Plant Floristics for the United States, Alaska, Canada, Greenland, and the U.S. Caribbean and Pacific Islands. *Mem. Torrey Bot. Club.*

²   Convention on International Trade in Endangered Species of Wild Fauna and Flora. Reprinted in *Extinction is Forever,* edited by G. T. Prance and T. S. Elias (Bronx, New York: New York Botanical Garden, 1977), pp. 401-407. See also *Federal Register* 42: 10461-10488, 1977, for U.S. regulations pertaining to the Trade Convention.

³   Convention on Nature Protection and Wildlife Preservation in the Western Hemisphere. [Pan American Union Treaty of October 12, 1940]. Reprinted in *Extinction is Forever,* edited by G. T. Prance and T. S. Elias, *op. cit.,* pp. 396-400.

⁴   G. Lucas and H. Synge. *The IUCN Plant Red Data Book* (Morges, Switzerland: International Union for Conservation of Nature and Natural Resources, 1978). See also G. Lucas and H. Synge, The IUCN Threatened Plants Committee and its work throughout the world, *Environmental Conserv.* 4(3): 179-187, 1977.

⁵   Endangered Species Act of 1973, Public Law 93-205, December 28, 1973. Reprinted in *Extinction is Forever,* edited by G. T. Prance and T. S. Elias, *op. cit.,* Appendix 4, pp. 417-437. See also amendments to the Act in Public Laws 94-325, 94-359, 95-212, 95-362, and 96-159. For a discussion of the Amendments, see "President Signs Endangered Species Act Amendments," *Endangered Species Techn. Bull.* 3(10): 1ff, 1978, and "Endangered Species Act Extended and Amended," *ibid* 5(1): 1-4, 1980. [Reprinted, this volume, as "The 1978 and 1979 Amendments to the Endangered Species Act: A Discussion"] (Note: The implications of these amendments have not been fully considered in these Guidelines.)

⁶   Among the *Federal Register* publications of general interest to date are *F.R.* July 1, 1975, notice of review for over 3000 taxa recommended by Smithsonian Institution for listing; *F.R.* June 16, 1976, proposed endangered status for 1700 taxa, and *F.R.* April 26, 1978, determination of 11 endangered and 2 threatened taxa, which includes responses to comments received on the major proposal. For a discussion of these publications see B. MacBryde's paper "Plant Conservation in the United States Fish and Wildlife Service" in *Extinction is Forever,* edited by G. T. Prance and T. S. Elias, *op. cit.,* pp. 62-74, and his papers in *Nature Conserv. News* 29: 9-11, 1979, and in this volume.

⁷   Smithsonian Institution, *Report on Endangered and Threatened Plant Species of the United States.* Committee on Merchant Marine and Fisheries, Serial No. 94-A. 94th Congress, 1st Session, House Document No. 94-51. (Washington, D.C.: U.S. Government Printing Office, 1975).

⁸   Edward S. Ayensu and Robert A. DeFilipps. *Endangered and Threatened Plants of the United States.* (Washington, D.C.: Smithsonian Institution Press and World Wildlife Fund, Inc., 1978).

⁹   W. R. Powell, editor. *Inventory of Rare and Endangered Vascular Plants of California.* (Berkeley, Ca.: California Native Plant Society, 1974).

¹⁰   J. T. Kartesz and R. Kartesz. *The Biota of North America. Part I. Vascular Plants. Vol. 1. Rare Plants.* (McKeesport, Pa.: Biota North America Committee, 1977).

¹¹   For a description of Michigan's Endangered Species Act of 1974 see W. H. Wagner *et al.,* Michigan's endangered and threatened species program, *Michigan Bot.* 16: 111-122, 1977; and J. H. Beaman, Commentary on endangered and threatened plants of Michigan, *op. cit.,* pp. 99-110.

¹²   Ronald L. McGregor, *Rare Native Vascular Plants of Kansas.* (Lawrence, Kansas: Technical Publication of the State Biological Survey of Kansas, No. 5, 1977).

¹³   A. E. Radford, L. J. Otte, and D. K. Strady Otte. Ecological Diversity Classification and Inventory. In Radford *et al., Natural Heritage Classification and Information Systems.* (Chapel Hill, North Carolina: University of North Carolina Student Stores, 1978).

¹⁴   R. E. Daubenmire. *Plants and the Environment:* A Textbook of Plant Autecology. 3rd Ed. (New York: John Wiley and Sons, 1974). [See also: P. S. White, Pattern, process, and natural disturbance in vegetation, *Bot. Rev.* 45: 229-299. 1979.]

¹⁵   E. A. Ackermann. *The Köppen Classification of Climates in North America.* (Cambridge: Harvard University Press, 1941). A map of the Köppen System is reprinted in *Climatic Atlas of the United States,* edited by S. S. Vishner. (Cambridge: Harvard University Press, 1966), fig. 984, p. 370.

¹⁶   N. J. Rosenberg. *Microclimate: The Biological Environment.* (New York: John Wiley and Sons, 1974).

¹⁷   N. M. Fenneman, *Physiography of the Eastern United States.* New York: McGraw Hill, 1938), and *Physiography of the Western United States.* (New York: McGraw Hill, 1931).

¹⁸   Charles B. Hunt, *Natural Regions of the United States and Canada.* (San Francisco: W. H. Freeman and Company, 1974).

¹⁹   Soil Survey Staff, *Soil Taxonomy: A Basic System of Soil Classification for Making and Interpreting Soil Surveys.* Soil Conservation Service, U.S. Department of Agriculture. Agriculture Handbook No. 436 (Washington, D.C.: U.S. Government Printing Office, 1975).

²⁰   D. Mueller-Dombois and H. Ellenberg, *Aims and Methods of Vegetation Ecology.* (New York: John Wiley and Sons, 1974).

²¹   R. E. Daubenmire, *Plant Communities: A Textbook of Plant Synecology.* New York: Harper and Row Publishers, 1968.

[22]   A. W. Küchler, *Potential Natural Vegetation of the Conterminous United States.* Amer. Geogr. Soc. Spec. Publ. 36, 1964; for Alaska and Hawaii, see "Potential Natural Vegetation of Alaska and Hawaii," sheet 89, *National Atlas,* U.S. Geological Survey, 1976.

[23]   R. G. Bailey, *Ecoregions of the United States* [map], and *Description of the Ecoregions of the United States.* (Ogden, Utah: Forest Service, U.S. Department of Agriculture, 1976 and 1978.) (A map combining the Küchler and Bailey classification systems is available from: Forest Service—Recreation Management, U.S. Department of Agriculture. P.O. Box 2417, Washington, D.C. 20013.)

[24]   *Forest Cover Types of North America (Exclusive of Mexico).* Report of the Committee on Forest Types, Society of American Foresters. (Washington, D.C.: S.A.F., 1954).

[25]   G. A. Garrison *et al. Vegetation and Environmental Features of Forest and Range Ecosystems.* Forest Service, U.S. Department of Agriculture. Agriculture Handbook 475. (Washington, D.C.: U.S. Government Printing Office, 1977).

[26]   J. L. Harper, *Population Biology of Plants* (New York: Academic Press, 1977); J. P. Grime, *Plant Strategies and Vegetation Processes* (New York: John Wiley and Sons, 1979); and O. T. Solbrig, S. Jain, G. B. Johnson, and P. H. Raven, *Topics in Plant Population Biology* (New York: Columbia Univ. Press, 1979).

[27]   For a description of cultivation techniques and a philosophical and practical discussion of the role of cultivation in rare plant conservation see *Conservation of Threatened Plants,* edited by J. B. Simmons *et al.* (New York: Plenum Press, 1977), especially Section III, "Techniques of Cultivation"; also, *Survival or Extinction,* edited by H. Synge and H. Townsend, Royal Botanic Gardens, Kew, 1979.

[28]   IUCN. First Draft of a World Conservation Strategy, January, 1978. Unpublished.

[29]   Director, U.S. Fish and Wildlife Service, Department of the Interior, Washington, D.C. 20240.

[30]   Procedures for listing species are described in the *Federal Register* 45 (40, IV): 13009-13026, 27 Feb. 1980.

[31]   "Critical Habitat" is defined in the Endangered Species Act Amendments of 1978 as: " . . . the specific areas within the geographical area occupied by the species at the time it is listed . . . on which are found those basic physical or biological features . . . essential to the conservation of the species and . . . which may require special management considerations or protection; and . . . specific areas outside the geographical area occupied by the species at the time it is listed . . . upon a determination . . . that such areas are essential to the conservation of the species."
   "Critical Habitat" has been further explained in the 4 Jan. 1978, *Federal Register* as: "any air, land, or water area (exclusive of those existing man-made structures or settlements which are not necessary to the survival and recovery of a listed species) and constituent elements thereof, the loss of which would appreciably decrease the likelihood of the survival and recovery of a listed species or a distinct segment of its population. The constituent elements of critical habitat include, but are not limited to: physical structures and topography, biota, climate, human activity, and the quality and chemical content of land, water, and air. Critical habitat may represent any portion of the present habitat of a listed species and may include additional areas for reasonable population expansion."
   Criteria for designating critical habitats have been published by the U.S. Fish and Wildlife Service and the National Oceanographic and Atmospheric Administration in the *Federal Register* 45(40, IV): 13009-13026, 27 Feb. 1980.

[32]   Some additional considerations in determining critical habitat for plants are:
   1. The essential environmental elements that the taxon required during a large number of previous generations.
   2. Associated plants and animals such as pollinators or seed-dispersers. The necessary portions of the ranges of such taxa should be included in critical habitat, if such relationships are essential for the survival of the candidate taxon.
   3. The inclusion of a minimal gene pool diverse enough to cope with natural forces and maintain evolutionary potential. Consider configuration of remaining populations and any need for interaction between them.
   4. Consideration of whether long-term essential habitat may be larger than that seen today.
   5. Need for buffer zones.

[33]   If land acquisition is recommended, information such as the following will help in acquisition efforts: Especially significant or unique natural features, history including past and existing preservation efforts, organizations interested in area, archeological features, and other threatened or endangered species in area. For further discussion see: P. M. Hoose, *Building an Ark: Tools for the Preservation of Natural Diversity* (Island Press, Covelo, Calif., 1980).

[34]   Office of Endangered Species, U.S. Fish and Wildlife Service, Washington, D.C. 20240.

[35]   Endangered Flora Project, MNH 166, Smithsonian Institution, Washington, D.C. 20560.

[36]   International Union for Conservation of Nature and Natural Resources (IUCN), avenue du Mont Blanc, CH-1196 Gland, Switzerland.

[37]   Science Programs, The Nature Conservancy, 1800 N. Kent St., Arlington, Va. 22209.

[38] Technical information on availability of maps on such topics as political units, hydrologic units, census subdivisions, and federal land ownership may be obtained from Geography Program, Land Information and Analysis Office, U.S. Geological Survey, 710 National Center, Reston, Va. 22092.

NOTE ADDED IN PROOF

A new "Notice of Review" including 2999 candidate plant taxa was published by the U.S. Fish and Wildlife Service in the Dec. 15, 1980, issue of the *Federal Register* (vol. 45, pp. 82479–82569). This publication supersedes the lists in the Smithsonian Institution's 1975 report and 1978 book and the previous FWS *Federal Register* notices. The new notice also lists 797 vascular plant taxa no longer under federal review. Of these, 51 taxa are believed extinct, 193 are considered synonyms of non-candidate taxa, and 553 are considered more common or less vulnerable than previously thought.

Report on the Conservation Status of *Hudsonia montana*, a Candidate Endangered Species[1]

Larry E. Morse[2]

Scientific name of taxon: Hudsonia montana Nutt.
Common name of taxon: Mountain Golden Heather
Family: Cistaceae
State(s) where taxon occurs: North Carolina, U.S.A.
Current Federal Status: Proposed Endangered Species
Recommended Federal Status: Threatened Species
Author of Report: Larry E. Morse
Original date of report: 15 January 1979
Date of most recent revision: 8 Nov. 1979
Individual to whom further information and comments should be sent: The Author

---

[1] This status report was prepared by the Cooperative Parks Study Unit of the New York Botanical Garden as a test of the guidelines for status reports being developed, in part, by that Unit. The report draws heavily upon research conducted by the author while a graduate student at the Gray Herbarium, Harvard University, supplemented by further research conducted by him with support from the U.S. National Park Service through its cooperative research agreement with the New York Botanical Garden.
[2] Gray Herbarium, Harvard University, Cambridge, Mass. 02138, and Cooperative Parks Study Unit, New York Botanical Garden, Bronx, N.Y. 10458. Present address: The Nature Conservancy, 1800 N. Kent St., Arlington, Va. 22209.

Pages 283-308 *in:* Larry E. Morse and Mary Sue Henifin, eds., *Rare Plant Conservation Geographical Data Organization*. The New York Botanical Garden, Bronx, N.Y.

Contents

I. Species Information

1. Classification and nomenclature
A. Species
1. Scientific name
     a. Binomial—*Hudsonia montana* Nutt.
     b. Full bibliographic citation—Nuttall, Thomas. 1818. *The Genera of North American Plants*. Philadelphia. (Vol. 2, p. 5)
     c. Type specimen(s)—Table Rock, North Carolina, [1816], in Academy of Natural Sciences, Philadelphia; presumed isotypes at Philadelphia, Gray Herbarium (Harvard Univ.), the New York Botanical Garden, and the British Museum (Natural History).
2. Pertinent synonym—*Hudsonia ericoides* ssp. *montana* (Nutt.) Nickerson and J. Skog (p. 456 in: Skog, J. T., and N. H. Nickerson, 1972. Varia-

284

tion and speciation in the genus *Hudsonia. Ann. Missouri Bot. Gard.* 59: 454-464.)

   3. **Common names**—This species has no single, well-established common name; the name Mountain Golden Heather is descriptive and appropriate. Other species in the genus are known as Golden Heathers, Sand Heathers, Beach Heathers, and Hudsonias.
   4. **Taxon codes**—HUMO (SCS, 1971).
   5. **Size of genus**—Three species, all North American, of which all three occur in the United States, and two occur in North Carolina.

B. **Family classification**
   1. **Family name**—Cistaceae
   2. **Pertinent synonyms**—None
   3. **Common name for family**—Rockrose Family

C. **Major plant group**—Dicot

D. **History of knowledge of taxon**—*Hudsonia montana* was discovered by Thomas Nuttall on the summit Table Rock, North Carolina, in 1816, on his first extensive expedition to the Southern States (Pennell, 1936; Graustein, 1967). He described this species as new to science in 1818, giving its distribution as:

> On the highest summits of the mountains of North Carolina, forming extensive caespitose patches; abundant on the romantic summit of the Table-Rock, a singularly elevated and isolated portion of the Catawba ridge, in company with *Rhododendron Catawbiense,* &c.

   After Nuttall's 1816 discovery of *Hudsonia montana,* it has been collected at infrequent intervals to the present time. Most specimens are labelled Table Rock (or, erroneously, Table Mountain; see Small, 1895), but a few are from three other nearby localities. There is no evidence that *Hudsonia montana* presently occurs, or has occurred historically, anywhere besides these four sites, all on U.S. Forest Service lands within five miles of the summit of Table Rock, and all in Burke County, North Carolina.

   It has been widely assumed that Table Rock is the only locality for the species, and the failure of several botanists to find any plants of *Hudsonia montana* there in the 1960's and early 1970's led to the assumption the species was possibly extinct. However, plants were again found on Table Rock in recent years, and two other historically reported sites have now been relocated, as well as a fourth, previously unreported occurrence. These four known sites were studied by the author in site visits in 1975, 1977, and 1978, and three of them have also been studied extensively by the U.S. Forest Service under the leadership of Wildlife Biologist Ben A. Sanders. Ruby Harbison, and E. LaVerne Smith also visited three of the sites in 1978.

   Scientific studies of the genus *Hudsonia* include those of Nuttall (1818), Skog and Nickerson (1972), and the author (Morse, 1978 and 1979).

E. **Comments on current alternative taxonomic treatments**—Nickerson and Skog (1972) concluded that this taxon should be treated as a subspecies of *Hudsonia ericoides,* rather than a distinct species; they also included plants generally known as *Hudsonia tomentosa* in their broad concept of *Hudsonia ericoides,* a concept that has never been followed in any subsequent taxonomic or floristic work.

The Smithsonian Institution followed Nickerson and Skog's treatment for their Report to the Congress (1975) and the subsequent revision by Ayensu and DeFilipps (1978); this treatment is also followed in the *Federal Register* notices derived from the Smithsonian's Report (U.S. Fish and Wildlife Service, 1975 and 1976). However, this taxon has been recognized recently at the species level by Hardin *et al.* (1977) in a work on the rare and endangered plants of North Carolina, and in general species checklists by Shetler and Skog (1978) and by Kartesz (in prep.; pers. comm.), as well as in a horticultural discussion of the genus (Bailey Hortorium, 1976) and its treatment in *The IUCN Plant Red Data Book* (Lucas and Synge, 1979).

The author has recently studied this genus (Morse, 1979) and concluded that *Hudsonia montana* is a species distinct from *H. eriocoides* and *H. tomentosa,* readily identified by its larger leaves, long-acuminate sepals, and 20-30 stamens.

2. Present legal or other formal status
A. International
  1. **Present designated or proposed protection or regulation**—None.
  2. **Other current formal status recommendations**—Listed as Endangered by the IUCN (Lucas and Synge, 1979).
  3. **Review of past status**—Not applicable.
B. National
  1. **United States**
     a. **Present designated or proposed legal protection or regulation**—*U.S. Fish and Wildlife Service:* Under the synonymous name *Hudsonia ericoides* ssp. *montana,* this taxon was proposed as an endangered species 16 June 1976, following a notice of consideration as an endangered species 1 July 1975 (U.S. Fish and Wildlife Service, 1975 and 1976). Section 7(c) of the Endangered Species Act of 1973, as amended in 1978, requires a biological assessment for certain federal agency actions that might affect a proposed (or listed) species. The present proposal of *Hudsonia montana* expires on 10 Nov. 1979.

     U.S. Forest Service: Hudsonia montana is considered a "sensitive species" because of its proposed federal status and/or its listing in the state of North Carolina (W. Zeedyk, pers. comm., 1977; B. A. Sanders, pers. comm., 1978).
     b. **Other current formal status recommendations**—Listed as "Endangered" by the Smithsonian Institution (1975) and by Ayensu and DeFilipps (1978), under the synonymous name *Hudsonia ericoides* ssp. *montana.*
     c. **Review of past status**—Not applicable.
C. State
  1. **North Carolina**
     a. **Present designated or proposed legal protection or regulation**—None.
     b. **Other current formal status recommendations**—Listed as an "Endangered Endemic Species" by Hardin *et al.* (1977) in the state of North Carolina.

Considered a "Special Species" by the North Carolina Natural Heritage Program.

   c. **Review of past status**—Not applicable.

3. Description

  A. **General nontechnical description**—A low, needle-leaved shrub with yellow flowers and long-stalked fruit capsules. Usually growing in clumps 4-8" across and about 6" high, and sometimes seen in larger patches a foot or two across. The plants have the general aspect of a big moss or a low juniper, but their branching is more open, their leaves are about 1/4" long, and the plant is often somewhat yellow-green in color, especially in shade. The leaves from previous years persist scale-like on the older branches. The flowers appear in early or mid-June, and are yellow, nearly an inch across, with five blunt-tipped petals and 20-30 stamens. The fruit capsules are on 1/2" stalks, roundish, and with three projecting points at the tips; they often persist after opening, and may be seen at any time of the year.

  B. **Technical description**—Leaves acicular, 3-7 mm long, spreading-ascending, at first villous, soon glabrate. Flowers solitary, pedicellate at the ends of short spur shoots or terminating normal branches. Calyx 6-7 mm long, lobes lanceolate, villous, shorter ones acute, the longer ones acuminate with a tip to 2.5 mm long; petals ca. 2 × as long as the calyx; ovary pubescent to base. (Radford *et al.*, 1968) Chromosome number: Diploid, n = 10 (Morse, 1979).

    When in flower or fruit, *Hudsonia montana* is readily distinguished by its elongated sepal tips and its high stamen number (20-30). Vegetative specimens resemble those of *H. ericoides,* and cannot always be identified with certainty; *H. montana* tends to have leaves 5-8 mm long, slightly longer than the 3-6 mm leaves of *H. ericoides. H. montana* leaves are also slightly less pubescent, with a larger proportion of the trichomes straight rather than wavy.

    Seedlings have the aspect of upright vegetative branches of the mature plants, with two small cotyledons just below the first leaves, and about half an inch above the soil level.

  C. **Local field characters**—The capsule is distinctive and diagnostic—about 1/4" long, rounded at the base, with three projecting points at the tip, on a stalk about 1/2" long, and opening to reveal a 3-valved ovary from which 1-3 seeds are dispersed. Old capsules can be found on most plants.

    The vegetative form is also distinctive: a needle-leaved plant in low open clumps, not densely compacted like a moss or a selaginella. The aspect is close to that of a delicate, low juniper, but no junipers grow in the same habitat.

    The flowers are conspicuous, yellow, nearly an inch across, and borne singly on long stalks. They superficially resemble flowers of *Hypericum,* the St. John's-Wort.

    Hudsonia montana is usually found on exposed quartzite ledges, and occasionally along paths or in small openings in oak and pine woods on quartzite ridgetops. Occasional plants persisting in shade under overtopping shrubs can be easily overlooked.

No other species of *Hudsonia* are known from the southern Appalachians, except *H. tomentosa* at one site in West Virginia (Rydberg, 1927).

D. **Identifying characteristics of material which is in interstate or international trade or commerce**—No interstate or international trade or commerce known.

E. **Photographs and/or line drawings**—Color photographs in Rickett (1967, p. 215) and Justice and Bell (1968, p. 118).

Line drawing in Hardin *et al.* (1967), p. 112; and in Morse (1979, p. 155).

Other color and black-and-white photographs are available from the author on request or from the slide collection of the Library of the New York Botanical Garden.

4. Significance

A. **Natural**—This is probably a relict stand of a primitive *Hudsonia* species once more widespread in a period of warmer and drier climate, and could become ecologically more important if a warmer climate returned to the area. Its xerophytic adaptations are unusual in the Southern Appalachian region.

The role of this species in succession on the exposed sandy ledges in the area is not fully understood; the plants appear to occupy a fairly stable ecotone, yet are sometimes seen overtopped by larger shrubs.

B. **Human**—When in flower in June, this species is of local aesthetic value in the Table Rock area, which is frequently visited by hikers, and is a subject of wildflower photography. No horticultural, medicinal, or other economic uses of *Hudsonia montana* are presently known; the related *Hudsonia tomentosa* has been studied in Canada for possible use in revegetation.

5. Geographical Distribution

A. **Geographical range**—Known from elevations of 2800-4000 feet (850-1200 m) on Table Rock and three nearby sites in Burke Co., North Carolina, U.S.A.; no paleontological sites or extirpated historical sites are known.

B. **Precise occurrences**

   1. *Populations currently or recently known extant:*

      a. Table Rock (USA: North Carolina: Burke Co., "Tablerock Mtn" of U.S.G.S. 7.5-minute Linville Falls, N.C., 1:24000 topographic map quadrangle, 1956; approximate location, lat. 35° 53' 25" N., long. 81° 53' 02" W.)

      Discovered by Thomas Nuttall in 1816; specimens collected irregularly to present. Reportedly extirpated in 1960's by tourists and climbers, but verified extant in 1970's and last observed extant in 1978. Some specimens presumably from Table Rock are labelled "Table Mountain," an error attributed to Asa Gray by Small (1895).

      b. The "Campground Site" (USA: North Carolina: Burke Co., exact location publicity-sensitive but on Forest Service land within five miles of the summit of Tablerock Mountain.)

      Discovered by E. J. Alexander in 1923; last observed extant in 1978.

      c. The "Lookout Site" (USA: North Carolina: Burke Co., exact location publicity-sensitive but on Forest Service land within five miles of Tablerock Mountain.)

Discovered by Morse in 1975; last observed extant in 1978.

    d. The "Flat Ledge Site" (USA: North Carolina: Burke Co., exact location publicity-sensitive but on Forest Service land within five miles of the summit of Tablerock Mountain.)

Presumably the site discovered by Sargent in 1950, and possibly seen by Nuttall in 1816; rediscovered and observed extant by Morse and by E. L. Smith in 1978.

  2. Populations known or assumbed extirpated:
No such sites known; the Table Rock site was believed extirpated until rediscovered by J. T. and L. E. Skog in 1976.

  3. Historically known populations where current status not known:
No such sites known.

  4. Locations not yet investigated believed likely to support other possibly extant natural occurrences:
Other areas of similar habitat in the Linville Gorge area should be investigated, particularly quartzite ledges supporting *Leiophyllum*-dominated heath balds mergining into a pine/oak canopy.

  5. Reports having ambiguous or incomplete locality information:
Several nineteenth-century specimens have merely "North Carolina" or "Mountains of North Carolina" as geographical data; however, in all such cases examined, the collectors are known (from other specimen labels) to have visited Table Rock. Hence these vague references are not regarded as indicating the existence of other locations.

  6. Locations known or suspected to be erroneous reports:
Reports of *Hudsonia montana* from Chesterfield Co., S.C., from central New Hampshire, and from St. Pierre et Miquelon (near Newfoundland) are all traceable to misidentified specimens of the vegetatively similar *H. ericoides,* a widespread species.

A specimen in the herbarium of the University of Minnesota dated 1884 has a recopied label suggesting Statesville, N.C., as a locality; this is presumably merely the address of the collector, M. E. Hyams, who visited Table Rock in 1879 and perhaps on other occasions.

The epithet *"montana"* in this species' scientific name refers to these plants' montane habitat, not to the State of Montana.

C. Status and location of presently cultivated material—No cultivated material known; the Plant Records Center of the American Horticultural Society has no current records of presently cultivated material of *Hudsonia montana* (S. H. Davis, pers. comm., 1978).

D. Biogeographical and phylogenetic history—*Hudsonia montana* appears to be the most primitive member of its genus, and may be a relict species surviving from the early diversification of *Hudsonia* following its divergence from *Helianthemum*. Its survival in sandy soils of an unusual bedrock type in the Southern Appalachians may indicate an affinity to the Coastal Plain flora (including *Hudsonia ericoides*), since several other species of the Table Rock area are disjunct from the Coastal Plain, including *Leiophyllum buxifolium* and *Xerophyllum asphodeleoides.*

6. General environment and habitat description
A. Concise statement of general environment and habitat—Exposed quartzite

ledges partially dominated by *Leiophyllum buxifolium* and other ericads, with scattered *Pinus* spp. outward from forest border, and mosses, *Selaginella,* and lichens on the rocks nearest the brink of the ledge. These areas have a sandy soil derived from the underlying quartzite, with some loam content, and often a thin litter of pine needles. Some sites have a veneer of fine gravel over the soil. The general aspect is that of a microclimatically maintained ecotone between bare rock and pine/oak forest, with *Hudsonia montana* as a local dominant in the more open areas of the heath bald vegetation.

A few plants of *Hudsonia montana* have been noted on exposed sandy soil in sunny clearings along trails near such ledges.

B. Physical characteristics

    1. Climate

        a. Köppen Climate Classification—Type Dfa, cold snowy forest (humid microthermal) climate moist in all seasons, with a hot summer, warmest month over 22° C. (Ackermann, 1941).

        b. Regional macroclimate—Monthly averages for temperature, precipitation, wind velocity, daylength, etc. are given by Schwartz (1977) for Asheville, N.C., about 50 miles southwest and 2250 feet in elevation, about 1000′ lower than the *Hudsonia montana* sites.

        Recorded rainfall in the area is about 50-55 inches/year, but rainfall is estimated to be greater locally in the mountainous areas (Bryant and Reed, 1970).

        c. Local microclimate—No quantitative information available. The characteristic ledge habitats are generally exposed to direct sunlight, although in some cases lightly shaded part of the day by stunted pine trees. The exposure of the ledges suggests severe winter conditions.

        Bryant and Reed (1970) report winter temperatures as low as −20°F on the highest peaks in the Grandfather Mountain area.

    2. Air and water quality requirements—Not known.

    3. Physiographic province—Mapped as part of the Blue Ridge Province by Hunt (1967), and lying along the crest of the escarpment at the contact of the Blue Ridge and Piedmont Plateau provinces (Bryant and Reed, 1970).

    4. Physiographic and topographic characteristics—All known sites are on the Table Rock Thrust Sheet within the Grandfather Mountain Window, an area along the Blue Ridge Escarpment near Morganton, North Carolina. The bedrock is Chilhowee quartzite, forming flat to gently sloping ledges atop steep cliffs of the Linville Gorge. Elevation range for the known sites is approximately 2800-4000 feet, and each lies near a drop of at least 20 feet, in one case perhaps 200 feet (Bryant and Reed, 1970).

        The populations lie within the watersheds of Roses Creek, Irish Creek, and the Linville River, all tributaries of the Catawba River.

    5. Edaphic factors—The soils in all sites are presumably derived from the underlying Chilhowee quartzite, and are shallow, with quartzite gravel interspersed. In some sites the gravel also forms a veneer on the surface, suggesting deflation. The soils are sandy, with low to moderate loam content, and usually a sparse litter of pine needles. Although no quantitative soil analyses have been made, the abundance of Ericaceae and pines suggests acid soil.

6. **Dependence on dynamic aspects**—The ecotonal setting is clearly dependent on the steep cliffs resulting from erosion of the Chilhowee and underlying formations by the Linville River to form the Linville Gorge. This erosion appears slow enough at present that ecological communities can develop and mature along the ledges atop these cliffs to form stable, self-perpetuating assemblages. In many places the forest cover extends completely to the edge of the bluff, with no ledge community at all.

7. **Other unusual physical features of environment and habitat**—None.

C. **Biological characteristics**

1. **Vegetation physiognomy and community structure**—An open heath-bald vegetation type, with scattered shrubs (mostly Ericaceae), few herbs or forbs, and much bare ground. Occasional larger shrubs or stunted pines overtop portions of the sites, but in most places the shrubs are exposed to full sunlight and to winter weather. Radford (1978) presents a more detailed account of the vegetation of the *Leiophyllum* bald on the summit of Table Rock.

2. **Regional vegetation type**—Mapped as Appalachian Oak Forest by Küchler (1964), and as Type 2214, Appalachian Oak Forest, by Bailey (1976).

3. **Frequently associated species**—*Leiophyllum buxifolium, Gaylussacia baccata, Rhododendron minus, Selaginella tortipila,* and *Pinus pungens.* Other conspicuous species present in at least one site include *Kalmia latifolia, Vaccinium* sp., *Xerophyllum asphodelioides,* and *Cladonia* sp. E. L. Smith (pers comm., 1978) describes a crust of a low moss, possibly *Grimmea,* seen on the soil in some sites.

4. **Dominance and frequency**—*Hudsonia montana* is a co-dominant in one of the sites, accounting for about 20% cover in a small local area there. In other sites it is scarce or scattered, with individual plants typically several decimeters or meters apart.

5. **Successional phenomena**—Persists for some time in partial shade of pines, but appears less healthy than in open areas. Never encountered under closed canopies.

6. **Dependence on dynamic biotic features**—None clearly evident, although local soil disturbance appears necessary for seedling establishment in areas of gravel veneers or moss crusts.

   The forest adjacent to the *Hudsonia* ledges is a pine/oak mixture suggestive of the New Jersey Pine Barrens and other areas where fire is known to be significant in maintaining certain vegetation types; at present this forest is mostly a closed canopy, but future fires might open clearings which could be colonized by *Hudsonia montana* much as *H. ericoides* colonizes openings in the New Jersey Pine Barrens. Some local evidence of past fire is present in the area; Forest Service staff regard this area as having a low frequency of fires.

   Supporting this idea is the observation by Sanders (pers. comm., 1978) that a *Hudsonia montana* clump top-killed by a campire resprouted from the root collar, and seedlings also germinated in the bare ground resulting from the campfire.

7. **Other endangered species**—*Shortia galacifolia,* also proposed for endangered species status, is reported from the Linville River west of Table Rock

(Davies, 1955). *Carex misera,* another candidate endangered plant species, has been seen in the vicinity of the "Campground" site (B. A. Sanders, pers. comm., 1979), as has *Liatris helleri,* under review as a threatened species (R. Harbison, pers. comm., 1979). A number of plant species on the North Carolina state list are found in this area.

7. Population biology

A. General summary—Populations generally consist of a few dozen mature clones 4-8 dm in diameter and presumably several years or decades old, and at most a few seedlings and juvenile plants. This structure suggests good survivorship once clones become well established, with poor establishment or high mortality of seedlings or juveniles. The populations appear able to endure periodic stresses affecting establishment, and presumably accumulate viable seeds in the soil. (Seeds four years old have been germinated under laboratory conditions.) The breeding system of *Hudsonia montana* is unknown, but by comparison with *H. ericoides* may be assumed to involve occasional cross-pollination by bees and bee-like flies, although depending predominantly on self-fertilization induced by the closing of the flowers in the afternoon.

B. Demography

1. Known populations—Four populations of *Hudsonia montana* are known, all within a five-mile radius of the summit of Table Rock. Recent field studies suggest there may be as many as 1000-2000 individuals altogether, including seedlings, although fewer than 200 individuals were located in earlier counts. The number of viable seeds in the soil is unknown.

2. Demographic details

a. Table Rock Population:

1. Area—About 20 × 20 meters
2. Number and size of plants—21 plants observed 1978, including 13 mature clones, 5 juveniles, and 3 seedlings; also one dead clone noted.
3. Density—scattered.
4. Presence of dispersed seeds—Not determined; opened seed capsules present on some plants.
5. Evidence of reproduction—Seedlings present; some mature clones rooting from branches while growing outward from overtopping shrubs.
6. Evidence of expansion/contraction—Insufficient evidence; Nuttall (1818) reported the species as "abundant" on Table Rock. An 1879 label by Canby (at NY) notes Table Rock as "the only locality and even there not plentiful," and a label by M. E. Hyams (at OS) reports the species as "scarce" the same year. In 1951, M. T. Hall was able to sample 29 plants while collecting small branches for a population study. Several botanists visiting Table Rock in the 1960's and 1970's have failed to locate the species; in a careful search by four people in 1978, 13 mature *Hudsonia* plants, 5 juveniles, and 3 first-year seedlings were located in one area about 20 by 20 meters on Table Rock.

b. "Campground" Population

1. Area—Several small ledges typically 10 × 30 meters each, along about 1/4 mile of mountain crest. B. A. Sanders (pers. comm., 1978) reports seeing *Hudsonia montana* here "on many ledges," adding his belief that at this site "there are more than 1,500 plants, including seedlings." These newly discovered plants are not included in the subpopulation tallies presented below.
2. Number and sizes of plants—Tabulated below by subpopulations, each on a distinct ledge:

Size	Subpopulation			Total
	A	B	C	
Mature (2-6 dm)	16	20	40	76
Huge (1 m.)	1	0	0	1
Juvenile or Seedlings	9	0	0	9
Dead Mature	0	(1)	0	(1)
Dead Seedlings	(1)	0	0	(1)
Totals	26	20	40	86

3. Density—Scattered, to several plants per square meter.
4. Presence of dispersed seeds—Not directly determined; open fruit capsules and presence of seedlings suggests seed is present in soil.
5. Evidence of reproduction—Seedlings and juvenile plants are present, and large clones are rooting vegetatively in soil at margins.
6. Evidence of expansion/contraction—The tallies presented above were made 16 Oct. 1977, and about the same number of plants were noted although not counted on 23 May 1978. Wilbur noted the abundance at this site as "occasional" in 1964, and Morse saw at least 105 plants on 17 July 1975. Sanders has estimated the total population at this site as more than 1500, including seedlings, based on observations in August, 1978.

   The population at this site thus appears stable in an overall sense, although locally depleted on certain ledges where mature plants have been destroyed and replaced by seedlings of uncertain fate. Sanders' estimate does not distinguish size classes, nor note the extent to which ramets of obviously common genetic origin were separately counted. However, his recent studies suggest there are subpopulations at this site that were not included in the previous estimates.

c. "Lookout" Population:

1. Area—About 10 × 20 yards
2. Number and size of plants—16 mature or juvenile plants seen 15

July 1975; 5 mature plants seen 22 May 1978.
3. Density—scattered.
4. Presence of dispersed seed—Not determined; plants have open seed capsules, but no seedlings were noted in the area.
5. Evidence of reproduction—None evident in 1978; some juvenile plants seen in 1975.
6. Evidence of expansion/contraction—Population has declined significantly since 1975; no successful reproduction was evident in 1978.

d. *"Flat Ledge" Population*
1. Area—About 8 × 15 meters on a ledge, and along a nearby trail for about 40 meters.
2. Number and size of plants—About 80 clumps, mostly mature, including a few fruiting juveniles. Eight seedlings noted along the trail, but only two seen within the ledge site itself. Two dead clumps were noted on the ledge.
3. Density—Scattered to locally closely spaced, in a few cases contiguous making tallying of individuals difficult.
4. Presence of dispersed seed—No seed immediately found in soil under plants, but opened seed capsules are present on most plants.
5. Evidence of reproduction—Few seedlings found, except along a trail at edge of area. Vegetative cloning extensive; some of the "plants" tallied may in fact be ramets of common genetic origin.
6. Evidence of expansion/contraction—Not readily determined; population appears stable and productive, but having relatively few seedlings and juvenile plants.

C. Phenology
1. **Patterns**—Open flowers are present for about two weeks in June, with leaves expanding a few weeks earlier. Fruit mature in August, opening subsequently to release seeds. The fruit capsules and old leaves persist on the plants an additional year or longer.

Emerging seedlings were seen in late June, and well-developed seedlings in mid-July.
2. **Relation to climate and microclimate**—Leaf production and flowering is later with increasing altitude, as is general for Southern Appalachian plants. In the related *Hudsonia ericoides,* flowering is delayed about a week for shaded branches, and accelerated about a week for buds close to warm sand.

D. Reproductive ecology
1. **Types of reproduction**—The plants being flowering in about their third year, and root vegetatively at the edges once they form well-rounded clumps, after perhaps ten years. Large, well-rooted clones may become fragmented into separate, self-maintaining plants.

Details of breeding system are unknown, but the flowers are similar in structure to those of other *Hudsonia* species, which are self-compatible and become self-pollinated in the late afternoon as the flower closes, forcing the anthers and their loose pollen against the pistil. Cross-pollination as well as self-pollination by insects also occurs in the other two species of *Hudsonia.*

While the plants may be apomictic, this is considered unlikely in view of the production of hybrids of *Hudsonia ericoides* and *H. tomentosa* when these two species come into parapatric contact, such as on Cape Cod (Morse, 1979). *Hudsonia montana* plants have well-formed pollen (unpublished), suggesting sexual rather than apomictic reproduction.

2. Pollination
    a. Mechanisms—By insects and by selfing, as discussed in the preceding section.

    b. Specific known pollinators—None specifically known.

    c. Other suspected pollinators—Various beeds, bombyliid flies, and syrphid flies frequently visit other species of *Hudsonia,* and are present in the Table Rock area. However, no specific observations of insect visitation to *Hudsonia montana* flowers are known. There is no reason to believe any specialized pollinators are dependent upon this species.

    d. Vulnerability of pollinators—Unknown.

3. Seed dispersal
    a. General mechanisms—Although not yet directly observed, the seeds are presumably shaken from the seed capsules and fall a short distance from the plants; opened capsules persist for a year or more on the shrubs. While most of the seeds may accumulate in the soil under the shrubs, the possibility of passive secondary dispersal by locally abundant ants should be considered.

    In other *Hudsonia* species, the capsule itself is the dispersal unit, but in *H. montana* the capsules remain firmly attached to the plants, merely opening to release the seeds.

    b. Specific agents—Unknown.

    c. Responses of mechanisms—Unknown.

    d. Dispersal patterns—Not known in detail; seedlings have been located only within a few meters of fruiting plants.

4. Seed
    a. Amount of seed production—A few capsules to several hundred capsules per clump, with each capsule usually containing 2 or 3 seeds.

    b. Seed viability and longevity—Unknown; seeds of *H. montana* four years old germinated under laboratory conditions (Morse, 1979).

    c. Dormancy—Unknown.

    d. Germination requirements—Unknown.

    e. Percent germination—Unknown.

5. Seedling ecology
—Seedlings have been noted occasionally in habitats of the mature plants, usually on slightly disturbed sandy soil with some humus content, and often a light litter of pine needles. Slight disturbance of the soil crust appears necessary for seedling establishment.

    B. A. Sanders (pers. comm., 1978) reports seedling densities of several per square meter in such slightly disturbed areas, with some seedlings as far as 3 meters from the nearest extant mature plants.

    Seedlings have also been seen on bare ground resulting from trailside erosion, trampling of ledge vegetation, campfires, and natural erosion. Soil disturbance by ants or other animals may be a factor in seedling establishment, since the undisturbed soil is often covered by a crust or a pebble pavement.

6. **Survival and mortality**—Little quantitative information; survival of well-established clones (those over 4″ in diameter) appears much better than for seedlings or immature clones, if one assumes the population structure to be stable. Alternatively, establishment of populations could be cyclic, with the present observations being in a period of maturity.

7. **Overall assessment of reproductive success**—Two of the four known populations presently appear stable, and the two others to be persisting but declining. Reproduction through seedlings and juvenile plants appears inadequate to maintain population levels in view of the increased mortality to older plants presently incurred from impacts of human recreational activities.

8. Population ecology

A. **General summary**—*Hudsonia montana* occurs primarily in exposed ledge habitats, and is quickly overtopped by larger shrubs or trees. No specific obligate relationships are known.

B. **Positive and neutral interactions**—None known; a fungus *(Laccaria trulisata)* has been tentatively identified as a mycorrhizal symbiont of *Hudsonia tomentosa* (Narciso and Collins, 1979).

C. **Negative interactions**

1. **Herbivores, predators, pests, parasites and diseases**—None specifically known; two species of true bugs *(Polymerus rostratus* and *P. vaccini)* are known to feed on *Hudsonia ericoides* (Henry, 1978). Small weevils have been noted in fruit capsules of *H. ericoides* (Morse, 1979).

2. **Competition**

   a. **Intraspecific**—Little information; most populations are too sparse for intraspecific competition to be evident. Root competition may be more significant than crown competition, in view of the shallow soils and xeric nature of the species habitats.

   b. **Interspecific**—Overtopping and shading by taller shrubs and trees appears to be a major factor limiting *Hudsonia montana* to the more open areas of the ledges.

3. **Toxic and allelopathic interactions**—None known; the possibility of allelopathic effects of *Hudsonia tomentosa* on surrounding vegetation is considered by Narciso and Collins (1979).

D. **Hybridization**

1. **Naturally occurring**—Not expected and not known, since no other species of *Hudsonia* occur within 100 miles of the Table Rock area to which *Hudsonia montana* is endemic.

2. **Artificially induced**—None known.

3. **Potential in cultivation**—Unknown, but parapatric hybrids are locally abundant where *Hudsonia ericoides* and *H. tomentosa* occur together (Nickerson and Skog, 1972; Morse, 1978, 1979).

E. **Other factors of population ecology**—None known.

9. Current land ownership and management responsibility

A. **General nature of ownership**—United States Government.

B. Specific landowner—U.S. Forest Service; all known sites for *Hudsonia montana* are on U.S. Forest Service lands within Unit 26 of the North Fork Catawba River Planning Unit of Pisgah National Forest, in North Carolina. The U.S. Forest Service holds both surface and subsurface rights to these areas.

C. Management responsibility—Pisgah National Forest.

D. Easements, conservation restrictions, etc.—Two of the four known sites for *Hudsonia montana* are partially within the Linville Gorge Wilderness, and additional *Hudsonia* habitat would be included in this Wilderness if currently proposed extensions are approved.

10. Management practices and experience
   A. Habitat management
      1. Review of past management and land-use experiences
         a. *Hudsonia montana*—No experience.
         b. Related taxa—Prevention of trampling and soil compaction have been shown significant in the conservation of *Hudsonia tomentosa* in dune areas (MacDonnell, 1979).
         c. Other ecologically similar taxa—Not reviewed.
      2. Performance under changed conditions—Seedlings, juveniles, and mature plants of *H. montana* have been noted in a few places along trails, but are vulnerable there to continue trampling and trailside erosion.
      3. Current management policies and actions—Consultation of maps and other records on file at the U.S. Forest Service office in Asheville, N.C., on 17 Oct. 1977 showed the *Hudsonia montana* sites to have no extant timber sales or leases, no prospecting permits, no active wildlife management programs, no fisheries development plans and no habitat improvement programs.

        Collection of specimens of *Hudsonia montana* is presently regulated by the U.S. Forest Service, and visitor entry into some of the habitat areas is restricted by an Entry Permit System implemented for the Linville Gorge Wilderness.

        Staff of the U.S. Forest Service are currently conducting research on the status and occurrence of this species in the Linville Gorge area.

      4. Future land use—The area is primarily valuable for human recreation, and increases in this use can be anticipated, including hiking, camping, climbing, and hunting. Little commercially valuable timber occurs near the *Hudsonia montana* sites, and the steep and rugged terrain should deter commercial loggers.

        Placement and maintenance of trails, control of access by vehicles to present trailheads, and promulgation and enforcement of policies and regulations regarding visitor use and special use permits in Pisgah National Forest have effects relating to frequency and intensity of visitor access and use of sites supporting *Hudsonia montana*.

   B. Cultivation
      1. Controlled propagation techniques—No experience; propagation by seed and/or by layering may be possible.
      2. Ease of transplanting—No experience.

3. **Pertinent horticultural knowledge**—Hortus Third (Bailey Hortorium, 1976) reports that *"Hudsonias* may be colonized in dry places in sandy soil or along the seashore. They are difficult to grow and short-lived. Propagated by seeds and probably by cuttings."

Seeds of *Hudsonia montana* can be germinated by placing them on a damp piece of paper towel in a covered dish, holding the seed with tweezers, and nicking the seed coat gently with a scalpel; germination follows within two days (Morse, 1979).

4. **Status and location of presently cultivated material**—No cultivated material known; the Plant Records Center of the American Horticultural Society has no current records of presently cultivated material of *Hudsonia montana* (S. H. Davis, pers. comm., 1978).

11. Evidence of threats to survival

A. Present or threatened destruction, modification, or curtailment of habitat or range

1. **Past threats**—Visitor impacts in Table Rock and Linville Gorge area may have been differently distributed before the visitor permit system was initiated; however, no specific information on past impacts of human recreation on *Hudsonia* habitat are known. Nor is any information available to indicate whether the construction of the trail and observation tower on Table Rock resulted in any specific destruction of *Hudsonia* habitat or loss of *Hudsonia montana* plants.

2. **Existing threats**—Present human recreational use of the Linville Gorge area, including Table Rock, is resulting in trampling of vegetation, compaction of soil, and destruction of *Hudsonia montana* plants and seeds by campfires.

Trampling and soil compaction by human visitors is the single most noticeable change in the habitat at the "Campground Site" from 1975 to 1978, and effects of trampling are also evident at the other three sites, especially the "Lookout Site" where the trail is wider and the ledge area used as a lookout or scenic overlook now much more heavily trampled than in 1975. Broken and dead *Leiophyllum* shrubs are frequent on the ledges; many are obviously damaged by trampling, but some may have died from natural causes. Even the relatively inaccessible "Flat Ledge Site" shows some effects of trampling, and two broad but little utilized trails cut through the extant population on Table Rock.

Although trampling seems to stimulate seed germination by disturbing the soil, these seedlings are presumably quite vulnerable to destruction from further trampling in these areas if present visitor-use patterns continue.

Campfire circles are also present on several of the ledges in the area of the "Campground Site," an area presently used by campers who do not hold overnight Wilderness permits. One such campfire was built directly in a sizeable population of *Hudsonia montana,* damaging or destroying some of the plants. One top-killed clone was recently observed to have resprouted from the root collar (B. A. Sanders, pers. comm., 1978), and seedlings were also observed in the bare soil resulting from the campfire. However, prolonged fire is known to alter soil properties (Fenn, Gogue, and Burge, 1976).

Accelerated erosion along the steep bank of the trail at the "Lookout Site" may have contributed to the decline in the number of mature plants of *Hudsonia* present there from 1975 to 1978, although also providing some seedbed areas. Along the trail near the "Flat Ledge" site, the lesser amount of erosion appears to provide areas colonized by *Hudsonia* seedlings with enough stability for these to develop into mature plants.

A prospecting permit has been applied for by Minatome for an area about 2 miles (3 km) from the *Hudsonia* habitat, but does not appear to pose a threat to the species (B. A. Sanders, pers. comm., 1978).

   3. Potential threats—Anticipated increases in recreational use of the area would be expected to increase the intensity of the threats cited above.

The use of the steep to sheer cliffs of the Linville Gorge area for climbing is becoming increasingly popular, and leading to damage of vegetation on and near these cliffs, especially on Table Rock, the Chimneys, and Shortoff Mountain. While no damage to *Hudsonia* sites was specifically attributable to climbing, some sites are atop cliffs that may be attractive to climbers. Organized climbing instruction activities by a special permit holder are presently conducted in areas where *Hudsonia montana* is not known to occur.

A viewing platform has been proposed for the summit of Table Rock, but away from the presently known *Hudsonia* site there.

The possibility of impacts from prospecting or mineral development in the habitat area needs further investigation.

B. Overutilization for commercial, sporting, scientific, or educational use— No significant existing or potential threats known.

Specimen collecting by botanists may have had some minor effect on *Hudsonia* populations from time to time, but the amount of collecting in any one year appears to have been only a small proportion of the plants present. When in bloom, the plants may be vulnerable to wildflower collecting, but no such cases are known.

There is no evidence of collecting or digging of *Hudsonia montana* for transplanting or for commercial purposes.

C. Disease, predation, or grazing—No significant existing or potential threats known.

D. Inadequacy of existing regulatory mechanisms
   1. Past threats—None known.
   2. Existing threats—Present regulation of visitor activities appears inadequate to prevent significant damage to *Hudsonia montana* habitat and plants, especially to areas outside the boundaries of the Linville Gorge Wilderness.
   3. Potential threats—Increase in visitor quotas may lead to greater visitor impacts on *Hudsonia montana* habitats unless mitigating regulations are implemented.

No vehicle access is permitted at any of the sites, although a jeep trail approaches within a few minutes' walk of the "Campground Site."

E. Other natural or manmade factors

1. **Past threats**—Not known.
2. **Existing threats**—Small total area of suitable habitat renders species vulnerable to extirpation or loss of genetic variability if natural population fluctuations are accentuated by impacts of human activities.
3. **Potential threats**—No additional threats are presently anticipated.

II. Summary and Recommendations

12. **General assessment of vigor, trends, and status**—Habitat modification and direct damage to *Hudsonia montana* plants resulting from inadequately regulated visitor use of certain ledges in the Linville Gorge area has depleted the numbers and decreased the habitable area of *Hudsonia montana* to a precarious level, and one of the four known populations could be completely extirpated by a few careless footsteps. The other three populations are presently larger, yet still vulnerable to further depletion and possible extirpation from impacts of recreational activities should present usage patterns continue or increase. From their present unnaturally depleted population levels, the populations could also be extirpated by a natural reduction of numbers as part of natural population fluctuations, a threat that will continue until the present unnatural habitat restrictions are mitigated and reversed.

13. **Priority of listing or status change**
    A. **Recommendation to U.S. Fish and Wildlife Service**—On the basis of the evidence presented here, I have recommended to the U.S. Fish and Wildlife Service (letter, 16 Sept. 1978) that the plant species *Hudsonia montana,* also known as *Hudsonia ericoides* subsp. *montana,* be listed as a threatened species under the provisions of the Endangered Species Act of 1973. I see no great urgency for listing, since the appropriate Forest Service staff are already aware of this species on their lands, and have initiated inventory, monitoring, and management activities.

    Delisting should be considered only when the populations have stabilized reproductively, as shown by findings of monitoring studies, and the threats from impacts of recreational visitation have been reduced substantially through enforced regulations.
    B. **Recommendations to Other U.S. Federal Agencies**
    1. **U.S. Forest Service**—The Forest Service has the opportunity and the responsibility to minimize future impacts on this species and to mitigate damage from past and present abuses. Should this species be listed as an endangered or threatened species by the federal government, the U.S. Forest Service should also develop a specific management and recovery plant in consultation with the U.S. Fish and Wildlife Service and appropriate specialists, including specialists in visitor impact management. The needs of this species should also be addressed in the next revision of the management plan for the North Fork Catawba Unit of the Pisgah National Forest, North Carolina, and the need for local regulations should be considered.

        Since known impacts to this species are primarily from visitor activities, special attention should be given to the need for revision of policies and regulations regarding visitor use in the Linville Gorge area, including group climbing activities, camping and fire-building on ledges, and trail development and maintenance. Reduction of visitor use of the southeastern portion of the Linville Gorge area may prove necessary, and could be accomplished by terminating the present access road further downhill, thus requiring a longer hike to the ridgetops.

Direct and indirect impacts on this species of possible mining developments in this area should also be considered.

Trail realignment at the place here called the "Lookout Site" should be specifically considered, and steps taken to mitigate past erosion problems along the present trail at that site. However, there may be no need to close off access to this viewpoint entirely, merely to route through traffic around it.

The plan for development of a viewing platform on Table Rock should be reviewed to verify no *Hudsonia* habitat would be significantly affected by construction or use of the facility.

Further research on the ecology and population biology of this species should be encouraged to provide additional background for development and revision of impact assessments and management and recovery plans.

C. **Other status recommendations**
1. **Counties and local areas**—No need for further regulation at the county or township levels of government is seen.
2. **State**—A change of status from endangered to threatened in the State of North Carolina should be considered in view of current information on the populations and nature and degree of threats.

No need for further regulation at the county or township levels of government is seen.
3. **Other Nations**—Not pertinent, since no material of *Hudsonia montana* is known in international trade, and no localities of this species are known outside the United States.
4. **International**—Unless evidence of commercial use of *Hudsonia montana* is developed, there should be no need for listing of this species on the Trade Convention. Listing under the Pan-American convention is not seen as necessary unless this action would provide additional protection not available through the U.S. Endangered Species Act and through U.S. Forest Service regulations.

14. **Recommended critical habitat**
A. **Concise statement**—Based on the information reviewed in this statement and the field studies of the author, the following area is recommended for designation as the Critical Habitat of *Hudsonia montana* under the provisions of the Endangered Species Act of 1973:
1. Table Rock, from the 3600-foot contour line to the summit.
2. The Chimneys, from the 3000-foot contour to the summit, and areas of adjacent ridgetop within a one-kilometer radius of the summit.
3. Shortoff Mountain, from the 2800-foot contour to the summit.

These areas are generally the flat ledges atop the steep cliffs characteristic of the Linville Gorge, along with the forested summits of the high

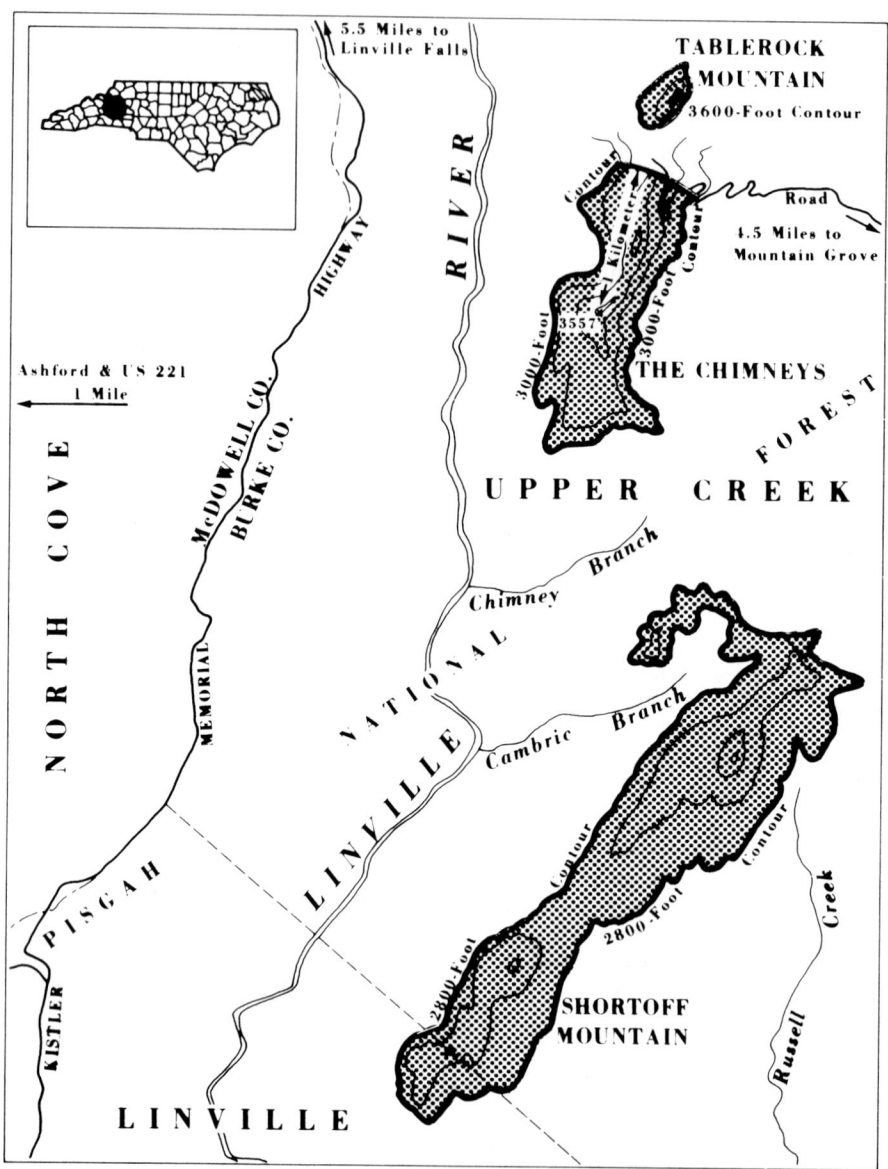

Map showing areas where appropriate habitat for *Hudsonia montana* is known; the species is endemic to the mapped area. (Courtesy U.S. Fish and Wildlife Service.)

ridges on the east side of the Gorge (map appended).

B. Legal description of boundaries—Not developed.

C. Latitude and longitude—The recommended critical habitat area lies within a circle of radius five miles, centered on the summit of Table Rock, location 35° 53′ 25″ N., 81° 53′ 02″ W.

302

D. Publicity-sensitivity of critical habitat area—The area of proposed critical habitat is of low publicity sensitivity. However, publication of the specific localities of *Hudsonia montana* populations could increase the possibility of their accidental or deliberate destruction.

15. **Conservation/recovery recommendations**
    A. **General conservation recommendations**
       1. **Recommendations regarding present or anticipated activities**—The U.S. Forest Service should continue to develop a management and recovery plan for this species, as described in section 12-B above.
       2. **Areas recommended for protection**—All known sites for *Hudsonia montana* are already on public lands; designation as a special area is not necessary and may prove detrimental through increased publicity and visitor curiosity.

          The Linville Gorge Ledges should be reviewed for possible designation as a National Natural Landmark having unusual geologic and biological features.
       3. **Habitat management recommendations**—Study should be made to determine whether current fire-management policy is appropriate for this area. Visitor abuse of some areas may require trail closure or fencing. Trail realignment and erosion control may be appropriate for protection of the "Lookout Site."
       4. **Publicity sensitivity**—Low to moderate.
       5. **Other recommendations**—Seed should be collected and cultivation attempted in order to maintain the species for future human uses should all extant sites be extirpated; this may be an appropriate student project.
    B. **Monitoring activities and further studies recommended**
       1. Area should be explored for further populations, particularly during flowering season when the plants are most conspicuous (June). Forest Service personnel in local area should be instructed in recognition of the species and provided color photographs showing flowers, vegetative stems, and general habit, then instructed to report possible sightings to a designated individual for verification.
       2. Pollination ecology should be examined to determine extent of outbreeding and dependence, if any, on specific kinds of insects. This could be an appropriate short-term project for a graduate student or advanced ungraduate in the area.
       3. Techniques for propagation and cultivation should be developed, perhaps using material from *Hudsonia ericoides* for initial experimentation.
       4. Monitoring of all sites should be done frequently enough to assure compliance with regulations and to note changes in populations and possible additional threats to them. Volunteers may be available for this work; otherwise, the U.S. Forest Service should conduct these studies.

16. Interested parties:

Office of Endangered Species
   (Attn: E. LaVerne Smith)
U. S. Fish and Wildlife Service
Washington, D.C. 20240

Regional Office
U. S. Fish and Wildlife Service
P. O. Box 95067
Atlanta, Ga. 30347

National Forests in North
   Carolina
   (Attn: Ben A. Sanders)
U. S. Forest Service
P. O. Box 2750
Asheville, N.C. 28802

U.S. Forest Service
   (Attn: Levester Pendergrass)
Southeast Regional Office
1720 Peachtree Road, N.W.
Atlanta, Ga. 30309

North Carolina Natural Heritage
   Program
Dept. Natural Resources and
   Community Development
P. O. Box 27689
Raleigh, N.C. 27611

Endangered Flora Program
   (Attn: Edward S. Ayensu)
Smithsonian Institution
Washington, D.C. 20560

International Union for the
   Conservation of Nature and
   Natural Resources
Threatened Plants Committee
   (Attn: G. Ll. Lucas)
c/o The Herbarium
Royal Botanic Gardens
Kew, Richmond, Surrey TW9 3AB
England

Science Programs
   (Attn: R. E. Jenkins)
The Nature Conservancy
1800 North Kent St.
Arlington, Va. 22209

Dr. J. Dan Pittillo
Dept. Biology
Western Carolina University
Cullowhee, N.C. 28723

Dr. Albert E. Radford
Dept. Botany
Univ. North Carolina
Chapel Hill, N.C. 27514

Dr. James W. Hardin
Dept. Botany
North Carolina State Univ.
Raleigh, N.C. 27607

Dr. Larry E. Morse
New York Botanical Garden
Bronx, N.Y. 10458

Dr. Norton H. Nickerson
Dept. Biology
Tufts University
Medford, Mass. 02155

Ms. Ruby Harbison
111 York St.
Morganton, N.C. 28655

III. Information Sources

17. Sources of information
  A. Publications
    1. **References cited in report**—List appended
    2. **Other technical publications**—None known
  B. Museum collections consulted—Specimens from many U.S. and Canadian herbaria were consulted, including those of Harvard University, The New York Botanical Garden, the Academy of Natural Sciences (Philadelphia), the Smithsonian Institution (U.S. National Herbarium), Tufts University, the University of New Hampshire, the University of North Carolina, North Carolina State University, the University of Michigan, Michigan State University, the Field Museum of Natural History (Chicago), and the University of Minnesota, and about twenty smaller collections (Morse, 1979).
  C. Fieldwork
    Site visits by the author:
    16-19 July 1975 —General fieldwork including popualtion estimates; notes, photographs, maps, sketches, and herbarium specimens.
    16 Oct. 1977 —Primarily population counts and studies of visitor impacts.
    21-22 May 1978 —Cytological samples and population assessments.
    21 June 1978 —Review of impacts and population status, with B. A. Sanders and others from U.S. Forest Service.

    Site visits by others:
    2 June 1978 —E. LaVerne Smith; photographs
    30 Aug. 1978 —Ben A. Sanders, E. LaVerne Smith, and others; reconnaissance and study of population status at "Campground Site" in vicinity of Table Rock.
  D. Knowledgeable individuals—Information from the following individuals was considered and reviewed in the preparation of this status report:
  Dr. Judith T. Skog
  Dept. Biology
  George Mason College of the
      Univ. Virginia
  Fairfax, Va. 22030

  Dr. John A. Churchill
  6851 Castle Drive
  Birmingham, Mich. 48010

  Dr. Carroll E. Wood, Jr.
  Arnold Arboretum
  Harvard University
  Cambridge, Mass. 02138

R. D. Whetstone
Dept. Botany
Univ. North Carolina
Chapel Hill, N.C.

M. P. Lee
28 Eastwood Road
Biltmore, N.C. 28803

As well as the following, for whom addresses are provided under "Interested Parties":

Ruby Harbison
James W. Hardin
Norton H. Nickerson
Levester Pendegrass
J. Dan Pitillo
Albert E. Radford
Ben A. Sanders
E. LaVerne Smith

E. Other information sources—
Materials on file were consulted at:

Office of Endangered Species
U.S. Fish and Wildlife Service, Washington, D.C.

Endangered Flora Program
Smithsonian Institution
Washington, D.C.

National Forests in North Carolina
U.S. Forest Service
Asheville, N.C.

8. Summary of materials on file—Nearly all items cited or referred to in this report, as well as author's field notes, photos, and population status reports.

V. Authorship

9. Initial authorship
Larry E. Morse*
Research Associate
New York Botanical Garden
Bronx, New York 10458
(212-220-8700)

---

* Present address: The Nature Conservancy, 1800 N. Kent St., Arlington, Va. 22209.

20. **Maintenance of status report**—Should this species be listed as an Endangered or a Threatened Species by the U.S. Fish and Wildlife Service, the Service, through its Office of Endangered Species in Washington, should maintain the primary file of information on it, encourage others to provide it new information, and distribute new findings, as received, to the others on the current version of the list of interested parties, above.

Should the species be determined to be neither endangered nor threatened, the files might be transferred to an appropriate agency within the State of North Carolina.

V. New Information

21. **Record of revisions to this status report**—*For each batch of revisions made to this status report, append a record of the date of distribution or publication, the name of the person making the revisions, and the topic(s) affected.*

A. **8 Nov. 1979**, Topics 1.A.12.c, 1.A.2,1.E, 2.A.2,2.B.1.a, 3,6.C.7, 7.A, 7.B., 7.D.4.b, 7.D.5, 8.B, 8.C.1, 8.C.3, 10.A.3, 10.B, 13.B.1, 14.A, 14.D, 17.D, all revised by Larry E. Morse, with additional minor revisions to conform with revised edition of Henifin *et al.*'s guidelines.

Literature Cited

Ackermann, E. A. 1941. *The Köppen Classification of Climate in North America.* Harvard Univ. Press, Cambridge, Mass.

Ayensu, E. S., and R. A. DeFilipps. 1978. *Endangered and Threatened Plants of the United States.* Smithsonian Institution and World Wildlife Fund, Inc., Washington, D.C.

Bailey, R. G. 1976. *Ecoregions of the United States.* U.S. Forest Service, Ogden, Utah.

The L. H. Bailey Hortorium. 1976. *Hortus Third.* Macmillan, New York.

Bryant, B., and J. C. Reed, Jr. 1970. *Geology of the Grandfather Mountain Window and Vicinity, North Carolina and Tennessee.* [U.S.] Geol. Surv. Prof. Pap. no. 615. 190 pp.

Davies, P. A. 1955. Distribution and abundance of *Shortia galacifolia. Rhodora* 57: 189-201.

Fenn, D. B., G. J. Gogue, and R. E. Burge. 1976. Effects of campfires on soil properties. *Ecol. Serv. Bull.* [U.S. Nat. Park Serv.] no. 5. 16 pp.

Graustein, J. E. 1967. *Thomas Nuttall, Naturalist.* Harvard Univ. Press, Cambridge, Mass.

Hardin, J. L., R. L. Kologiski, J. R. Massey, J. F. Matthews, J. D. Pitillo, and A. E. Radford. 1977. Vascular plants, pp. 56-142 in *Endangered and Threatened Plants and Animals of North Carolina,* ed. J. E. Cooper, S. S. Robinson, and J. B. Funderburg, N. C. St. Mus. Nat. Hist., Raleigh, N.C.

Henry, T. J. 1978. Description of a new *Polymerus,* with notes on two other little-known Mirids from the New Jersey Pine-Barrens (Hemiptera: Miridae). *Proc. Entomol. Soc. Wash.* 80: 543-547.

Heritage Conservation and Recreation Service [HCRS]. 1978. *Draft Natural Heritage Classification Systems.* [144 pp.]

Hunt, C. B. 1967. *Physiography of the United States.* Freeman, San Francisco.

Justice, W. S., and C. R. Bell. 1968. *Wild Flowers of North Carolina.* Univ. N. C. Press, Chapel Hill, N. C.

Lucas, G., and H. Synge. 1979. *The IUCN Plant Red Data Book.* International Union for Conservation of Nature and Natural Resources, Morges, Switzerland.

McDonnell, M. J. 1979. *The vascular flora of Plum Island, Essex County, Massachusetts, with an analysis of the impact of human trampling on the coastal dune vegetation.* M. S. Thesis, Dept. Botany and Plant Path., Univ. New Hampshire, Durham, N.H. 199 pp.

Morse, L. E. 1978. Studies of mixed-species populations of *Hudsonia* on Cape Cod, Massachusetts. *Ohio J. Sci.* 78 (suppl.): 15. (Abstract only).

_____. 1979. *Systematics and ecological biogeography of the genus* Hudsonia *(Cistaceae), the sand heathers.* Ph.D. thesis, Dept. Biology, Harvard Univ., Cambridge, Mass. 274 pp.

Narciso, J., and R. P. Collins. 1979. The beach heather community on Cape Cod. *New England Wild Fl. Notes* (Spring 1979), pp. 1-2.

Nuttall, T. 1818. *The Genera of North American Plants.* Philadelphia. 2 vols.

Pennell, F. W. 1936. Travels and scientific publications of Thomas Nuttall. *Bartonia* 18: 1-51.

Radford, A. E. 1976. *Vegetation, Habitats, Floras, Natural Areas in the Southeastern United States: Field Data and Information.* Univ. N. C. Student Stores, Chapel Hill, N. C.

Radford, A. E., H. E. Ahles, and C. R. Bell. 1968. *Manual of the Vascular Flora of the Carolinas.* Univ. N. C. Press, Chapel Hill.

Rickett, H. W. 1967. *Wildflowers of the United States. Vol. 2. The Southeast.* McGraw-Hill, New York.

Rydberg, P. A. 1926. Botanizing in the higher Allegheny Mountains. I. West Virginia. *J. New York Bot. Gard.* 27: 1-6.

Schwartz, G. 1977. *The Climate Advisor.* Climate Guide Publications, Flushing, New York.

Shetler, S. G., and L. E. Skog. 1978. *A Provisional Checklist of Species for Flora North America (Revised).* Missouri Botanical Garden, St. Louis. 199 pp.

Skog, J. T., and N. H. Nickerson. 1972. Variation and speciation in the genus *Hudsonia. Ann. Missouri Bot. Gard.* 59: 454-464.

Small, J. K. 1895. Review: Synoptical Flora of North America. *Bull. Torrey Bot. Club* 22: 472-475.

Smithsonian Institution. 1975. *Report on Endangered and Threatened Plant Species of the United States.* House Document no. 94-51, Committee on Merchant Marine and Fisheries, Washington.

U. S. Fish and Wildlife Service. 1975. Review of status of over 3000 vascular plants and determination of "critical habitat." *Federal Register* 40(127): 27824-27924 (1 July 1975).

_____. 1976. Proposed endangered status for some 1700 U.S. vascular plant taxa. *Federal Register* 41(117): 24524-24572 (16 June 1976).

Note added in proof. On May 29, 1980, the U.S. Fish and Wildlife Service proposed that *Hudsonia montana* be listed as a threatened species (*Federal Register* 45(105): 36332-36335).

Hudsonia montana was determined to be a threatened species, with critical habitat designated, on Oct. 20, 1980 (*Federal Register* 45 (204): 69360-63363).

Planning Field Work on Rare or Endangered Plant Populations

Mary Sue Henifin,[1] Larry E. Morse,[2] Steve Griffith,[3] and Janet E. Hohn[4]

Current site-specific field information is essential to informed management of local populations of rare endangered plant species, as well as to meaningful synthesis of information on rarity and threat throughout a plant species' range. Since information needs and interests of field workers differ, no single data-collection form can suffice to solicit pertinent information efficiently for every purpose. Instead, we offer suggestions for developing data-collection procedures for individual situations. One level of information, such as location, plant identification, and land ownership, needs to be collected only once per site, and revised only if the information changes. Another category of data should be recorded on repeat visits; among these latter topics are population counts, phenological observations, and assessment of threats. The information we recommend collecting is selected to support development of species-wide status reports.

Current field information is the major basis for identifying which plant taxa are rare or endangered, and making up-to-date decisions on their needs for protection and management. Since the information needed to make these decisions differs from that obtained in traditional biological studies, care must be taken to plan field work in ways that assure pertinent information is obtained. Such planning is especially important for projects such as contract field inventories of federal lands, or field efforts of state native plant societies or Natural Heritage Programs.

Comprehensive *status reports,* such as those following the guidelines presented by Henifin *et al.* (this volume), are necessary for synthesizing information and making decisions on a species-wide basis. Such status reports incorporate all pertinent information available from the literature and from museum collections as well as from past and current field studies. The *field report,* on the other hand, concentrates on one local population, organizing field observations pertaining to this site. Species-level status reports depend on these field reports for significant parts of their conclusions regarding distribution, rarity, threat patterns, and management recommendations.

[1] New York Botanical Garden, Bronx, N.Y. 10458; present address: Div. Environmental Sciences, Columbia University School of Public Health, New York, N.Y. 10032
[2] New York Botanical Garden, Bronx, N.Y. 10458; present address: The Nature Conservancy, 1800 N. Kent St., Arlington, Va. 22209.
[3] 1561 Sheldon Drive, Denver, Colo. 80229.
[4] Endangered Species Office, U.S. Fish and Wildlife Service, 500 N.E. Multnomah St., Portland, Ore. 97232

Pages 309-312 *in:* Larry E. Morse and Mary Sue Henifin, eds., *Rare Plant Conservation: Geographical Data Organization for,* The New York Botanical Garden, Bronx, N.Y.

For more widespread species, it would usually be impractical and unjustifiable to gather such detailed information on every population. However, for rare or endangered species such detail is often possible, since these plants usually have only a few populations, and a limited geographical distribution. Current field information of this kind is also essential to the conservation of disjunct, peripheral, or otherwise interesting populations of more widespread species.

When organizing field information on rare or endangered plant taxa, there are several important topics to consider. No particular data-collection format can serve all users efficiently, because specific needs differ depending on the group to be using the information, the personnel doing the fieldwork, and the nature of the terrain being studied. For example, different types of habitat information may be more important in different geographical areas, or federal agencies may need more specific data on threats to determine appropriate land-management actions, or native plant societies or similar groups may want field forms that are easy for amateurs to use. Rather than a specific data-collection format, we offer the following ideas and suggestions for use in developing forms and procedures for particular cases.

Five major kinds of information should be recorded and documented in a field report form:

1. Exact locality information for known populations of rare or endangered plant taxa.

2. Status of extant populations, and threats to them or their habitats.

3. Ecological observations necessary to the delimitation of population areas and critical habitats.

4. Management actions recommended for the conservation of the population and its habitat.

5. Types of habitats and potential localities to be searched.

Exact locality information is perhaps the most important field data that can be generated. Care should be taken to develop field data forms that encourage the recording of specific locality information. Requiring a map with the location of the plant population marked as well as range-township-section and/or latitude/longitude will help insure that specific locality information is obtained.

Habitat information on each population is also of great help in determining what other areas may possibly support occurrences of the species. The habitat of each population can be synthesized to indicate the habitat type or types where the taxon may be found, making it possible to concentrate future inventory work in habitats where the species is most likely to occur.

In doing field work, there are two types of information to consider: those data that only need to be gathered once because they remain static (for example exact locality, elevation, and soil type) and information which changes over time. Distinguishing between these two different types of information allows more efficient use of time in the field.

Assessing trends in rare or endangered plant populations is perhaps the most difficult task for the field botanist. Determination of population trends must be based on information gathered over time. By comparing population numbers on a yearly basis one can begin to determine the impact of threats and whether a population is increasing or decreasing. Often field personnel change from field season to field season. Also, different field personnel may gather information on the species in different segments of its range. For these reasons it is important for all field personnel to gather information in a standard manner that will allow across-time and across-geographical-range comparisons. It is not enough to state that a plant taxon is "abundant" at a particular location or "rare." Exact numbers of the taxon for that particular population or an estimate of numbers (with the method used for making the estimate indicated) should be given as well as the area the population covers. This information is essential for documenting population trends.

Observations on the population biology of a species may be particularly important in determining appropriate management actions for its conservation. Documenting a plant population's response to particular management actions over time can help land managers refine their actions to aid the recovery of threatened or endangered species.

Vouchering of information is particularly important. *In most cases, collection of specimens of threatened and endangered plant species must be strongly discouraged;* photographs or other methods of documentation should be used. A description of field methodologies can also help others evaluate the accuracy of reported information.

Keeping the above points in mind, it should be possible to design field report forms which will generate valuable data. Such data will greatly aid attempts to learn more about the biology of rare and endangered plant species and to determine management actions necessary to ensure their conservation.

We suggest the information categories in the Appendix be included in such field report forms.

Appendix I.
Suggested Information Categories for Field Reports

General Information
>Name, address, and phone number of observer
>Scientific name of plant taxon
>Date(s) of field information
>Indication of whether or not a voucher specimen, fragment, and/or photograph obtained, and where deposited or available.

Information to be collected only once for a given plant population, unless specified details change

Geographical information
>Location searched (Give state, county, nearest landmark, nearest road or trail; attach map or map map sketch to report, indicating type and scale of map.)
>Date Searched
>Presence or absence of taxon
>Key characters verified; reference work(s) or specialist(s) consulted
>Number of populations at locality (Indicate each on map.)
>Ownership of land; include owner's address if private
>Management activities affecting area, past and present
>Recommended critical habitat (Note in detail on map.)

General habitat description for each population
>Topography (relief, elevation, geologic formation, slope, exposure, etc.)
>Soils (texture, parent material, bedrock type, thickness of litter, structure and porosity, nutrient status, etc.)
>Microclimate and microhabitat characteristics (localized conditions influencing distribution of individuals)
>Vegetation type (*e.g.,* coniferous forest, grassland, alpine, riverbottom)
>Associated plants

Dependence on disturbance, successional stage, or other phenomena (*e.g.,* tree falls, fire, insect invasions)

Hydrology

Other information, where not previous presented

Other persons who know about locality and its management or natural history

Other known or possible locations for the taxon

Information on propagation techniques, horticultural knowledge, or stored seed

Any known ecological, evolutionary, or human significance for this taxon

Conservation, recovery, and monitoring measures recommended

Population biology information to be recorded each time a population is visited

Number of individuals in each population and area population covers

Size/age classes present (define terms used) and number (or percentage) of individuals in each class

Phenology—Estimate of percentage for each population in vegetative, flowering, fruiting, or senescent stages

Number of seedlings observed

Observations relating to survivorship and nature of mortality at each life stage (predation on seedlings, competition, etc.)

Evidence of herbivores, predators, diseases, and/or pests

Evidence of disturbance by exotic plants, animals, microbes

Human impacts observed or suspected (trampling, damage by vehicles, wild flower collecting, etc.)

Other threatening factors and their severity (land development, grazing, etc.) Give both existing and potential threats

Assessment of the vigor and status of individuals and the population

Also, general information recorded earlier should be reviewed and revised as necessary on each site visit.

Other useful population biology information that may be recorded each time population is visited

Types of reproduction noted (seeds, vegetative, etc.)

Pollinators (wind, water, insect, etc.) Distinguish between those visiting plant and other suspected pollinators

Observations on seed dispersal (general mechanisms, dispersal patterns, amount of seed, germination requirements)

Seedling ecology (morphology, microhabitat, localized conditions restricting establishment)

Other species of this genus at or near site, and hybrids observed, if any

Evidence of symbiotic or parasitic relationships

Response of taxon to disturbance

The 1978 and 1979 Amendments to the Endangered Species Act: A Discussion*

U.S. Fish and Wildlife Service

The 1978 Amendments

On Friday, November 10th [1978] President Carter signed "The Endangered Species Act Amendments of 1978," reauthorizing administration of the Endangered Species Act of 1973 and, among other things, establishing a cabinet-level committee authorized to exempt Federal agencies from compliance with some of the Act's protective provisions.

(Section 7 of the Act requires all Federal agencies to insure that their actions do not jeopardize the continued existence of Endangered or Threatened species, or result in the adverse modification of their Critical Habitats.)

Although congratulating Congress for working hard to resolve this difficult issue, the President expressed some misgivings in approving the compromise approach to handling irresolvable conflicts under the section 7 mandate. "While I believe that this new exemption process is not necessary, I hope that as the committee carries out its responsibilities, it will make the utmost efforts to protect the existence of the species inhabiting this planet."

The President emphasized his belief that the Act has worked without such an exemption process "because all agencies have made efforts to resolve conflicts and, where necessary, to pursue alternate courses of action. This consultation and cooperation should continue under these new amendments, minimizing the number of requests for exemptions." Upon signing the bill, Carter directed committee members to be "exceedingly cautious in considering exemptions," and asked that national security exemptions be exercised "only in grave circumstances posing a clear and immediate threat to national security." In the words of the President, "Destruction of the life of an endangered or threatened species should never be undertaken lightly, no matter how insignificant the species may appear today."

Congress Sought More Flexibility. President Carter's approval followed a flurry of congressional activity during which House and Senate conferees worked through

---

*Reprinted by permission of the U.S. Fish and Wildlife Service from the October 1978 and January 1980 issues of the *Endangered Species Technical Bulletin* (GPO number 281-326); these articles were originally entitled "President Signs Endangered Species Amendments" and "Endangered Species Act Amended and Extended." The editor, Dona Finnley, included the following note with the first article: "Although we would like to present a detailed analysis of the 1978 amendments, Public Law 95-632 has brought many changes which have not yet been subjected to legal interpretation. We have therefore refrained from premature attempts to explain new or revised provisions. Detailed definitions of new terms, procedures for implementing the amendments, and other pertinent interpretations will be provided in the form of solicitors' opinions or proposed regulations, as appropriate, at the earliest possible time."

the last hours of the 95th Congress to hammer out the compromise bill. Motivated primarily by the recent Supreme Court ruling upholding the applicability of the 1973 Act to the nearly completed Tellico dam, a number of members of Congress believed the legislation should be changed to provide for human and economic as well as biological considerations in resolving conflicts under section 7. The Senate on July 19 approved a bill to create a special committee to consider exemptions for Federal actions, while on October 14 the House voted out yet a separate administrative mechanism to rule on conflicts (creating not only a committee, but also a review board to determine the appropriateness of exemption applications).

In submitting its final report on H.R. 14104, subsequent to oversight hearings on the 1973 legislation, the House Committee on Merchant Marine and Fisheries stated that ''. . . the evidence developed at these hearings suggests that the consultation process can resolve many if not most of the conflicts that might develop under the Act. . . . It is clear, nevertheless, that there will continue to be some federally authorized activities which cannot be modified in a manner which will avoid a conflict with a listed species.'' The report concluded that ''the bill attempts to retain the basic integrity of the Endangered Species Act, while introducing some flexibility which will permit exemptions from the Act's stringent requirements.''

As discussed in the following sections, the amendments not only provide for an exacting, two-tiered review process to consider exemptions under section 7, they also affect the consultation process, listing, Critical Habitat determinations, cooperative agreements with the States, enforcement and penalties, recovery planning, captive-held raptors, and public hearing/notice procedures.

Consultation Process Method. Report language from both the House and Senate indicates that their intent in the wording of the new section 7(a) was not to diminish

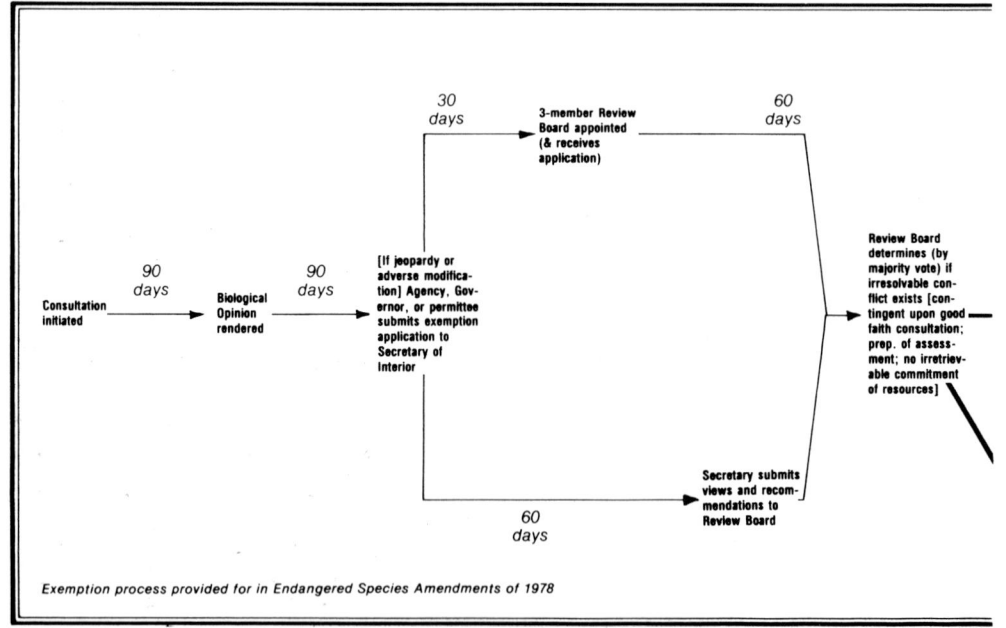

Exemption process provided for in Endangered Species Amendments of 1978

the agencies' mandate to consult with the Secretary, but rather to strengthen the consultation process. According to the new provisions, consultation is to be concluded within 90 days (formerly 60 days by Service regulations) after initiated, or at a time agreed to by the Secretary and the involved agency.

A new section 7(b) requires the Secretary's biological opinion (rendered at the conclusion of consultation) to detail how the agency's action would affect the listed species or its Critical Habitat, and to suggest "reasonable and prudent alternatives" that would avoid jeopardy to the species or adverse modification of its Critical Habitat.

Under a new section 7(c), each Federal agency is now required—with respect to actions for which no contract for construction has been entered into and no construction has begun on the date of enactment of the amendments—to request information from the Secretary regarding the presence of any listed or proposed species within the area of the proposed action. If such species are present, then the agency must prepare a biological assessment within 180 days (or a time mutually agreed to by the agency and the Secretary) identifying species likely to be affected by its action.

Once consultation has been initiated, the amendments stipulate that no irreversible or irretrievable commitment of resources may be made by the agency which forecloses the implementation of alternative measures to avoid jeopardy or adverse impacts on the species or its Critical Habitat.

The Exemption Process. Following consultation, the amendments provide for an elaborate review process through which Federal agencies (and permit or license applicants) may be exempted from the requirements of section 7. Should the Service's biological opinion result in a finding of jeopardy to the species or modification of its

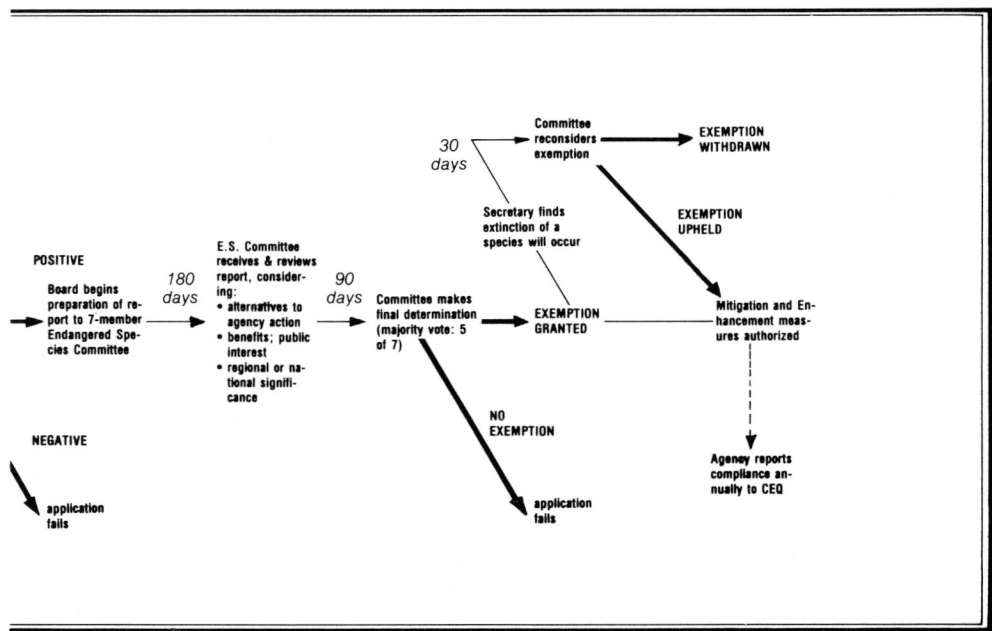

Critical Habitat, the involved agency, the Governor of the State in which the action was to occur, or the permit or license applicant may—within 90 days of issuance of the biological opinion—submit an application to the Secretary of the Interior requesting exemption from the Act's protective requirements.

As outlined in the accompanying flowchart, the first step in the exemption process is the formulation of a Review Board. A board is to be established upon the receipt of the exemption application, to consist of one member appointed by the Secretary of the Interior within 15 days, one member from the affected State to be appointed by the President within 30 days, after consideration of any recommendations by the Governor(s) of the affected State(s), and an Administrative Law Judge selected by the Civil Service Commission.

The Review Board must then consider the application within 60 days after its appointment, making a full review of the consultation carried out and determining, by majority vote, (1) whether an irresolvable conflict exists and (2) whether the exemption applicant has:

● carried out its consultation responsibilities in good faith and has made "a reasonable and responsible effort to develop and fairly consider modifications or reasonable and prudent alternatives to the proposed agency action" which would avoid jeopardy to the species or adverse modification to its Critical Habitat;
● conducted a biological assessment, if required; and
● refrained from making an irreversible or irretrievable commitment of resources.

The Secretary of the Interior is also called upon to submit to the board his views and recommendations concerning the exemption request within 60 days after his receipt of the application.

Any finding by the board that the applicant has failed to meet any of the exemption criteria given above will be considered final action under the Administrative Procedures Act.

If, however, the Review Board makes a positive finding concerning the applicant's eligibility, then the board will proceed to prepare a report for the cabinet-level Endangered Species Committee, to be presented within 180 days following the board's findings. As provided under section 7(g)(7), the report must address the following issues:

"(A) the availability of reasonable and prudent alternatives to the agency action, and the nature and extent of the benefits of the agency action and of alternative courses of action consistent with conserving the species or the critical habitat;

"(B) a summary of the evidence concerning whether or not the agency action is in the public interest and is of national or regional significance;

"(C) appropriate reasonable mitigation and enhancement measures which should be considered by the Committee."

The Endangered Species Committee, which is to make a final decision on whether or not to exempt a Federal agency action from the requirements of section 7(a) is to consist of the following seven members:

The Secretary of the Interior (as Chairman)
The Secretary of the Agriculture
The Secretary of the Army
The Chairman of the Council of Economic Advisors
The Administrator of the Environmental Protection Agency
The Administrator of the National Oceanic and Atmospheric Administration

A State representative, as appointed by the President after consideration of recommendations from the Governor(s) of the affected State(s).

Five members of the Committee must be present to constitute a quorum, and the Committee shall meet at the call of the Chairman or five of its members. An exemption may be granted by the Committee, by majority vote of at least five of its members (voting in person), if it determines within 90 days after receipt of the Review Board's report that:

"(i) there are no reasonable and prudent alternatives to the agency action;

"(ii) the benefits of such action clearly outweigh the benefits of alternative courses of action consistent with conserving the species or its critical habitat, and such action is in the public interest; and

"(iii) the action is of regional or national significance. . . .''

At the time of an exemption determination, the Committee must also establish "reasonable mitigation and enhancement measures, including, but not limited to, live propagation, transplantation, and habitat acquisition and improvement, as are necessary and appropriate to minimize the adverse effects of the agency action upon the endangered species, threatened species, or critical habitat concerned."

Once granted, an exemption shall be considered permanent with respect to all Endangered and Threatened species for the purposes of completing an agency action, if any required biological assessment has been conducted. If, however, the Secretary of the Interior finds that the project or action would result in the extinction of the species, the exemption shall not be permanent, and the Committee must reconsider the exemption (within 30 days following the Secretary's finding) and determine whether to uphold the exemption order.

Three exceptions to the exemption review process are provided for:

● the Secretary of State may prohibit exemption consideration for actions that would violate an international treaty obligations of the United States (by submitting such findings to the Committee in writing within 60 days after the receipt of an exemption application).

● the Secretary of Defense can exempt actions from the provisions of section 7 if he finds that the actions are necessary for national defense.

● the President may grant exemptions for declared major disaster areas.

The Commitee's final decision is subject to judicial review, and any person wishing to appeal may bring such action to the U.S. Court of Appeals.

Once an exemption order is granted, the applicant shall pay for and carry out any mitigation and enhancement measures specified by the Committee. The applicant must also submit an annual report to the Council on Environmental Quality, describing its compliance with the prescribed mitigation and enhancement measures. (These reports will then be published by CEQ in the *Federal Register*.)

To insure implementation of the new exemption process, the amendments also call for the promulgation of regulations by the Secretary ". . . which set forth the form and manner in which applications for exemption shall be submitted to the Secretary and the information to be contained in such applications."

Tellico and Grayrocks to be Reviewed. Two projects, Tennessee Valley Authority's Tellico dam and the Grayrocks Dam and Reservoir Project, have been pinpointed for early review. Within 30 days of enactment of the amendments, the Endangered Species Commitee is to proceed to consider the exemption of the two projects (the latter proposed, and the former nearly completed but enjoined by the Supreme Court ruling) from the provisions of section 7.

The Committee is directed to exempt the projects—within 90 days after enactment of the amendments—if it determines that (1) there are no reasonable and prudent alternatives to the projects and (2) the benefits of the projects outweigh the benefits of alternative courses of action consistent with conserving the affected species or their Critical Habitat, and the projects are in the public interest. If no decision is made by the Committee within the 90-day period, the projects shall be automatically exempted.

New Definitions, Listing Requirements.   Several new definitions are provided in the amendments. A "species" which may be considered for protection under the Act is now limited to ". . . any subspecies of fish or wildlife or plant, and any distinct population segment of any species of *vertebrate* fish or wildlife which interbreeds when mature." (emphasis added)

Critical Habitat has been defined for the first time, revising the Service's definition (by regulation) to include: "the specific areas within the geographical area occupied by the species at the time it is listed . . . on which are found those physical or biological features (I) essential to the conservation of the species and (II) which may require special management consideration or protection; and . . . specific areas outside the geographical area . . . upon a determination by the Secretary that such areas are essential for the conservation of the species."

Economic Impact to be Considered.   An amendment to section 4(b) now requires the Secretary to consider the economic impact of specifying any particular area as Critical Habitat. The language reads:

"(4) In determining the critical habitat of any endangered or threatened species, the Secretary shall consider the economic impact, and any other relevant impacts of specifying any particular area as critical habitat, and he may exclude any such area from the critical habitat if he determines that the benefits of such exclusion outweigh the benefits of specifying the area as part of the critical habitat, unless he determines, based on the best scientific and commercial data available, that the failure to designate such area as critical habitat will result in the extinction of the species."

A number of other provisions have been incorporated in section 4. To "the maximum extent prudent," Critical Habitat must now be determined at the time a species is listed. A more involved public notification process is now required prior to the listing of species or Critical Habitat determination. The amendments specify three additional steps to be taken during the listing process:
- notice to local government units whose boundaries include or are adjacent to the proposed Critical Habitat
- publication of the substance of proposed regulations in affected area newspapers
- publication of the substance of regulations in appropriate scientific journals.

A public hearing must now be held before designation of Critical Habitat. Also, where a species is listed but no Critical Habitat is to be determined, a public meeting shall be held when requested.

All listings and Critical Habitat determinations proposed after enactment of the amendments must be finalized within two years or be withdrawn. (Listings already proposed must be finalized within one year of the date of enactment.) The amendments also call for the periodic review—at least once every five years—of listed species.

Finally, to the maximum extent practicable, the Secretary must include in the proposed and final Critical Habitat rulings a description of activities which may adversely modify the habitat or which may be affected by the designation.

Agreements to Cover Plants.   The amendments, under section 6(c), now permit the Service to enter into cooperative agreements with the States for the conservation of Endangered and Threatened plants.

The use of Land and Water Conservation funds for the acquisition of habitat for Endangered or Threatened plants has also been authorized (under Section 5(a)).

Recovery Planning.   Under a new subsection 4(g), the Secretary is directed to develop and implement recovery plans for the conservation and survival of all Endangered or Threatened species, ''. . . unless he finds that such a plan will not promote the conservation of the species.'' (The services of public and private agencies may be utilized for this purpose.)

Penalties/Enforcement.   With regard to civil penalties, section 11(a) has been amended to make violations by commercial importers or exporters of fish, wildlife, and plants a liability offense subject to a fine of $10,000. (Violators are now subject to such penalties without a ''knowledge requirement.'') The maximum fine for noncommercial offenders under the strict liability provisions has been lowered from $1,000 to $500.

Under the criminal penalty provisions, ''knowingly'' replaces ''willfully.'' Heads of Federal agencies are now authorized to modify, suspend, or revoke permits or licenses to import or export animals or plants or to operate quarantine stations for any person convicted of a criminal violation of the Act.

In both the criminal and civil enforcement sections, the requirement that a person ''commit an act'' has been removed, thereby allowing prosecution of offenses of omission.

Finally, persons taking Endangered or Threatened species on the good faith belief that they were acting to protect themselves or others from bodily harm are protected from civil and criminal penalties under the Act.

Exceptions for Raptors, Antiquities.   Captive-held raptors and their progeny may be exempted from the Act's permit requirements, if held on the effective date of the amendments. (The Secretary is authorized to require documents, records, inventories, and other proof of eligibility.)

Also, antique articles (except scrimshaw) made from parts of products of listed species before 1830 are now exempted from the Act's provisions. The article must not have been modified with any part of a listed species after December 28, 1973, must be accompanied by appropriate documentation, and must be imported through a designated port of entry. Individuals wishing to reclaim such articles confiscated since enactment of the 1973 Act may apply for the return of their items within one year of the date of enactment of the 1978 amendments.

Appropriations Reauthorized.   Administration of the Endangered Species Act of 1973 by the Departments of Interior and Commerce is once again official, with appropriations now authorized for another year and a half. For Fiscal Year 1979, the Secretary of the Interior may utilize up to $23,000,000 in carrying out his responsibilities under the Endangered Species Act, while up to $12,500,000 is authorized for the 6-month period ending March 31, 1980.

Also, to assist the review boards and the Endangered Species Committee in carrying out their functions, the Secretary is authorized an additional $900,000 for the same 18-month period.

The 1979 Amendments

On December 28, 1979, President Carter signed into law, for the second consecutive year, substantial amendments to the Endangered Species Act of 1973. Pointing to reauthorization of the Act (for a 3-year period) as one of his highest legislative priorities, the President also approved revisions to the listing, Section 7 consultation, and exemption provisions under the Act, and—perhaps more significantly—the creation of a new commission to advise on scientific policy under the Convention on International Trade in Endangered Species of Wild Fauna and Flora.

Calling the Act "one of the most far-reaching and progressive laws ever enacted by any nation to protect wildlife and plant resources," the President said, "I look forward to and will continue to support the implementation of a vigorous endangered species program."

Listing and related provisions. Among other things, Public Law 96-159, in the words of President Carter, "strengthens our endangered species protection program by including plant as well as animal species in the emergency listing and international cooperation provisions."

Other revisions follow:

● A summary of proposed regulations (rather than the complete text) and, where applicable, a map of the proposed Critical Habitat, must be published in local newspapers within or adjacent to the habitat.

● Public meetings and hearings on Critical Habitat proposals are to be held separately (with a hearing to be held if requested within 15 days of a public meeting).

● The time period for which emergency listing and Critical Habitat designations are effective (now applicable to both animals and plants) has been extended from 120 to 240 days.

● A new provision requires the development and notice (with opportunity for public comment) of guidelines for the handling of petitions for listing, for priority systems for listing, and for priority systems for developing and implementing recovery plans.

● A "status review" is now required prior to the preparation of proposals for listing.

● Foreign nations—with the help of Department personnel—are encouraged under a new subsection to develop programs for the conservation of listed plants.

Section 7 Consultations/Exemptions. Necessary changes in language have been made throughout the Act to revise the jeopardy standard under Section 7 from "would jeopardize" to "is likely to jeopardize."

320

Other new provisions:

● All Federal agencies are required to "confer" with the Secretary on any action likely to jeopardize a proposed species. (The intent is for agencies to begin informal discussions at an early stage). No "irreversible or irretrievable commitment of resources" requirement is imposed for *proposed* species impacts.

● Biological assessments (as required prior to filing for an exemption from Section 7) must be conducted in cooperation with the Secretary and under the supervision of the appropriate Federal agency. (Completion of an adequate assessment then qualifies the applicant for a possible permanent exemption under the Act.)

● An exemption application from a permit or license applicant must be filed within 90 days of final agency action (such as permit denial, which may follow issuance of a biological opinion).

● With regard to exemption applications initiated subsequent to the issuance of negative biological opinions by both the Secretary of Interior and Commerce for the same agency action (such as that involving sea turtles, for which jurisdiction is shared), the two Departments will jointly convene a review board.

● Regarding exemptions under Section 7, threshold requirements shall apply to both the Federal agency and the exemption applicant (i.e., the applicant must carry out all consultation requirements, conduct any necessary biological assessment, and refrain from making a commitment of resources in order to qualify for exemption consideration, regardless of the applicant's identity.)

● An exemption granted by the Endangered Species Committee shall be permanent with respect to all listed species regardless of whether the species was included in the biological assessment (and only if the assessment was conducted), *unless* a listed species not so identified will become extinct. (In this case, the Committee, which must meet within 30 days of such a finding by the Secretary, has 60 additional days to determine permanence.)

New Commission/Scientific Authority. Under a new section, the Secretary of the Interior (acting through the Fish and Wildlife Service) has been designated as both the U.S. Management Authority and the U.S. Scientific Authority for purposes of the Convention on International Trade in Endangered Species of Wild Fauna and Flora (CITES). (Director Greenwalt has placed the Scientific Authority function under the Service's Associate Director for Research, while the Management Authority function will continue under the Associate Director for Federal Assistance.)

While abolishing the existing Endangered Species Scientific Authority (ESSA) —established previously under Executive Order 11911 as the U.S. Scientific Authority—within 90 days of enactment, the new law also creates an independent International Convention Advisory Commission (ICAC) to advise on scientific policy under CITES. Similar in structure to the existing ESSA, ICAC is to be composed of at least six (and possibly seven) members: with one member each appointed by:

—the Secretary of the Interior
—the Secretary of Agriculture
—the Secretary of Commerce
—the Director of the National Science Foundation.

One member (to serve a 2-year term) shall also be appointed by the Secretary of the Interior from among officers and employees of State fish and wildlife agencies, and the Secretary of the Smithsonian Institution is invited to appoint a seventh member.

A Chairman is to be elected annually by the members. All members must be scientifically qualified.

Speaking of the new Commission upon signing the 1979 amendments, President Carter said, "that scientific integrity of the Convention will be preserved by the Commission's advice on the effects of trade, the listing of species on Convention appendices, and the interpretation and implementation of the Convention." Under the new provisions, the Commission will make recommendations (by majority vote) on all matters pertaining to the responsibilities of the Scientific Authority under the terms of the Convention.

As set forth in the new law, the new Commission will "to the extent practicable, ascertain the views of, and utilize the expertise of, the governmental and nongovernmental scientific communities, State agencies responsible for the conservation of wild fauna and flora, humane groups, zoological and botanical institutions, recreational and commercial interests, the conservation community, and others as appropriate" in discharging its responsibilities.

The public will have an opportunity to comment on all Management and Scientific Authority decisions, and the Scientific Authority must also provide, upon publication of final notices, an explanation of its reasons for any decision not consistent with the Commission's recommendations.

Until such time as the Chairman and members are appointed (or no longer than 90 days after enactment), the current ESSA will carry out the functions of the newly-created Commission.

Appropriations. Reauthorizing administration of the Endangered Species Act for an additional three years, the 1979 amendments allow appropriations to implement the Act's provisions (except as authorized under Section 6 and as discussed below for portions of Section 7) not to exceed the following amounts:

Interior:	$23,000,000 for FY 1980
	25,000,000 for FY 1981
	27,000,000 for FY 1982
Commerce:	$ 2,500,000 for FY 1980
	3,000,000 for FY 1981
	3,500,000 for FY 1982

Additional appropriations of $600,000 are also authorized to support Endangered Species Committee and review board functions (under Section 7) for each of Fiscal Years 1980, 1981, and 1982.

For the first time, the amendments also authorize appropriations for the Department of Agriculture to facilitate enforcement of the Act and the Convention with regard to the importation and exportation of terrestrial plants. Amounts not to exceed $1,500,000 for FY 1980, $1,750,000 for FY 1981, and $1,850,000 for FY 1982 have been allowed.

Scrimshaw. Finally, the deadline for the sale of scrimshaw has been extended under the amendments. Section 10(f) now provides for one last renewal of certificates of exemption, allowing the sale for three more years *only* of whale parts and products held in stock prior to 1973.

322

The Endangered Species Act of 1973*

An Act

To provide for the conservation of endangered and threatened species of fish, wildlife, and plants, and for other purposes.

Be it enacted, by the Senate and House of Representatives of the United States of America in Congress assembled. That this Act may be cited as the "Endangered Species Act of 1973."

Table of Contents

---

* P.L. 93-205 (87 Stat. 884), including the amendments of:

June 30, 1976	P.L. 94-325 (90 Stat. 724)
July 12, 1976	P.L. 94-359 (90 Stat. 911)
December 19, 1977	P.L. 95-212 (91 Stat. 1493)
November 10, 1978	P.L. 95-632 (92 Stat. 3751)

This unofficial reprint of the Endangered Species Act of 1973, as amended, was edited by the Office of Legislative Services, U.S. Fish and Wildlife Service, December, 1978. It does not include the December, 1979, amendments (P.L. 96-159).

SEC. 2. (a) FINDINGS.—The Congress finds and declares that—

(1) various species of fish, wildlife, and plants in the United States have been rendered extinct as a consequence of economic growth and development untempered by adequate concern and conservation;

(2) other species of fish, wildlife, and plants have been so depleted in numbers that they are in danger of or threatened with extinction;

(3) these species of fish, wildlife, and plants are of esthetic, ecological, educational, historical, recreational, and scientific value to the Nation and its people;

(4) the United States has pledged itself as a sovereign state in the international community to conserve to the extent practicable the various species of fish or wildlife and plants facing extinction, pursuant to—

(A) migratory bird treaties with Canada and Mexico;

(B) the Migratory and Endangered Bird Treaty with Japan;

(C) the Convention on Nature Protection and Wildlife Preservation in the Western Hemisphere;

(D) the International Convention for the Northwest Atlantic Fisheries;

(E) the International Convention for the High Seas Fisheries of the North Pacific Ocean;

(F) the Convention on International Trade in Endangered Species of Wild Fauna and Flora; and

(G) other international agreements.

(5) encouraging the States and other interested parties, through Federal financial assistance and a system of incentives, to develop and maintain conservation programs which meet national and international standards is a key to meeting the Nation's international commitments and to better safeguarding, for the benefit of all citizens, the Nation's heritage in fish and wildlife.

(b) PURPOSES.—The purposes of this Act are to provide a means whereby the ecosystems upon which endangered species and threatened species depend may be conserved, to provide a program for the conservation of such endangered species and threatened species, and to take such steps as may be appropriate to achieve the purposes of the treaties and conventions set forth in subsection (a) of this section.

(c) POLICY.—It is further declared to be the policy of Congress that all Federal departments and agencies shall seek to conserve endangered species and threatened species and shall utilize their authorities in furtherance of the purposes of this Act.

SEC. 3. For the purposes of this Act—

(1) The term "alternative courses of action" means all alternatives and thus is not limited to original project objectives and agency jurisdiction.

(2) The term "commercial activity" means all activities of industry and trade, including, but not limited to, the buying or selling of commodities and activities conducted for the purpose of facilitating such buying and selling: *Provided, however,* That it does not include exhibition of commodities by museums or similar cultural or historical organizations.

(3) The terms "conserve", "conserving", and "conservation" mean to use and the use of all methods and procedures which are necessary to bring any endangered species or threatened species to the point at which the measures provided pursuant to this Act are no longer necessary. Such methods and procedures include, but are not limited to, all activities associated with scientific resources management such as research, census, law enforcement, habitat acquisition and maintenance, propagation, live trapping, and transplantation, and, in the extraordinary case where population pressures within a given ecosystem cannot be otherwise relieved, may include regulated taking.

(4) The term "Convention" means the Convention on International Trade in Endangered Species of Wild Fauna and Flora, signed on March 3, 1973, and the appendices thereto.

(5) (A) The term "critical habitat" for a threatened or endangered species means—

(i) the specific areas within the geographical area occupied by the species, at the time it is listed in accordance with the provisions of section 4 of this Act, on which are found those physical or biological features (I) essential to the conservation of the species and (II) which may require special management considerations or protection; and

(ii) specific areas outside the geographical area occupied by the species at the time it is listed in accordance with the provisions of section 4 of this Act, upon a determination by the Secretary that such areas are essential for the conservation of the species.

(B) Critical habitat may be established for those species now listed as threatened or endangered species for which no critical habit has heretofore been established as set forth in subparagraph (A) of this paragraph.

(C) Except in those circumstances determined by the Secretary, critical habit shall not include the entire geographical area which can be occupied by the threatened or endangered species.

(6) The term "endangered species" means any species which is in danger of extinction throughout all or a significant portion of its range other than a species of the Class Insecta determined

by the Secretary to constitute a pest whose protection under the provisions of this Act would present an overwhelming and overriding risk to man.

(7) The term "Federal agency" means any department, agency, or instrumentality of the United States.

(8) The term "fish or wildlife" means any member of the animal kingdom, including without limitation any mammal, fish, bird (including any migratory, nonmigratory, or endangered bird for which protection is also afforded by treaty or other international agreement), amphibian, reptile, mollusk, crustacean, arthropod or other invertebrate, and includes any part, product, egg, or offspring thereof, or the dead body or parts thereof.

(9) The term "foreign commerce" includes, among other things, any transaction—

(A) between persons within one foreign country;

(B) between persons in two or more foreign countries;

(C) between a person within the United States and a person in a foreign country; or

(D) between persons within the United States, where the fish and wildlife in question are moving in any country or countries outside the United States.

(10) The term "import" means to land on, bring into, or introduce into, or attempt to land on, bring into, or introduce into, any place subject to the jurisdiction of the United States, whether or not such landing, bringing, or introduction constitutes an importation within the meaning of the customs laws of the United States.

(11) The term "irresolvable conflict" means, with respect to any action authorized, funded, or carried out by a Federal agency, a set of circumstances under which, after consultation as required in section 7(a) of this Act, completion of such action would (A) jeopardize the continued existence of an endangered or threatened species, or (B) result in the adverse modification or destruction of a critical habitat.

(12) The term "permit or license applicant" means, when used with respect to an action of a Federal agency for which exemption is sought under section 7, any person whose application to such agency for a permit or license has been denied primarily because of the application of section 7(a) to such agency action.

(13) The term "person" means an individual, corporation, partnership, trust, association, or any other private entity, or any officer, employee, agent, department, or instrumentality of the Federal Government, of any State or political subdivision thereof, or of any foreign government.

(14) The term "plant" means any member of the plant kingdom, including seeds, roots and other parts thereof.

(15) The term "Secretary" means, except as otherwise herein provided, the Secretary of the Interior or the Secretary of Commerce as program responsibilities are vested pursuant to the provisions of Reorganization Plan Numbered 4 of 1970; except that with respect to the enforcement of the provisions of this Act and the Convention which pertain to the importation or exportation of terrestrial plants, the term means the Secretary of Agriculture.

(16) The term "species" includes any subspecies of fish or wildlife or plants, and any distinct population segment of any species of vertebrate fish or wildlife which interbreeds when mature.

(17) The term "State" means any of the several States, the District of Columbia, the Commonwealth of Puerto Rico, American Samoa, the Virgin Islands, Guam, and the Trust Territory of the Pacific Islands.

(18) the term "State agency" means any State agency, department, board, commission, or other governmental entity which is responsible for the management and conservation of fish, plant, or wildlife resources within a State.

(19) The term "take" means to harass, harm, pursue, hunt, shoot, wound, kill, trap, capture, or collect, or to attempt to engage in any such conduct.

(20) The term "threatened species" means any species which is likely to become an endangered species within the foreseeable future throughout all or a significant portion of its range.

(21) The term "United States", when used in a geographical context, includes all States.

DETERMINATION OF ENDANGERED SPECIES AND THREATENED SPECIES

SEC. 4. (a) GENERAL.—(1) The Secretary shall by regulation determine whether any species is an endangered species or a threatened species because of any of the following factors:

(1) the present or threatened destruction, modification, or curtailment of its habitat or range;

(2) overutilization for commercial, sporting, scientific, or educational purposes;

(3) disease or predation;

(4) the inadequacy of existing regulatory mechanisms; or

(5) other natural or manmade factors affecting its continued existence.

At the time any such regulation is proposed, the Secretary shall also by regulation, to the maximum extent prudent, specify any habitat of such species which is then considered to be critical habitat. The requirement of the preceding sentence shall not apply with respect to any species which was listed prior to enactment of the Endangered Species Act Amendments of 1978.

(2) With respect to any species over which program responsibilities have been vested in the Secretary of Commerce pursuant to Reorganization Plan Numbered 4 of 1970—

(A) in any case in which the Secretary of Commerce determines that such species should—

(i) be listed as an endangered species or a threatened species, or

(ii) be changed in status from a threatened species to an endangered species,

he shall so inform the Secretary of the Interior, who shall list such species in accordance with this section;

(B) in any case in which the Secretary of Commerce determines that such species should—

(i) be removed from any list published pursuant to subsection (c) of this section, or

(ii) be changed in status from an endangered species to a threatened species,

he shall recommend such action to the Secretary of the Interior, and the Secretary of the Interior, if he concurs in the recommendation, shall implement such action; and

(C) the Secretary of the Interior may not list or remove from any list any such species, and may not change the status of any such species which are listed, without a prior favorable determination made pursuant to this section by the Secretary of Commerce.

(b) BASIS FOR DETERMINATIONS.—(1) The Secretary shall make determinations required by subsection (a) of this section on the basis of the best scientific and commercial data available to him and after consultation, as appropriate, with the affected States, interested persons and organizations, other interested Federal agencies, and, in cooperation with the Secretary of State, with the country or countries in which the species concerned is normally found or whose citizens harvest such species on the high seas; except that in any case in which such determinations involve resident species of fish or wildlife, the Secretary of the Interior may not add such species to, or remove such species from, any list published pursuant to subsection (c) of this section, unless the Secretary has first—

(A) published notice in the Federal Register and notified the Governor of each State within which such species is then known to occur that such action is contemplated;

(B) allowed each such State 90 days after notification to submit its comments and recommendations, except to the extent that such period may be shortened by agreement between the Secretary and the Governor or Governors concerned; and

(C) published in the Federal Register a summary of all comments and recommendations received by him which relate to such proposed action.

(2) In determining whether or not any species is an endangered species or a threatened species, the Secretary shall take into consideration those efforts, if any, being made by any nation or any political subdivision of any nation to protect such species, whether by predator control, protection of habitat and food supply, or other conservation practices, within any area under the jurisdiction of any such nation or political subdivision, or on the high seas.

(3) Species which have been designated as requiring protection from unrestricted commerce by any foreign country, or pursuant to any international agreement, shall receive full consideration by the Secretary to determine whether each is an endangered species or a threatened species.

(4) In determining the critical habitat of any endangered or threatened species, the Secretary shall consider the economic impact, and any other relevant impacts, of specifying any particular area as critical habitat, and he may exclude any such area from the critical habitat if he determines that the benefits of such exclusion outweigh the benefits of specifying the area as part of the critical habitat, unless he determines, based on the best scientific and commercial data available, that the failure to designate such area as critical habitat will result in the extinction of the species.

(c) LISTS.—(1) The Secretary of the Interior shall publish in the Federal Register, and from time to time he may by regulation revise, a list of all species determined by him or the Secretary of Commerce to be endangered species and a list of all species determined by him or the Secretary of Commerce to be threatened species. Each list shall refer to the species contained therein by scientific and common name or names, if any, and shall specify with respect to each such species over what portion of its range it is endangered or threatened and specify any critical habitat within such range.

(2) The Secretary shall, within 90 days of the receipt of the petition of an interested person under subsection 553(e) of title 5, United States Code, conduct and publish in the Federal Register a review of the status of any listed or unlisted species proposed to be removed from or added to either of the lists published pursuant to paragraph (1) of this subsection, but only if he makes and publishes a finding that such person has presented substantial evidence which in his judgment warrants such a review. Such review and finding shall be made and published prior to the initiation of any procedures under subsection (b)(1).

(3) Any list in effect on the day before the date of the enactment of this Act of species of fish or wildlife determined by the Secretary of the Interior, pursuant to the Endangered Species Conservation Act of 1969 to be threatened with extinction shall be republished to conform to the classification for endangered species or threatened species, as the case may be, provided for in this Act, but until such republication, any such species so listed

shall be deemed an endangered species within the meaning of this Act. The republication of any species pursuant to this paragraph shall not require public hearing or comment under section 553 of title 5, United States Code.

(4) The Secretary shall—

(A) conduct, at least once every five years, a review of all species included in a list which is published pursuant to paragraph (1) and which is in effect at the time of such review; and

(B) determine on the basis of such review whether any such species should—

(i) be removed from such list;

(ii) be changed in status from an endangered species to a threatened species; or

(iii) be changed in status from a threatened species to an endangered species.

Each determination under subparagraph (B) shall be made in accordance with the provisions of subsections (a) and (b).

(d) PROTECTIVE REGULATIONS.—Whenever any species is listed as a threatened species pursuant to subsection (c) of this section, the Secretary shall issue such regulations as he deems necessary and advisable to provide for the conservation of such species. The Secretary may by regulation prohibit with respect to any threatened species any act prohibited under section 9(a)(1), in the case of fish or wildlife, or section 9(a)(2), in the case of plants, with respect to endangered species; except that with respect to the taking of resident species of fish or wildlife, such regulations shall apply in any State which has entered into a cooperative agreement pursuant to section 6(a) of this Act only to the extent that such regulations have also been adopted by such State.

(e) SIMILARITY OF APPEARANCE CASES.—The Secretary may, by regulation, and to the extent he deems advisable, treat any species as an endangered species or threatened species even though it is not listed pursuant to section 4 of this Act if he finds that—

(A) such species so closely resembles in appearance, at the point in question, a species which has been listed pursuant to such section that enforcement personnel would have substantial difficulty in attempting to differentiate between the listed and unlisted species;

(B) the effect of this substantial difficulty is an additional threat to an endangered or threatened species; and

(C) such treatment of an unlisted species will substantially facilitate the enforcement and further the policy of this Act.

(f) REGULATIONS.—(1) Except as provided in paragraphs (2) and (3) of this subsection and subsection (b) of this section, the provisions of section 553 of title 5, United States Code (relating to rulemaking procedures), shall apply to any regulation promulgated to carry out the purposes of this Act.

(2) (A) Except as provided in subparagraph (B), in the case of any regulation proposed by the Secretary to carry out the purposes of this Act—

(i) the Secretary shall publish general notice of the proposed regulation (including the complete text of the regulation) in the Federal Register not less than 60 days before the effective date of the regulation; and

(ii) if any person who feels that he may be adversely affected by the proposed regulation files (within 45 days after the date of publication of general notice) objections thereto and requests a public hearing thereon, the Secretary may grant such request, but shall, if he denies such request, publish his reasons therefore in the Federal Register.

(B) In the case of any regulation proposed by the Secretary to carry out the purposes of this section with respect to the determination and listing of endangered or threatened species and their critical habitats in any State (other than regulations to implement the Convention), the Secretary—

(i) shall publish general notice of the proposed regulation (including the complete text of the regulation), not less than 60 days before the effective date of the regulation—

(I) in the Federal Register, and

(II) if the proposed regulation specifies any critical habitat, in a newspaper of general circulation within or adjacent to such habitat;

(ii) shall offer for publication in appropriate scientific journals the substance of the Federal Register notice referred to in clause (i)(I);

(iii) shall give actual notice of the proposed regulation (including the complete text of the regulation), and any environmental assessment or environmental impact statement prepared on the proposed regulation, not less than 60 days before the effective date of the regulation to all general local governments located within or adjacent to the proposed critical habitat, if any; and

(iv) shall—

(I) if the proposed regulation does not specify any critical habitat, promptly hold a public meeting on the proposed regulation within or adjacent to the area in which the endangered or threatened species is located, if request therefore is filed with the Secretary by any person within 45 days after the date of publication of general notice under clause (i)(I), and

(II) if the proposed regulation specifies any critical habitat, promptly hold a public meeting on the proposed regulation within the area in which such habitat is located in each State, and, if requested, hold a public hearing in each such State.

If a public meeting or hearing is held on any regulation, the regulation may not take effect before the 60th day after the date on which the meeting or hearing is concluded and if more than one public meeting or hearing is held, before the 60th day after the date on which the last such meeting or hearing is concluded. Any accidental failure to provide actual notice under clause (ii) to all general local governments required to be given notice shall not invalidate the proposed regulation.

(C) Neither subparagraph (A) or (B) of this paragraph nor section 553 of title 5, United States Code, shall apply in the case of any of the following regulations and any such regulation shall, at the discretion of the Secretary, take effect immediately upon publication of the regulation in the Federal Register:

(i) Any regulation appropriate to carry out the purposes of this Act which was originally promulgated to carry out the Endangered Species Conservation Act of 1969.

(ii) Any regulation (including any regulation implementing section 6(g)(2)(B)(ii) of this Act) issued by the Secretary in regard to any emergency posing a significant risk to the well-being of any species of fish or wildlife, but only if (I) at the time of publication of the regulation in the Federal Register the Secretary publishes therein detailed reasons why such regulation is necessary, and (II) in the case such regulation applies to resident species of fish and wildlife, the requirements of subsection (b) (A), (B), and (C) of this section have been complied with. Any regulation promulgated under the authority of this clause (ii) shall cease to have force and effect at the close of the 120-day period following the date of publication unless, during such 120-day period, the rulemaking procedures which would apply to such regulation without regard to this subparagraph are complied with.

(3) The publication in the Federal Register of any proposed or final regulation which is necessary or appropriate to carry out the purposes of this Act shall include a summary by the Secretary of the data on which such regulation is based and shall show the relationship of such data to such regulations.

(4) Any proposed or final regulation which specifies any critical habitat of any endangered species or threatened species shall be based on the best scientific data available, and the publication in the Federal Register of any such regulation shall, to the maximum extent practicable, be accompanied by a brief description and evaluation of those activities (whether public or private) which, in the opinion of the Secretary, if undertaken may adversely modify such habitat, or may be impacted by such designation.

(5) A final regulation adding a species to any list published pursuant to subsection (c) shall be published in the Federal Register not later than two years after the date of publication of notice of the regulation proposing such listing under paragraph (B)(i)(I). If a

final regulation is not adopted within such two-year period, the Secretary shall withdraw the proposed regulation and shall publish notice of such withdrawal in the Federal Register not later than 30 days after the end of such period. The Secretary shall not propose a regulation adding to such a list any species for which a proposed regulation has been withdrawn under this paragraph unless he determines that sufficient new information is available to warrant the proposal of a regulation. No proposed regulation for the listing of any species published before the date of the enactment of the Endangered Species Act Amendments of 1978 shall be withdrawn under this paragraph before the end of the one-year period beginning on such date of enactment.

(g) RECOVERY PLANS.—The Secretary shall develop and implement plans (hereinafter in this subsection referred to as "recovery plans") for the conservation and survival of endangered species and threatened species listed pursuant to this section, unless he finds that such a plan will not promote the conservation of the species. The Secretary, in developing and implementing recovery plans, may procure the services of appropriate public and private agencies and institutions, and other qualified persons. Recovery teams appointed pursuant to this subsection shall not be subject to the Federal Advisory Committee Act.

LAND ACQUISITION

SEC. 5. (a) PROGRAM.—The Secretary, and the Secretary of Agriculture with respect to the National Forest System, shall establish and implement a program to conserve fish, wildlife, and plants, including those which are listed as endangered species or threatened species pursuant to section 4 of this Act. To carry out such a program, the appropriate Secretary—

(1) shall utilize the land acquisition and other authority under the Fish and Wildlife Act of 1956, as amended, the Fish and Wildlife Coordination Act, as amended, and the Migratory Bird Conservation Act, as appropriate; and

(2) is authorized to acquire by purchase, donation, or otherwise, lands, waters, or interest therein, and such authority shall be in addition to any other land acquisition authority vested in him.

(b) ACQUISITIONS.—Funds made available pursuant to the Land and Water Conservation Fund Act of 1965, as amended, may be used for the purpose of acquiring lands, waters, or interests therein under subsection (a) of this section.

COOPERATION WITH THE STATES

SEC. 6. (a) GENERAL.—In carrying out the program authorized by this Act, the Secretary shall cooperate to the maximum extent practicable with the States. Such cooperation shall include consultation with the States concerned before acquiring any land or water, or interest therein, for the purpose of conserving any endangered species or threatened species.

(b) MANAGEMENT AGREEMENTS.—The Secretary may enter into agreements with any State for the administration and management of any area established for the conservation of endangered species or threatened species. Any revenues derived from the administration of such areas under these agreements shall be subject to the provisions of section 401 of the Act of June 15, 1935 (49 Stat. 383; 16 U.S.C. 715s).

(c) COOPERATIVE AGREEMENTS.—(1) In furtherance of the purposes of this Act, the Secretary is authorized to enter into a cooperative agreement in accordance with this section with any State which establishes and maintains an adequate and active program for the conservation of endangered species and threatened species. Within one hundred and twenty days after the Secretary receives a certified copy of such a proposed State program, he shall make a determination whether such program is in accordance with this Act. Unless he determines, pursuant to this paragraph, that the State program is not in accordance with this Act, he shall enter into a cooperative agreement with the State for the purpose of assisting in implementation of the State program. In order for a State program to be deemed an adequate and active program for the conservation of endangered species and threatened species, the Secretary must find, and annually thereafter reconfirm such finding, that under the State program—

(A) authority resides in the State agency to conserve resident species of fish or wildlife determined by the State agency or the Secretary to be endangered or threatened;

(B) the State agency has established acceptable conservation programs, consistent with the purposes and policies of this Act, for all resident species of fish or wildlife in the State which are deemed by the Secretary to be endangered or threatened, and has furnished a copy of such plan and program together with all pertinent details, information, and data requested to the Secretary;

(C) the State agency is authorized to conduct investigations to determine the status and requirements for survival of resident species of fish and wildlife;

(D) the State agency is authorized to establish programs, including the acquisition of land or aquatic habitat or interests therein, for the conservation of resident endangered or threatened species of fish or wildlife;

(E) provision is made for public participation in designating resident species of fish or wildlife as endangered or threatened or that under the State program—

(i) the requirements set forth in subparagraph (C), (D), and (E) of this paragraph are complied with, and

(ii) plans are included under which immediate attention will be given to those resident species of fish and wildlife which are determined by the Secretary or the State agency to be endangered or threatened and which the Secretary and the State agency agree are most urgently in need of conservation programs; except that a cooperative agreement entered into with a State whose program is deemed adequate and active pursuant to clause (i) and this clause shall not affect the applicability of prohibitions set forth in or authorized pursuant to section 4(d) or section 9(a)(1) with respect to the taking of any resident endangered or threatened species.

(2) In furtherance of the purposes of this Act, the Secretary is authorized to enter into a cooperative agreement in accordance with this section with any State which establishes and maintains an adequate and active program for the conservation of endangered species and threatened species of plants. Within one hundred and twenty days after the Secretary receives a certified copy of such a proposed State program, he shall make a determination whether such program is in accordance with this Act. Unless he determines, pursuant to this paragraph, that the State program is not in accordance with this Act, he shall enter into a cooperative agreement with the State for the purpose of assisting in implementation of the State program. In order for a State program to be deemed an adequate and active program for the conservation of endangered species of plants and threatened species of plants, the Secretary must find, and annually thereafter reconfirm such finding, that under the State program—

(A) authority resides in the State agency to conserve resident species of plants determined by the State agency or the Secretary to be endangered or threatened;

(B) the State agency has established acceptable conservation programs, consistent with the purposes and policies of this Act, for all resident species of plants in the State which are deemed by the Secretary to be endangered or threatened, and has furnished a copy of such plan and program together with all pertinent details, information, and data requested to the Secretary;

(C) the State agency is authorized to conduct investigations to determine the status and requirements for survival of resident species of plants; and

(D) provision is made for public participation in designating resident species of plants as endangered or threatened; or
that under the State program—

(i) the requirements set forth in subparagraphs (C) and (D) of this paragraph are complied with, and

(ii) plans are included under which immediate attention will be given to those resident species of plants which are determined by the Secretary of the State agency to be endangered or threatened and which the Secretary and the State agency agree are most urgently in need of conservation programs; except that a cooperative agreement entered into with a State whose program is deemed adequate and active pursuant to clause (i) and this clause shall not affect the applicability of prohibitions set forth in a authorized pursuant to section 4(d) or section 9(a)(1) with respect to the taking of any resident engangered or threatened species.

(d) ALLOCATION OF FUNDS.—(1) The Secretary is authorized to provide financial assistance to any State, through its respective State agency, which has entered into a cooperative agreement pursuant to subsection (c) of this section to assist in development of programs for the conservation of endangered and threatened species. The Secretary shall make an allocation of appropriated funds to such States based on consideration of—

(A) the international commitments of the United States to protect endangered species or threatened species;

(B) the readiness of a State to proceed with a conservation program consistent with the objectives and purposes of this Act;

(C) the number of endangered species and threatened species within a State;

(D) the potential for restoring endangered species and threatened species within a State; and

(E) the relative urgency to initiate a program to restore and protect an endangered species or threatened species in terms of survival of the species.

So much of any appropriated funds allocated for obligation to any State for any fiscal year as remains unobligated at the close thereof is authorized to be made available to that State until the close of the succeeding fiscal year. Any amount allocated to any State which is unobligated at the end of the period during which it is available for expenditure is authorized to be made available for expenditure by the Secretary in conducting programs under this section.

(2) Such cooperative agreements shall provide for (A) the actions to be taken by the Secretary and the States; (B) the benefits that are expected to be derived in connection with the conservation of endangered or threatened species; (C) the estimated cost of these actions; and (D) the share of such costs to be borne by the Federal Government and by the States; except that—

(i) the Federal share of such program costs shall not exceed 66 2/3 per centum of the estimated program cost stated in the agreement; and

(ii) the Federal share may be increased to 75 per centum whenever two or more States having a common interest in one or more endangered or threatened species, the conservation of which may be enhanced by cooperation of such States, enter jointly into an agreement with the Secretary.

The Secretary may, in his discretion, and under such rules and regulations as he may prescribe, advance funds to the State for financing the United States pro rata share agreed upon in the cooperative agreement. For the purposes of this section, the non-Federal share may, in the discretion of the Secretary, be in the form of money or real property, the value of which will be determined by the Secretary, whose decision shall be final.

(e) REVIEW OR STATE PROGRAMS.—Any action taken by the Secretary under this section shall be subject to his periodic review at no greater than annual intervals.

(f) CONFLICTS BETWEEN FEDERAL AND STATE LAWS.—Any State law or regulation which applies with respect to the importation or exportation of, or interstate or foreign commerce in, endangered species or threatened species is void to the extent that it may effectively (1) permit what is prohibited by this Act or by any regulation which implements this Act, or (2) prohibit what is authorized pursuant to an exemption or permit provided for in this Act or in any regulation which implements this Act. This Act shall not otherwise be construed to void any State law or regulation which is intended to conserve migratory, resident, or introduced fish or wildlife, or to permit or prohibit sale of such fish or wildlife. Any State law or regulation respecting the taking of an endangered species or threatened species may be more restrictive than the exemptions or permits provided for in this Act or in any regulation which implements this Act but not less restrictive than the prohibitions so defined.

(g) TRANSITION.—(1) For purposes of this subsection, the term "establishment period" means, with respect to any State, the period beginning on the date of enactment of this Act and ending on whichever of the following dates first occurs: (A) the date of the close of the 120-day period following the adjournment of the first regular session of the legislature of such State which commences after such date of enactment, or (B) the date of the close of the 15-month period following such date of enactment.

(2) The prohibitions set forth in or authorized pursuant to sections 4(d) and 9(a)(1)(B) of this Act shall not apply with respect to the taking of any resident endangered species or threatened species (other than species listed in Appendix I to the Convention or otherwise specifically covered by any other treaty or Federal law) within any State—

(A) which is then a party to a cooperative agreement with the Secretary pursuant to section 6(c) of this Act (except to the extent that the taking of any such species is contrary to the law of such State); or

(B) except for any time within the establishment period when—

(i) the Secretary applies such prohibition to such species at the request of the State, or

(ii) the Secretary applies such prohibition after he finds, and publishes his finding, that an emergency exists posing a significant risk to the well-being of such species and that the prohibition must be applied to protect such species. The Secretary's finding and publication may be made without regard to the public hearing or comment provisions of section 553 of title 5, United States Code, or any other provision of this Act; but such prohibition shall expire 90 days after the date of its imposition unless the Secretary further extends such prohibition by publishing notice and a statement of justification of such extension.

(h) REGULATIONS.—The Secretary is authorized to promulgate such regulations as may be appropriate to carry out the provisions of this section relating to financial assistance to States.

(i) APPROPRIATIONS.—For the purposes of this section, there are authorized to be appropriated not to exceed the following sums:

(1) $10,000,000 through the period ending September 30, 1977.

(2) $16,000,000 for the period beginning October 1, 1977, and ending September 30, 1981.

INTERAGENCY COOPERATION

SEC. 7 (a) CONSULTATION.—The Secretary shall review other programs administered by him and utilize such programs in furtherance of the purposes of this Act. All other Federal agencies shall, in consultation with and with the assistance of the Secretary, utilize their authorities in furtherance of the purposes of this Act by carrying out programs for the conservation of endangered species and threatened species listed pursuant to section 4 of this Act. Each Federal agency shall, in consultation with and with the assistance of the Secretary, insure that any action authorized, funded, or carried out by such agency (hereinafter in this section referred to as an "agency action") does not jeopardize the continued existence of any endangered species or threatened species or result in the destruction or adverse modification of habitat of such species which is determined by the Secretary, after consultation as appropriate with the affected States, to be critical, unless such agency has been granted an exemption for such action by the Committee pursuant to subsection (h) of this section.

(b) SECRETARY'S OPINION.—Consultation under subsection (a) with respect to any agency action shall be concluded within 90 days after the date on which initiated or within such other period of time as is mutually agreeable to the Federal agency and the Secretary. Promptly after the conclusion of consultation, the Secretary shall provide to the Federal agency concerned a written statement setting forth the Secretary's opinion, and a summary of the information on which the opinion is based, detailing how the agency action affects the species or its critical habitat. The Secretary shall suggest those reasonable and prudent alnteratives which he believes would avoid jeopardizing the continued existence of any endangered or threatened species or adversely modifying the critical habitat of such species, and which can be taken by the Federal agency or the permit or license applicant in implementing the agency action.

(c) BIOLOGICAL ASSESSMENT.—To facilitate compliance with the requirements of subsection (a), each Federal agency shall, with respect to any agency action of such agency for which no contract for construction has been entered into and for which no construction has begun on the date of enactment of the Endangered Species Act Amendments of 1978, request of the Secretary information whether any species which is listed or proposed to be listed may be present in the area of such proposed action. If the Secretary advises, based on the best scientific and commercial data available, that such species may be present, such agency shall conduct a biological assessment for the purpose of identifying any endangered species or threatened species which is likely to be affected by such action. Such assessment shall be completed within 180 days after the date on which initiated (or within such other period as is mutually agreed to by the Secretary and such agency) and, before any contract for construction is entered into and before construction is begun with respect to such action. Such assessment may be undertaken as part of a Federal agency's compliance with the requirements of section 102 of the National Environmental Policy Act of 1969 (42 U.S.C. 4332).

(d) LIMITATION ON COMMITMENT OF RESOURCES.— After initiation of consultation required under subsection (a), the Federal agency and the permit or license applicant shall not make any irreversible or irretrievable commitment of resources with respect to the agency action which has the effect of foreclosing the formulation or implementation of any reasonable and prudent alternative measures which would avoid jeopardizing the continued existence of any endangered or threatened species or adversely modifying or destroying the critical habitat of any such species.

(e)(1) ESTABLISHMENT OF COMMITTEE.—There is established a committee to be known as the Endangered Species Committee (hereinafter in this section referred to as the "Committee").

(2) The Committee shall review any application submitted to it pursuant to this section and determine in accordance with subsection (h) of this section whether or not to grant an exemption from the requirements of subsection (a) of this section for the action set forth in such application.

(3) The Committee shall be composed of seven members as follows:

(A) The Secretary of Agriculture.
(B) The Secretary of the Army.
(C) The Chairman of the Council of Economic Advisors.

(D) The Administrator of the Environmental Protection Agency.

(E) The Secretary of the Interior.

(F) The Administrator of the National Oceanic and Atmospheric Administration.

(G) The President, after consideration of any recommendations received pursuant to subsection (g)(2)(B) shall appoint one individual from each affected State, as determined by the Secretary, to be a member of the Committee for the consideration of the application for exemption for an agency action with respect to which such recommendations are made, not later than 30 days after an application is submitted pursuant to this section.

(4) (A) Members of the Committee shall receive no additional pay on account of their service on the Committee.

(B) While away from their homes or regular places of business in the performance of services for the Committee, members of the Committee shall be allowed travel expenses, including per diem in lieu of subsistence, in the same manner as persons employed intermittently in the Government service are allowed expenses under section 5703 of title 5 of the United States Code.

(5) (A) Five members of the Committee or their representatives shall constitute a quorum for the transaction of any function of the Committee, except that, in no case shall any representative be considered in determining the existence of a quorum for the transaction of any function of the Committee if that function involves a vote by the Committee on any matter before the Committee.

(B) The Secretary of the Interior shall be the Chairman of the Committee.

(C) The Committee shall meet at the call of the Chairman or five of its members.

(D) All meetings and records of the Committee shall be open to the public.

(6) Upon request of the Committee, the head of any Federal agency is authorized to detail, on a nonreimbursable basis, any of the personnel of such agency to the Committee to assist it in carrying out its duties under this section.

(7) (A) The Committee may for the purpose of carrying out its duties under this section hold such hearings, sit and act at such times and places, take such testimony, and receive such evidence, as the Committee deems advisable.

(B) When so authorized by the Committee, any member or agent of the Committee may take any action which the Committee is authorized to take by this paragraph.

(C) Subject to the Privacy Act, the Committee may secure directly from any Federal agency information necessary to enable it to carry out its duties under this section. Upon request of the Chairman of the Committee, the head of such Federal agency shall furnish such information to the Committee.

(D) The Committee may use the United States mails in the same manner and upon the same conditions as a Federal agency.

(E) The Administrator of General Services shall provide to the Committee on a reimbursable basis such administrative support services as the Committee may request.

(8) In carrying out its duties under this section, the Committee may promulgate and amend such rules, regulations, and procedures, and issue and amend such orders as it deems necessary.

(9) For the purpose of obtaining information necessary for the consideration of an application for an exemption under this section the Committee may issue subpoenas for the attendance and testimony of witnesses and the production of relevant papers, books, and documents.

(10) Except in the case of a member designated pursuant to paragraph (3)(G) of this subsection, no member shall designate any person to serve as his or her representative unless that person is, at the time of such designation, holding a Federal office the appointment to which is subject to the advice and consent of the United States Senate. In no case shall any representative, including a representative of a member designated pursuant to paragraph (3)(G) of this subsection, be eligible to cast a vote on behalf of any member.

(f) REGULATIONS.—Not later than 90 days after the date of enactment of the Endangered Species Act Amendments of 1978, the Secretary shall promulgate regulations which set forth the form and manner in which applications for exemption shall be submitted to the Secretary and the information to be contained in such applications. Such regulations shall require that information submitted in an application by the head of any Federal agency with respect to any agency action include, but not be limited to—

(1) a description of the consultation process carried out pursuant to subsection (a) of this section between the head of the Federal agency and the Secretary; and

(2) a statement describing why such action cannot be altered or modified to conform with the requirements of subsection (a) of this section.

(g) APPLICATION FOR EXEMPTION AND CONSIDERATION BY REVIEW BOARD.—(1) A Federal agency, the Governor of the State in which an agency action will occur, if any, or a permit or license applicant may apply to the Secretary for an exemption for an agency action of such agency if, after consultation under subsection (a), the Secretary's opinion under subsection (b) indicates that the agency action may jeopardize the continued existence of any endangered or threatened species or destroy or adversely modify the critical habitat of such species. An application for an exemption shall be considered initially by a review board in the manner provided in this subsection, and shall be considered by the Endangered Species Committee for a final determination under subsection (h) after a report is made by the review board. The applicant for an exemption shall be referred to as the 'exemption applicant' in this section.

(2) (A) An exemption applicant shall submit a written application to the Secretary, in a form prescribed under subsection (f) of this section, not later than 90 days after the completion of the consultation process. Such application shall set forth the reasons why the exemption applicant considers that the agency action meets the requirements for an exemption under this subsection.

(B) Upon receipt of an application for exemption for an agency action under paragraph (1), the Secretary shall promptly notify the Governor of each affected State, if any, as determined by the Secretary, and request the Governors so notified to recommend individuals to be appointed to the review board to be established under paragraph (3) and to the Endangered Species Committee for consideration of such application.

(3) (A) A review board shall be established for purposes of considering an application for exemption and submitting a report to the Endangered Species Committee under this subsection as follows:

(i) One individual shall be appointed to the board by the Secretary not later than 15 days after an application is submitted pursuant to paragraph (2).

(ii) One individual shall be appointed to the board by the President, not later than 30 days after an application is submitted pursuant to paragraph (2) and after consideration of any recommendations received pursuant to paragraph (2)(B). An individual appointed by the President under this subparagraph shall be a resident of a State, if any, in which the agency action will be, or is being, carried out.

(iii) One administrative law judge shall be selected to serve on the board by the Civil Service Commission in the same manner as administrative law judges are selected under section 3344 of title 5 of the United States Code to be detailed to an agency which occasionally or temporarily is insufficiently staffed with administrative law judges. The use by the review board of such an administrative law judge shall be on a reimbursable basis.

(B) Members of a review board who are full-time officers or employees of the United States shall receive no additional pay on account of their service on the board. All other members shall be entitled to receive an amount not to exceed the daily equivalent of the annual rate of basic pay in effect for grade GS-18 of the General Schedule for each day during which they are engaged in the actual performance of duties vested in the board. While away from their homes or regular places of business in the performance of services for a review board, members of the board shall be allowed travel expenses, including per diem in lieu of subsistence, in the same manner as persons employed intermittently in the Government service are allowed expenses under section 5703 of title 5 of the United States Code.

(4) The Secretary shall submit the applicant to the review board immediately after its appointment under paragraph (3), and the Secretary shall submit to the review board, in writing, his views and recommendations with respect to the application within 60 days after receiving a copy of any application under paragraph (2).

(5) It shall be the duty of a review board appointed under paragraph (3) to make a full review of the consultation carried out under subsection (a), and within 60 days after its appointment or within such longer time as is mutually agreed upon between the exemption applicant and the Secretary, to make a determination, by a majority vote, (1) whether an irresolvable conflict exists and (2) whether such exemption applicant has—

(A) carried out its consultation responsibilities as described in subsection (a) in good faith and made a reasonable and responsible effort to develop and fairly consider modifications or reasonable and prudent alternatives to the proposed agency action which will avoid jeopardizing the continued existence of an endangered or threatened species or result in the adverse modification or destruction of a critical habitat;

(B) conducted any biological assessment required of it by subsection(c); and

(C) refrained from making any irreversible or irretrievable commitment of resources prohibited by subsection (d).

Any determination by the review board that an irresolvable conflict does not exist or that the exemption applicant has not met the requirements of subparagraph (A), (B), or (C) shall be considered final agency action for purposes of chapter 7 of title 5 of the United States Code.

(6) If the review board determines that an irresolvable conflict exists and makes positive determinations under subparagraphs (A), (B), and (C) of paragraph (5), it shall proceed to prepare the report to be submitted under paragraph (7).

(7) Within 180 days after making the determinations under paragraph (6), the review board shall submit to the Committee a report discussing—

(A) the availability of reasonable and prudent alternatives to the agency action, and the nature and extent of the benefits of the agency action and of alternative courses of action consistent with conserving the species or the critical habitat;

(B) a summary of the evidence concerning whether or not the agency action is in the public interest and is of national or regional significance;

(C) appropriate reasonable mitigation and enhancement measures which should be considered by the Committee.

(8) To the extent practicable within the time required for action under subsection (g) of this section, and except to the extent

inconsistent with the requirements of this section, the consideration of any application for an exemption under this section and the conduct of any hearing under this subsection shall be in accordance with sections 554, 555, and 556 (other than subsection (b)(3) of section 556) of title 5, United States Code.

(9) In carrying out its duties under this subsection, a review board may, and any member of a review board if so authorized by the review board, may—

    (A) sit and act at such times and places, take such testimony, and receive such evidence, as the review board deems advisable;

    (B) subject to the Privacy Act of 1974, request of any Federal agency or applicant information necessary to enable it to carry out such duties, and upon such request the head of such Federal agency shall furnish such information to the review board; and

    (C) use the United States mails in the same manner and upon the same conditions as a Federal agency.

(10) Upon request of a review board, the head of any Federal agency is authorized to detail, on a nonreimbursable basis, any of the personnel of such agency to the review board to assist it in carry [sic] out its duties under this section.

(11) The Administrator of the General Services Administration shall provide to a review board, on a reimbursable basis, such administrative support services as the review board may request.

(12) All meetings and records of review boards shall be open to the public.

(h) EXEMPTION.—(1) The Committee shall make a final determination whether or not to grant an exemption within 90 days of receiving the report of the review board under subsection (g)(7). The Committee shall grant an exemption from the requirements of subsection (a) for an agency action if, by a vote of not less than five of its members voting in person—

    (A) it determines on the record, based on the report of the review board and on such other testimony or evidence as it may receive, that—

        (i) there are no reasonable and prudent alternatives to the agency action;

        (ii) the benefits of such action clearly outweigh the benefits of alternative courses of action consistent with conserving the species or its critical habitat, and such action is in the public interest; and

        (iii) the action is of regional or national significance; and

    (B) it establishes such reasonable mitigation and enhancement measures, including, but not limited to, live propagation, transplantation, and habitat acquisition and improvement, as are necessary and appropriate to minimize the adverse effects of the agency action upon the endangered species, threatened species, or critical habitat concerned.

Any final determination by the Committee under this subsection shall be considered final agency action for purposes of chapter 7 of title 5 of the United States Code.

    (2) (A) Except as provided in subparagraph (B), an exemption for an agency action granted under subsection (h) shall constitute a permanent exemption with respect to all endangered or threatened species for the purposes of completing such agency action: *Provided*, That a biological assessment has been conducted under subsection (c).

    (B) An exemption shall not be permanent under subparagraph (A) if the Secretary finds, based on the best scientific and commercial data available, that such exemption would result in the extinction of the species. If the Secretary so finds, the Committee shall determine within 30 days after such finding whether to grant an exemption for the agency action notwithstanding the Secretary's finding.

    (i) REVIEW BY SECRETARY OF STATE.—Notwithstanding any other provision of this Act, the Committee shall be prohibited from considering for exemption any application made to it, if the Secretary of State, after a review of the proposed agency action and its potential implications, and after hearing, certifies, in writing, to the Committee within 60 days of any application made under this section that the granting of any such exemption and the carrying out of such action would be in violation of an international treaty obligation or other international obligation of the United States. The Secretary of State shall, at the time of such certification, publish a copy thereof in the Federal Register.

    (j) Notwithstanding any other provision of this Act, the Committee shall grant an exemption for any agency action if the Secretary of Defense finds that such exemption is necessary for reasons of national security.

    (k) SPECIAL PROVISIONS.—An exemption decision by the Committee under this section shall not be a major Federal action for purposes of the National Environmental Policy Act of 1969 (42 U.S.C. 4321 et seq.): *Provided*, That an environmental impact statement which discusses the impacts upon endangered species or threatened species or their critical habitats shall have been previously prepared with respect to any agency action exempted by such order.

    (l) COMMITTEE ORDERS.—(1) If the Committee determines under subsection (h) that an exemption should be granted with respect to any agency action, the Committee shall issue an order granting the exemption and specifying the mitigation and enhancement measures established pursuant to subsection (h) which shall be carried out and paid for by the exemption applicant in implementing the agency action. All necessary mitigation and enhancement measures shall be authorized prior to the implementing of the agency action and funded concurrently with all other project features.

(2) The applicant receiving such exemption shall include the costs of such mitigation and enhancement measures within the overall costs of continuing the proposed action. Notwithstanding the preceding sentence the costs of such measures shall not be treated as project costs for the purpose of computing benefit-cost or other ratios for the proposed action. Any applicant may request the Secretary to carry out such mitigation and enhancement measures. The costs incurred by the Secretary in carrying out any such measures shall be paid by the applicant receiving the exemption. No later than one year after the granting of an exemption, the exemption applicant shall submit to the Council on Environmental Quality a report describing its compliance with the mitigation and enhancement measures prescribed by this section. Such a report shall be submitted annually until all such mitigation and enhancement measures have been completed. Notice of the public availability of such reports shall be published in the Federal Register by the Council on Environmental Quality.

(m) NOTICE.—The 60-day notice requirement of section 11(g) of this Act shall not apply with respect to review of any final determination of the Committee under subsection (h) of this section granting an exemption from the requirements of subsection (a) of this section.

(n) JUDICIAL REVIEW.—Any person, as defined by section 3(13) of this Act, may obtain judicial review, under chapter 7 of title 5 of the United States Code, of any decision of the Endangered Species Committee under subsection (h) in the United States Court of Appeals for (1) any circuit wherein the agency action concerned will be, or is being, carried out, or (2) in any case in which the agency action will be, or is being, carried out outside of any circuit, the District of Columbia, by filing in such court within 90 days after the date of issuance of the decision, a written petition for review. A copy of such petition shall be transmitted by the clerk of the court to the Committee and the Committee shall file in the court the record in the proceeding, as provided in section 2112, of title 28, United States Code. Attorneys designated by the Endangered Species Committee may appear for, and represent the Committee in any action for review under this subsection.

(o) EXCEPTION ON TAKING.—Notwithstanding sections 4(d) and 9(a) of this Act or any regulations promulgated pursuant to such sections, any action for which an exemption is granted under subsection (h) of this section shall not be considered a taking of any endangered or threatened species with respect to any activity which is necessary to carry out such action.

(p) EXEMPTIONS IN PRESIDENTIALLY DECLARED DISASTER AREAS.—In any area which has been declared by the President to be a major disaster area under the Disaster Relief Act of 1974, the President is authorized to make the determinations required by subsections (g) and (h) of this section for any project for the repair or replacement of a public facility substantially as it existed prior to the disaster under section 401 or 402 of the Disaster Relief Act of 1974, and which the President determines (1) is necessary to prevent the recurrence of such a natural disaster and to reduce the potential loss of human life, and (2) to involve an emergency situation which does not allow the ordinary procedures of this section to be followed. Notwithstanding any other provision of this section, the Committee shall accept the determinations of the President under this subsection.

(q) AUTHORIZATION.—There is authorized to be appropriated to the Secretary to assist review boards and the Committee in carrying out their functions under subsections (e), (f), (g), and (h) of this section not to exceed $600,000 for fiscal year 1979, and not to exceed $300,000 for the period beginning October 1, 1979, and ending March 31, 1980. The Chairman of the Committee shall report to the Congress before the end of fiscal year 1979 with respect to the adequacy of the budget authority contained in this subsection.

INTERNATIONAL COOPERATION

SEC. 8. (a) FINANCIAL ASSISTANCE.—As a demonstration of the commitment of the United States to the worldwide protection of endangered species and threatened species, the President may, subject to the provisions of section 1415 of the Supplemental Appropriation Act, 1953 (31 U.S.C. 724), use foreign currencies accruing to the United States Government under the Agricultural Trade Development and Assistance Act of 1954 or any other law to provide to any foreign country (with its consent) assistance in the development and management of programs in that country which the Secretary determines to be necessary or useful for the conservation of any endangered species or threatened species listed by the Secretary pursuant to section 4 of this Act. The President shall provide assistance (which includes, but is not limited to, the acquisition, by lease or otherwise, of lands, waters, or interests, therein) to foreign countries under this section under such terms and conditions as he deems appropriate. Whenever foreign currencies are available for the provision of assistance under this section, such currencies shall be used in preference to funds appropriated under the authority of section 15 of this Act.

(b) ENCOURAGEMENT OF FOREIGN PROGRAMS.—In order to carry out further the provisions of this Act, the Secretary, through the Secretary of State, shall encourage—

(1) foreign countries to provide for the conservation of fish or wildlife including endangered species and threatened species listed pursuant to section 4 of this Act;

(2) the entering into of bilateral or multilateral agreements with foreign countries to provide for such conservation; and

(3) foreign persons who directly or indirectly take fish or wildlife in foreign countries or on the high seas for importation into the United States for commercial or other purposes to develop and carry out with such assistance as he may provide, conservation practices designed to enhance such fish or wildlife and their habitat.

(c) PERSONNEL.—After consultation with the Secretary of State, the Secretary may—

(1) assign or otherwise make available any officer or employee of his department for the purpose of cooperating with foreign countries and international organizations in developing personnel resources and programs which promote the conservation of fish or wildlife; and

(2) conduct or provide financial assistance for the educational training of foreign personnel, in this country or abroad, in fish, wildlife, or plant management, research and law enforcement and to render professional assistance abroad in such matters.

(d) INVESTIGATIONS.—After consultation with the Secretary of State and the Secretary of the Treasury, as appropriate, the Secretary may conduct or cause to be conducted such law enforcement investigations and research abroad as he deems necessary to carry out the purposes of this Act.

(e) CONVENTION IMPLEMENTATION.—The President is authorized and directed to designate appropriate agencies to act as the Management Authority or Authorities and the Scientific Authority or Authorities pursuant to the Convention. The agencies so designated shall thereafter be authorized to do all things assigned to them under the Convention, including the issuance of permits and certificates. The agency designated by the President to communicate with other parties to the Convention and with the Secretariat shall also be empowered, where appropriate, in consultation with the State Department, to act on behalf of and represent the United States in all regards as required by the Convention. The President shall also designate those agencies which shall act on behalf of and represent the United States in all regards as required by the Convention on Nature Protection and Wildlife Preservation in the Western Hemisphere.

PROHIBITED ACTS

SEC. 9. (a) GENERAL.—(1) Except as provided in sections 6(g)(2) and 10 of this Act, with respect to any endangered species of fish or wildlife listed pursuant to section 4 of this Act it is unlawful for any person subject to the jurisdiction of the United States to—

(A) import any such species into, or export any such species from the United States;

(B) take any such species within the United States or the territorial sea of the United States;

(C) take any such species upon the high seas;

(D) possess, sell, deliver, carry, transport, or ship, by any means whatsoever, any such species taken in violation of subparagraphs (B) and (C);

(E) deliver, receive, carry, transport, or ship in interstate or foreign commerce, by any means whatsoever and in the course of a commercial activity, any such species;

(F) sell or offer for sale in interstate or foreign commerce any such species; or

(G) violate any regulation pertaining to such species or to any threatened species of fish or wildlife listed pursuant to section 4 of this Act and promulgated by the Secretary pursuant to authority provided by this Act.

(2) Except as provided in sections 6(g)(2) and 10 of this Act, with respect to any endangered species of plants listed pursuant to section 4 of this Act, it is unlawful for any person subject to the jurisdiction of the United States to—

(A) import any such species into, or export any such species from, the United States;

(B) deliver receive, carry, transport, or ship in interstate or foreign commerce, by any means whatsoever and in the course of a commercial activity, any such species;

(C) sell or offer for sale in interstate or foreign commerce any such species; or

(D) violate any regulation pertaining to such species or to any threatened species of plants listed pursuant to section 4 of this Act and promulgated by the Secretary pursuant to authority provided by this Act.

(b) Species Held in Captivity or Controlled Environment.—(1) The provisions of this section shall not apply to any fish or wildlife held in captivity or in a controlled environment on the effective date of this Act if the purposes of such holding are not contrary to the purposes of this Act; except that this subsection shall not apply in the case of any fish or wildlife held in the course of a commercial activity. With respect to any act prohibited by this section which occurs after a period of 180 days from the effective date of this Act, there shall be a rebuttable presumption that the fish or wildlife involved in such act was not held in captivity or in a controlled environment on such effective date.

(2) (A) This section shall not apply to—

(i) any raptor legally held in captivity or in a controlled environment on the effective date of the Endangered Species Act Amendments of 1978; or

(ii) any progeny of any raptor described in clause (i);

until such time as any such raptor or progeny is intentionally returned to a wild state.

(B) Any person holding any raptor or progeny described in subparagraph (A) must be able to demonstrate that the raptor or progeny does, in fact, qualify under the provisions of this paragraph, and shall maintain and submit to the Secretary, on request, such inventories, documentation, and records as the Secretary may by regulation require as being reasonably appropriate to carry out the purposes of this paragraph. Such requirements shall not unnecessarily duplicate the requirements of other rules and regulations promulgated by the Secretary.

(c) VIOLATION OF CONVENTION.—(1) It is unlawful for any person subject to the jurisdiction of the United States to engage in any trade in any speciments contrary to the provisions of the Convention, or to possess any specimens traded contrary to the provisions of the Convention, including the definitions of terms in article I thereof.

(2) Any importation into the United States of fish or wildlife shall, if—

(A) such fish or wildlife is not an endangered species listed pursuant to section 4 of this Act but is listed in Appendix II to the Convention.

(B) the taking and exportation of such fish or wildlife is not contrary to the provisions of the Convention and all other applicable requirements of the Convention have been satisfied,

(C) the applicable requirements of subsections (d), (e), and (f) of this section have been satisfied, and

(D) such importation is not made in the course of a commercial activity,

be presumed to be an importation not in violation of any provision of this Act or any regulation issued pursuant to this Act.

(d) IMPORTS AND EXPORTS.—(1) It is unlawful for any person to engage in business as an importer or exporter of fish or wildlife (other than shellfish and fishery products which (A) are not listed pursuant to section 4 of this Act as endangered species or threatened species, and (B) are imported for purposes of human or animal consumption or taken in waters under the jurisdiction of the United States or on the high seas for recreational purposes) or plants without first having obtained permission from the Secretary.

(2) Any person required to obtain permission under paragraph (1) of this subsection shall—

(A) keep such records as will fully and correctly disclose each importation or exportation of fish, wildlife, or plants made by him and the subsequent disposition made by him with respect to such fish, wildlife, or plants;

(B) at all reasonable times upon notice by a duly authorized representative of the Secretary, afford such representative access to his places of business, an opportunity to examine his inventory of imported fish, wildlife, or plants and the records required to be kept under subparagraph (A) of this paragraph, and to copy such records; and

(C) file such reports as the Secretary may require.

(3) The Secretary shall prescribe such regulations as are necessary and appropriate to carry out the purposes of this subsection.

(e) REPORTS.—It is unlawful for any person importing or exporting fish or wildlife (other than shellfish and fishery products which (1) are not listed pursuant to section 4 of this Act as endangered or threatened species, and (2) are imported for purposes of human or animal consumption or taken in waters under the jurisdiction of the United States or on the high seas for recreational purposes) or plants to fail to file any declaration or report as the Secretary deems necessary to facilitate enforcement of this Act or to meet the obligations of the Convention.

(f) DESIGNATION OF PORTS.—(1) It is unlawful for any person subject to the jurisdiction of the United States to import into or export from the United States any fish or wildlife (other than shellfish and fishery products which (A) are not listed pursuant to section 4 of this Act as endangered species or threatened species, and (B) are imported for purposes of human or animal consumption or taken in waters under the jurisdiction of the United States or on the high seas for recreational purposes) or plants, except at a port or ports designated by the Secretary of the Interior. For the purpose of facilitating enforcement of this Act and reducing the costs thereof, the Secretary of the Interior, with approval of the Secretary of the Treasury and after notice and opportunity for public hearing, may, by regulation, designate ports and change such designations. The Secretary of the Interior, under such terms and conditions as he may prescribe, may permit the importation or exportation at nondesignated ports in the interest of the health or safety of the fish or wildlife or plants, or for other reasons if, in his discretion, he deems it appropriate and consistent with the purpose of this subsection.

(2) Any port designated by the Secretary of the Interior under the authority of section 4(d) of the Act of December 5, 1969 (16 U.S.C. 666cc-4(d)), shall, if such designation is in effect on the day before the date of the enactment of this Act, be deemed to be a port designated by the Secretary under paragraph (1) of this subsection until such time as the Secretary otherwise provides.

(g) VIOLATIONS.—It is unlawful for any person subject to the jurisdiction of the United States to attempt to commit, solicit another to commit, or cause to be committed, any offense defined in this section.

EXCEPTIONS

SEC. 10. (a) PERMITS.—The Secretary may permit, under such terms and conditions as he may prescribe, any act otherwise prohibited by section 9 of this Act for scientific purposes or to enhance the propagation or survival of the affected species.

(b) HARDSHIP EXEMPTIONS.—(1) If any person enters into a contract with respect to a species of fish or wildlife or plant before the date of the publication in the Federal Register of notice of consideration of that species as an endangered species and the subsequent listing of that species as an endangered species pursuant to section 4 of this Act will cause undue economic hardship to such person under the contract, the Secretary, in order to minimize such hardship, may exempt such person from the application of section 9(a) of this Act to the extent the Secretary deems appropriate if such person applies to him for such exemption and includes with such application such information as the Secretary may require to prove such hardship; except that (A) no such exemption shall be for a duration of more than one year from the date of publication in the Federal Register of notice of consideration of the species concerned, or shall apply to a quantity of fish or wildlife or plants in excess of that specified by the Secretary; (B) the one-year period for those species of fish or wildlife listed by the Secretary as endangered prior to the effective date of this Act shall expire in accordance with the terms of section 3 of the Act of December 5, 1969 (83 Stat. 275); and (C) no such exemption may be granted for the importation or exportation of a specimen listed in Appendix I of the Convention which is to be used in a commercial activity.

(2) As used in this subsection, the term "undue economic hardship" shall include, but not be limited to:

(A) substantial economic loss resulting from inability caused by this Act to perform contracts with respect to species of fish and wildlife entered into prior to the date of publication in the Federal Register of a notice of consideration of such species as an endangered species;

(B) substantial economic loss to persons who, for the year prior to the notice of consideration of such species as an endangered species, derived a substantial portion of their income from the lawful taking of any listed species, which taking would be made unlawful under this Act; or

(C) curtailment of subsistence taking made unlawful under this Act by persons (i) not reasonably able to secure other sources of subsistence; and (ii) dependent to a substantial extent upon hunting and fishing for subsistence; and (iii) who must engage in such curtailed taking for subsistence purposes.

(3) The Secretary may make further requirements for a showing of undue economic hardship as he deems fit. Exceptions granted under this section may be limited by the Secretary in his discretion as to time, area, or other factor of applicability.

(c) NOTICE AND REVIEW.—The Secretary shall publish notice in the Federal Register of each application for an exemption or permit which is made under this section. Each notice shall invite the submission from interested parties, within thirty days after the date of the notice, written data, views, or arguments with respect to the application; except that such thirty-day period may be waived by the Secretary in an emergency situation where the health or life of an endangered animal is threatened and no reasonable alternative is available to the applicant, but notice of any such waiver shall be published by the Secretary in the Federal Register within ten days following the issuance of the exemption or permit. Information received by the Secretary as a part of any application shall be available to the public as a matter of public record at every state of the proceeding.

(d) PERMIT AND EXEMPTION POLICY.—The Secretary may grant exceptions under subsections (a) and (b) of this section only if he finds and publishes his finding in the Federal Register that (1) such exceptions were applied for in good faith, (2) if granted and exercised will not operate to the disadvantage of such endangered species, and (3) will be consistent with the purposes and policy set forth in section 2 of this Act.

(e) ALASKA NATIVES.—(1) Except as provided in paragraph (4) of this subsection the provisions of this Act shall not apply with respect to the taking of any endangered species or threatened species, or the importation of any such species taken pursuant to this section, by—

(A) any Indian, Aleut, or Eskimo who is an Alaskan Native who resides in Alaska; or

(B) any non-native permanent resident of an Alaskan native village;

if such taking is primarily for subsistence purposes. Non-edible byproducts of species taken pursuant to this section may be sold in interstate commerce when made into authentic native articles of handicrafts and clothing; except that the provisions of this subsection shall not apply to any non-native resident of an Alaskan native village found by the Secretary to be not primarily dependent upon the taking of fish and wildlife for consumption or for the creation and sale of authentic native articles of handicrafts and clothing.

(2) Any taking under this subsection may not be accomplished in a wasteful manner.

(3) As used in this subsection—

(i) The term "subsistence" includes selling any edible portion of fish or wildlife in native villages and towns in Alaska for native consumption within native villages or towns; and

(ii) The term "authentic native articles of handicrafts and clothing" means items composed wholly or in some significant respect of natural materials, and which are produced, decorated, or fashioned in the exercise of traditional native handicrafts without the use of pantographs, multiple carvers, or other mass copying devices. Traditional native handicafts include, but are not limited to, weaving, carving, stitching, sewing, lacking, beading, drawing, and painting.

(4) Notwithstanding the provisions of paragraph (1) of this subsection, whenever the Secretary determines that any species of fish

or wildlife which is subject to taking under the provisions of this subsection is an endangered species or threatened species, and that such taking materially and negatively affects the threatened or endangered species, he may prescribe regulations upon the taking of such species by any such Indian, Aleut, Eskimo, or non-Native Alaskan resident of an Alaskan native village. Such regulations may be established with reference to species, geographical description of the area included, the season for taking, or any other factors related to the reason for establishing such regulations and consistent with the policy of this Act. Such regulations shall be prescribed after a notice and hearings in the affected judicial districts of Alaska and as otherwise required by section 103 of the Marine Mammal Protection Act of 1972, and shall be removed as soon as the Secretary determines that the need for their impositions has disappeared.

(f)(1) As used in this subsection—

(A) The term "pre-Act endangered species part" means—

(i) any sperm whale oil, including derivatives thereof, which was lawfully held within the United States on December 28, 1973, in the course of a commercial activity; or

"(ii) any finished scrimshaw product, if such product or the raw material for such product was lawfully held within the United States on December 28, 1973, in the course of a commercial activity.

(B) The term "scrimshaw product" means any art form which involves the etching or engraving of designs upon, or the carving of figures, patterns, or designs from, any bone or tooth of any marine mammal of the order Cetacea.

(2) The Secretary, pursuant to the provisions of this subsection, may exempt, if such exemption is not in violation of the Convention, any pre-Act endangered species part from one or more of the following prohibitions:

(A) The prohibition on exportation from the United States set forth in section 9(a)(1)(A) of this Act.

(B) Any prohibition set forth in section 9(a)(1) (E) or (F) of this Act.

(3) Any person seeking an exemption described in paragraph (2) of this subsection shall make application therefore to the Secretary in such form and manner as he shall prescribe, but no such application may be considered by the Secretary unless the application—

(A) is received by the Secretary before the close of the one-year period beginning on the date on which regulations promulgated by the Secretary to carry out this subsection first take effect;

(B) contains a complete and detailed inventory of all pre-Act endangered species parts for which the applicant seeks exemption;

(C) is accompanied by such documentation as the Secretary may require to prove that any endangered species part or product claimed by the applicant to be a pre-Act endangered species part is in fact such a part; and

(D) contains such other information as the Secretary deems necessary and appropriate to carry out the purposes of this subsection.

(4) if the Secretary approves any application for exemption made under this subsection, he shall issue to the applicant a certificate of exemption which shall specify—

(A) any prohibition in section 9(a) of this Act which is exempted;

(B) the pre-Act endangered species parts to which the exemption applies;

(C) the period of time during which the exemption is in effect, but no exemption made under this subsection shall have force and effect after the close of the three-year period beginning on the date of issuance of the certificate; and

(D) any term or condition prescribed pursuant to paragraph (5) (A) or (B), or both, which the Secretary deems necessary or appropriate.

(5) The Secretary shall prescribe such regulations as he deems necessary and appropriate to carry out the purposes of this subsection. Such regulations may set forth—

(A) terms and conditions which may be imposed on applicants for exemptions under this subsection (including, but not limited to, requirements that applicants register inventories, keep complete sales records, permit duly authorized agents of the Secretary to inspect such inventories and records, and periodically file appropriate reports with the Secretary); and

(B) terms and conditions which may be imposed on any subsequent purchaser of any pre-Act endangered species part covered by an exemption granted under this subsection;

to insure that any such part so exempted is adequatley accounted for and not disposed of contrary to the provisions of this Act. No regulation prescribed by the Secretary to carry out the purposes of this subsection shall be subject to section 4(f)(2)(A)(i) of this Act.

(6) (A) Any contract for the sale of pre-Act endangered species parts which is entered into by the Administrator of General Services prior to the effective date of this subsection and pursuant to the notice published in the Federal Register on January 9, 1973, shall not be rendered invalid by virtue of the fact that fulfillment of such contract may be prohibited under section 9(a)(1)(F).

(B) In the event that this paragraph is held invalid, the validity of the remainder of the Act, including the remainder of this subsection, shall not be affected.

(7) Nothing in this subsection shall be construed to—

(A) exonerate any person from any act committed in violation of paragraphs (1)(A), (1)(E), or (1)(F) of section 9(a) prior to the date of enactment of this subsection; or

(B) immunize any person from prosecution for any such act.

(g) In connection with any action alleging a violation of section 9, any person claiming the benefit of any exemption or permit under this Act shall have the burden of proving that the exemption or permit is applicable, has been granted, and was valid and in force at the time of the alleged violation.

(h) CERTAIN ANTIQUE ARTICLES.—(1) Sections 4(d), 9(a), and 9(c) do not apply to any article (other than scrimshaw) which—

(A) was made before 1830;

(B) is composed in whole or in part of any endangered species or threatened species listed under section 4;

(C) has not been repaired or modified with any part of any such species on or after the date of the enactment of this Act; and

(D) is entered at a port designated under paragraph (3).

(2) Any person who wishes to import an article under the exception provided by this subsection shall submit to the customs officer concerned at the time of entry of the article such documentation as the Secretary of the Treasury, after consultation with the Secretary of the Interior, shall by regulation require as being necessary to establish that the article meets the requirements set forth in paragraph (1)(A), (B), and (C).

(3) The Secretary of the Treasury, after consultation with the Secretary of the Interior, shall designate one port within each customs region at which articles described in paragraph (1)(A), (B), and (C) must be entered into the customs territory of the United States.

(4) Any person who imported, after December 27, 1973, and on or before the date of the enactment of the Endangered Species Act Amendments of 1978, any article described in paragraph (1) which—

(A) was not repaired or modified after the date of importation with any part of any endangered species or threatened species listed under section 4;

(B) was forfeited to the United States before such date of the enactment, or is subject to forfeiture to the United States on such date of enactment, pursuant to the assessment of a civil penalty under section 11; and

(C) is in the custody of the United States on such date of enactment;

may, before the close of the one-year period beginning on such date of enactment, make application to the Secretary for return of the article. Application shall be made in such form and manner, and contain such documentation, as the Secretary prescribes. If on the basis of any such application which is timely filed, the Secretary is satisfied that the requirements of this paragraph are met with respect to the article concerned, the Secretary shall return the article to the applicant and the importation of such article shall, on and after the date of return, be deemed to be a lawful importation under this Act.

(i)(1) TELLICO AND GRAYROCKS PROJECTS.—Notwithstanding any other provision of this Act, the Committee shall, within 30 days of the date of the enactment of the Endangered Species Act Amendments of 1978, proceed to consider the exemption of the Tellico Dam and Reservoir Project and the Grayrocks Dam and Reservoir Project from the requirements of section 7(a). For the purposes of such consideration, the Committee shall grant an exemption to such projects if the criteria of section 7(h)(1)(A)(i) and 7(h)(1)(A)(ii) are met. A decision on any such exemption shall be made within 90 days after the date of the enactment of the Endangered Species Act Amendments of 1978. If no decision is made within such 90-day period, such project shall be deemed to be exempted from the requirements of section 7(a).

(2) Following the rendering of a biological opinion by the United States Fish and Wildlife Service concerning the effect, if any, of the operation of the Missouri Basin Power Project on endangered species or their critical habitat, the responsible officers of the Rural Electrification Administration, the Secretary of the Interior, and the Secretary of the Army, shall require such modifications in the operation or design of the project as they may determine are required to insure that actions authorized, funded, or carried out by them, relating to the Missouri Basin Power Project do not jeopardize the continued existence of such endangered species or result in the destruction or adverse modification of habitat of such species which is or has been determined to be critical by the Secretary of the Interior, after consultation as appropriate with the affected States.

PENALTIES AND ENFORCEMENT

SEC. 11. (a) CIVIL PENALTIES.—(1) Any person who knowingly violates, and any person engaged in business as an importer or exporter of fish, wildlife, or plants who violates, any provision of this Act, or any provision of any permit or certificate issued hereunder, or of any regulation issued in order to implement subsection (a)(1) (A), (B), (C), (D), (E), or (F), (a)(2) (A), (B), or (C), (c), (d) (other than regulation relating to recordkeeping or filing of reports), (f) or (g) of section 9 of this Act, may be assessed a civil penalty by the Secretary of not more than $10,000 for each violation. Any person who knowingly violates, and any person engaged in business as an importer or exporter of fish, wildlife, or plants who violates, any provision of any other regulation issued under this Act may be assessed a civil penalty by the Secretary of not more than $5,000 for each such violation. Any person who otherwise violates any provision of this Act, or any regulation, permit, or certificate issued hereunder, may be assessed a civil penalty by the Secretary of not more than $500 for each such violation. No penalty may be assessed under this subsection unless such person is given notice and opportunity for a hearing with respect to such violation. Each violation shall be a separate offense. Any such civil penalty may be remitted or mitigated by the Secretary. Upon any failure to pay a penalty assessed under this subsection, the Secretary may request the Attorney General to institute a civil action in a district court of the United States for any district in which such person is found, resides, or transacts business to collect the penalty and such court shall have jurisdiction to hear and decide any such action. The court shall hear such action on the record made before the Secretary and shall sustain his action if it is supported by substantial evidence on the record considered as a whole.

(2) Hearings held during proceedings for the assessment of civil penalties authorized by paragraph (1) of this subsection shall be conducted in accordance with section 554 of title 5, United States Code. The Secretary may issue subpoenas for the attendance and testimony of witnesses and the production of relevant papers, books, and documents, and administer oaths. Witnesses summoned shall be paid the same fees and mileage that are paid to witnesses in the courts of the United States. In case of contumacy or refusal to obey a subpoena served upon any person pursuant to this paragraph, the district court of the United States for any district in which such person is found or resides or transacts business, upon application by the United States and after notice to such person, shall have jurisdiction to issue an order requiring such person to appear and give testimony before the Secretary or to appear and produce documents before the Secretary, or both, and any failure to obey such order of the court may be punished by such court as a contempt thereof.

(3) Notwithstanding any other provision of this Act, no civil penalty shall be imposed if it can be shown by a preponderance of the evidence that the defendant committed an act based on a good faith belief that he was acting to protect himself or herself, a member of his or her family, or any other individual from bodily harm, from any endangered or threatened species.

(b) CRIMINAL, VIOLATIONS.—(1) Any person who knowingly violates any provision of this Act, of any permit or certificate issued hereunder, or of any regulation issued in order to implement subsection (a)(1) (A), (B), (C), (D), (E), or (F); (a) (2) (A), (B), or (C), (c), (d) (other than a regulation relating to record-keeping, or filing of reports), (f), or (g) of section 9 of this Act shall, upon conviction, be fined not more than $20,000 or imprisoned for not more than one year, or both. Any person who knowingly violates any provision for any other regulation issued under this Act shall, upon conviction, be fined not more than $10,000 or imprisoned for not more than six months, or both.

(2) The head of any Federal agency which has issued a lease, license, permit, or other agreement authorizing a person to import or export fish, wildlife, or plants, or to operate a quarantine station for imported wildlife, or authorizing the use of Federal lands, including grazing of domestic livestock, to any person who is convicted of a criminal violation of this Act or any regulation, permit, or certificate issued hereunder may immediately modify, suspend, or revoke each lease, license, permit, or other agreement. The Secretary shall also suspend for a period of up to one year, or cancel, any Federal hunting or fishing permits or stamps issued to any person who is convicted of a criminal violation of any provision of this Act or any regulation, permit, or certificate issued hereunder. The United States shall not be liable for the payments of any compensation, reimbursement, or damages in connection with the modification, suspension, or revocation of any leases, licenses, permits, stamps, or other agreements pursuant to this section.

(3) Notwithstanding any other provision of this Act, it shall be a defense to prosecution under this subsection if the defendant committed the offense based on a good faith belief that he was acting to protect himself or herself, a member of his or her family, or any other individual, from bodily harm from any endangered or threatened species.

(c) DISTRICT COURT JURISDICTION.—The several district courts of the United States, including the courts enumerated in section 460 of title 28, United States Code, shall have jurisdiction over any actions arising under this Act. For the purpose of this Act, American Samoa shall be included within the judicial district of the District Court of the United States for the District of Hawaii.

(d) REWARDS.—Upon the recommendation of the Secretary, the Secretary of the Treasury is authorized to pay an amount equal to one-half of the civil penalty or fine paid, but not to exceed $2,500, to any person who furnishes information which leads to a finding of civil violation or a conviction of a criminal violation of any provision of this Act or any regulation or permit issued thereunder. Any officer or employee of the United States or of any State or local government who furnishes information or renders service in the performance of his official duties shall not be eligible for payment under this section.

(e) ENFORCEMENT.—(1) The provisions of this Act and any regulations or permits issued pursuant thereto shall be enforced by the Secretary, the Secretary of the Treasury, or the Secretary of the Department in which the Coast Guard is operating, or all such Secretaries. Each such Secretary may utilize by agreement, with or without reimbursement, the personnel, services, and facilities of any other Federal agency or any State agency for purposes of enforcing this Act.

(2) The judges of the district courts of the United States and the United States magistrates may, within their respective jurisdictions, upon proper oath or affirmation showing probable cause, issue such warrants or other process as may be required for enforcement of this Act and any regulation issued thereunder.

(3) Any person authorized by the Secretary, the Secretary of the Treasury, or the Secretary of the Department in which the Coast Guard is operating, to enforce this Act may detain for inspection and inspect any package, crate, or other container, including its contents, and all accompanying documents, upon importation or exportation. Such person may make arrests without a warrant for any violation of this Act if he has reasonable grounds to believe that the person to be arrested is committing the violation in his presence or view, and may execute and serve any arrest warrant, search warrant, or other warrant or civil or criminal process issued by any officer or court of competent jurisdiction for enforcement of this Act. Such person so authorized may search and seize, with or without a warrant, as authorized by law. Any fish, wildlife, property, or item so seized shall be held by any person authorized by the Secretary, the Secretary of the Treasury, or the Secretary of the Department in which the Coast Guard is operating pending disposition of civil or criminal proceedings, or the institution of an action in rem for forfeiture of such fish, wildlife, property, or item pursuant to paragraph (4) of this subsection; except that the Secretary may, in lieu of holding such fish, wildlife, property, or item, permit the owner or consignee to post a bond or other surety satisfactory to the Secretary, but upon forfeiture of any such property to the United States, or the abandonment or waiver of any claim to any such property, it shall be disposed of (other than by sale to the general public) by the Secretary in such a manner, consistent with the purposes of this Act, as the Secretary shall by regulation prescribe.

(4) (A) All fish or wildlife or plants taken, possessed, sold, purchased, offered for sale or purchase, transported, delivered, received, carried, shipped, exported, or imported contrary to the provisions of this Act, any regulation made pursuant thereto, or any permit or certificate issued hereunder shall be subject to forfeiture to the United States.

(B) All guns, traps, nets, and other equipment, vessels, vehicles, aircraft, and other means of transportation used to aid the taking, possessing, selling, purchasing, offering for sale or purchase, transporting, delivering, receiving, carrying, shipping, exporting, or importing of any fish or wildlife or plants in violation of this Act, any regulation made pursuant thereto, or any permit or certificate issued thereunder shall be subject to forfeiture to the United States upon conviction of a criminal pursuant to section 11(b)(1) of this Act.

(5) All provisions of law relating to the seizure, forfeiture, and condemnation of a vessel for violation of the customs laws, the disposition of such vessel or the proceeds from the sale thereof, and the remission or mitigation of such forfeiture, shall apply to the seizures and forfeitures incurred, or alleged to have been incurred, under the provisions of this Act, insofar as such provisions of law are applicable and not inconsistent with the provisions of this Act; except that all powers, rights, and duties conferred or imposed by the customs laws upon any officer or employee of the Treasury Department shall, for the purposes of this Act, be exercised or performed by the Secretary or by such persons as he may designate.

(f) REGULATIONS.—The Secretary, the Secretary of the Treasury, and the Secretary of the Department in which the Coast Guard is operating, are authorized to promulgate such regulations as may be appropriate to enforce this Act, and charge reasonable fees for expenses to the Government connected with permits or certificates authorized by this Act including processing applications and reasonable inspections, and with the transfer, board, handling, or storage of fish or wildlife or plants and evidentiary items seized and forfeited under this Act. All such fees collected pursuant to this subsection shall be deposited in the Treasury to the credit of the appropriation which is current and chargeable for the cost of furnishing the services. Appropriated funds may be expended pending reimbursement from parties in interest.

(g) CITIZEN SUITS.—(1) Except as provided in paragraph (2) of this subsection any person may commence a civil suit on his own behalf—

> (A) to enjoin any person, including the United States and any other governmental instrumentality or agency (to the extent permitted by the eleventh amendment to the Constitution), who is alleged to be in violation of any provision of this Act or regulation issued under the authority thereof; or
>
> (B) to compel the Secretary to apply, pursuant to section 6(g)(2)(B)(ii) of this Act, the prohibitions set forth in or authorized pursuant to section 4(d) or section 9(a)(1)(B) of this Act with respect to the taking of any resident endangered species or threatened species within any State.

The district courts shall have jurisdiction, without regard to the amount in controversy, or the citizenship of the parties, to enforce any such provision or regulation, as the case may be. In any civil suit commenced under subparagraph (B) the district court shall compel the Secretary to apply the prohibition sought if the court finds that the allegation that an emergency exists is supported by substantial evidence.

(2) (A) No action may be commenced under subparagraph (1)(A) of this section—

(i) prior to sixty days after written notice of the violation has been given to the Secretary, and to any alleged violator of any such provision or regulation;

(ii) if the Secretary has commenced action to impose a penalty pursuant to subsection (a) of this section; or

(iii) if the United States has commenced and is diligently prosecuting a criminal action in a court of the United States or a State to redress a violation of any such provision or regulation.

(B) No action may be commenced under subparagraph (1)(B) of this section—

(i) prior to sixty days after written notice has been given to the Secretary setting forth the reasons why an emergency is thought to exist with respect to an endangered species or a threatened species in the State concerned; or

(ii) if the Secretary has commenced and is diligently prosecuting action under section 6(g)(2)(B)(ii) of this Act to determine whether any such emergency exists.

(3) (A) Any suit under this subsection may be brought in the judicial district in which the violation occurs.

(B) In any such suit under this subsection in which the United States is not a party, the Attorney General, at the request of the Secretary, may intervene on behalf of the United States as a matter of right.

(4) The court, in issuing any final order in any suit brought pursuant to paragraph (1) of this subsection, may award costs of litigation (including reasonable attorney and expert witness fees) to any party, whenever the court determines such award is appropriate.

(5) The injunctive relief provided by this subsection shall not restrict any right which any person (or class of persons) may have under any statute or common law to seek enforcement of any standard or limitation or to seek any other relief (including relief against the Secretary or a State agency).

(h) COORDINATION WITH OTHER LAWS.—The Secretary of Agriculture and the Secretary shall provide for appropriate coordination of the administration of this Act with the administration of the animal quarantine laws (21 U.S.C. 101-105, 111-135b, and 612-614) and section 306 of the Tariff Act of 1930 (19 U.S.C. 1306). Nothing in this Act or any amendment made by this Act shall be construed as superseding or limiting in any manner the functions of the Secretary of Agriculture under any other law relating to prohibited or restricted importations or possession of animals and other articles and no proceeding or determination under this Act shall preclude any proceeding or be considered determinative of any issue of fact or law in any proceeding under any Act administered by the Secretary of Agriculture. Nothing in this Act shall be construed as superseding or limiting in any manner the functions and responsibilities of the Secretary of the Treasury under the Tariff Act of 1930, including, without limitation, section 527 of that Act (19 U.S.C. 1527), relating to the importation of wildlife taken, killed, possessed, or exported to the United States in violation of the laws or regulations of a foreign country.

ENDANGERED PLANTS

SEC. 12. The Secretary of the Smithsonian Institution, in conjunction with other affected agencies, is authorized and directed to review (1) species of plants which are now or may become endangered or threatened and (2) methods of adequately conserving such species, and to report to Congress, within one year after the date of the enactment of this Act, the results of such review including recommendations for new legislation or the amendment of existing legislation.

CONFORMING AMENDMENTS

SEC. 13. (a) Subsection 4(c) of the Act of October 15, 1966 (80 Stat. 928, 16 U.S.C. 668dd(c)), is further amended by revising the second sentence thereof to read as follows: "With the exception of endangered species and threatened species listed by the Secretary pursuant to section 4 of the Endangered Species Act of 1973 in States wherein a cooperative agreement does not exist pursuant to section 6(c) of that Act, nothing in this Act shall be construed to authorize the Secretary to control or regulate hunting or fishing of resident fish and wildlife on lands not within the system."

(b) Subsection 10(a) of the Migratory Bird Conservation Act (45 Stat. 1224, 16 U.S.C. 715i(a)) and subsection 401(a) of the Act of June 15, 1935 (49 Stat. 383, 16 U.S.C. 715s(a)), are each amended by striking out "threatened with extinction," and inserting in lieu thereof the following: "listed pursuant to section 4 of the Endangered Species Act of 1973 as endangered species or threatened species".

(c) Section 7(a)(1) of the Land and Water Conservation Fund Act of 1965 (16 U.S.C. 4601–9(a)(1)) is amended by striking out:

"THREATENED SPECIES.—For any national area which may be authorized for the preservation of species of fish or widlife that are threatened with extinction."

"and inserting in lieu thereof the following"

"ENDANGERED SPECIES AND THREATENED SPECIES.—For lands, waters, or interests therein, the acquisition of which is authorized under section 5(a) of the Endangered Species Act of 1973, needed for the purpose of conserving endangered or threatened species of fish or wildlife or plants."

(d) The first sentence of section 2 of the Act of September 28, 1962, as amended (76 Stat. 653, 16 U.S.C. 460k-1), is amended to read as follows:

"The Secretary is authorized to acquire areas of land, or interests therein, which are suitable for—

"(1) incidental fish and wildlife-oriented recreational development,

"(2) the protection of natural resources,

"(3) the conservation of endangered species or threatened species listed by the Secretary pursuant to section 4 of the Endangered Species Act of 1973, or

"(4) carrying out two or more of the purposes set forth in paragraphs (1) through (3) of this section, and are adjacent to, or within, the said conservation areas, except that the acquisition of any land or interest therein pursuant to this section shall be accomplished only with such funds as may be appropriated therefor by the Congress or donated for such purposes, but such property shall not be acquired with funds obtained from the sale of Federal migratory bird hunting stamps."

(e) The Marine Mammal Protection Act of 1972 (16 U.S.C. 1361-1407)is amended—

(1) by striking out "Endangered Species Conservation Act of 1969" in section 3(1)(B) thereof and inserting in lieu thereof the following: "Endangered Species Act of 1973";

(2) by striking out "pursuant to the Endangered Species Conservation Act of 1969" in section 101(a)(3)(B) thereof and inserting in lieu thereof the following: "or threatened species pursuant to the Endangered Species Act of 1973";

(3) by striking out "endangered under the Endangered Species Conservation Act of 1969" in section 102(b)(3) thereof and inserting in lieu thereof the following: "an endangered species or threatened species pursuant to the Endangered Species Act of 1973"; and

(4) by striking out "of the Interior such revisions of the Endangered Species List, authorized by the Endangered Species Conservation Act of 1969," in section 202(a)(6) thereof and inserting in lieu thereof the following: "such revisions of the endangered species list and threatened species list published pursuant to section 4(c)(1) of the Endangered Species Act of 1973".

(f) Section 2(1) of the Federal Environmental Pesticide Control Act of 1972 (Public Law 92-516) is amended by striking out the words "by the Secretary of the Interior under Public Law 91-135" and inserting in lieu thereof the words "or threatened by the Secretary pursuant to the Endangered Species Act of 1973".

SEC. 14. The Endangered Species Conservation Act of 1969 (sections 1 through 3 of the Act of October 15, 1966, and sections 1 through 6 of the Act of December 5, 1969; 16 U.S.C. 668aa—668cc-6), is repealed.

AUTHORIZATION OF APPROPRIATIONS

SEC. 15. Except as authorized in sections 6 and 7 of this Act, there are authorized to be appropriated—

(1) not to exceed $25,000,000 for the fiscal year ending September 30, 1977, and the fiscal year September 30, 1978, not to exceed $23,000,000 for the fiscal year ending September 30, 1979, and not to exceed $12,500,000 for the period beginning October 1, 1979, and ending March 31, 1980.

(2) not to exceed $5,000,000 for the fiscal year ending September 30, 1977, and the fiscal year ending September 30, 1978, not to exceed $2,500,000 for the fiscal year ending September 30, 1979, not to exceed $12,500,000 for the period beginning October 1, 1979, and ending March 31, 1980, to enable the Department of Commerce to carry out such functions and responsibilities as it may have been given under this Act.

EFFECTIVE DATE

SEC. 16. This Act shall take effect on December 28, 1973.

MARINE MAMMAL PROTECTION ACT OF 1972

SEC. 17. Except as otherwise provided in this Act, no provision of this Act shall take precedence over any more restrictive conflicting provision of the Marine Mammal Protection Act of 1972.

☆ U.S. GOVERNMENT PRINTING OFFICE 1979 –288-221/6089

Management Policies of the National Park Service Pertinent to Conservation of Threatened and Endangered Plants*

Threatened and Endangered Plants and Animals

The Service will identify all threatened and endangered species within park boundaries and their critical habitat requirements. As necessary, the Service shall control visitor use and access to such habitat, including closure to entry for other than official purposes. Active management programs, where necessary, may be carried out to perpetuate the natural distribution and abundance of threatened or endangered species and the ecosystem on which they depend, in accordance with existing Federal laws.

The Service will cooperate with the Fish and Wildlife Service, which is recognized as the lead agency in matters pertaining to threatened or endangered species, including delineation of critical habitat on parklands.

Plant and animal species considered to be rare or unique to a park shall be identified also and their distribution within the park mapped. Management actions for their protection and perpetuation shall be incorporated into the natural resources management plan.

Research and Collecting Permits

Scientists may use parks for studies that cannot be performed outside the parks. Those studies should contribute to better understanding of park resources and environments and of their use by people, and must not interfere with other public uses nor have a lasting or significant physical impact on park resources. Where manipulative research occurs on lands both in and outside of parks, the unmanipulated control area will be in the park. Superintendents will issue permits for all research conducted in park areas. Issuance of such permits will be based on scientific validity of the research proposal and on potential conflicts with other resource uses. Conditions to be included in the permit may include restrictions as to locations, timing, methods, and number of specimens to be collected. The permit must include agreement that the researcher submit to the park a research proposal, annual progress reports, a final report, and, as appropriate, copies of all theses, dissertations, and publications resulting from the research.

Research permits may include collection of plants, animals, rocks, and other natural objects when specimens of such objects are essential for conducting a *bona fide* research project substantiated by an approved research proposal. This proposal must be consistent with the scope of collections statement prepared for each park.

---

*Reprinted by permission of the U.S. National Park Service from their *Management Policies,* 2-78 edition, pages IV-11 and VII-20-21.

367

Collectors must comply with all applicable State and Federal laws regulating collecting and associated activities, including the provisions of the Antiquities Act of 1906, where vertebrate fossils are concerned. The collection of specimens for use in off-site educational programs and/or the development of general study collections will be discouraged in instances where specimens can be obtained outside the boundaries of parks.

Collecting for personal use or profit will not be permitted.

Collecting by Service employees in the performance of their authorized duties shall conform to all applicable rules governing collection of specimens and their disposition. Where objects are not obtainable from a park, or additional objects are needed to supplement existing Service collections, such may be acquired by gift, loan, exchange, purchase, etc., in conformance with legal authorization and procedure.

Limited collecting by students at environmental study areas and in science classes sponsored by public elementary and secondary schools, and colleges and universities, may be permitted in natural environment subzones at the discretion of the superintendent when the students' activities are closely supervised by responsible adult instructors and/or Service employees, and the collecting and capture and release are considered essential to the learning process. Such collecting will be restricted to common plants and invertebrates and will be carried out in accordance with the terms of a special use permit or memorandum of understanding between the institution and the Service.

The collecting of threatened or endangered plant and animal species will comply with these policies and also be in accordance with provisions of the Endangered Species Act of 1973, as amended, and will be strictly limited according to applicable rules of the U.S. Fish and Wildlife Service and National Park Service.

Collecting Without Permit

Collecting for individual private use, and not for profit or distribution to others, may be permitted for certain renewable resources such as flotsam and jetsam along beaches, or for berries, fruits, mushrooms, and similar edibles taken for consumption in the area. This must be in conformance with a General Management Plan or regulations for each park area, which will specify what items may be collected and under what terms.

U.S. Forest Service Policy on Threatened and Endangered Plants*

Objectives

The objectives of the Forest Service threatened and endangered species program for plants and animals are:

1. Determine, in conjunction with appropriate State and other Federal agencies, the occurrence and distribution of listed plant and animal species on lands affected by Forest Service programs.

2. Describe essential and critical habitats for plant and animal species identified on Federal lists. Prescribe management direction for those habitats in land management planning.

3. Determine the use, condition, and trend of essential and critical habitats of federally listed plant and animal species in cooperation with appropriate State and other Federal agencies.

4. Review Cooperative Forestry Assistance programs with appropriate State and Federal agencies to ensure that planned management activities conform with the intent of the Endangered Species Act.

5. Identify sensitive species with special habitat needs that may be influenced by management programs. Define the quantity and quality of habitat required to manage those species in viable populations.

Policy

The conservation of endangered and threatened species and their habitats will receive priority in management, including State and Private Forestry and Research activities.

1. Sensitive species, although not subject to the provisions of the Act, will receive special management, as needed, to prevent their placement on Federal lists.

2. Protect habitats of listed and sensitive species from adverse modification or destruction, and protect individual organisms from harm or harassment as appropriate.

3. Formulate and implement, in cooperation with other agencies, programs for the recovery of species, and the perpetuation of their habitats. This effort may include reintroduction of a species into designated historic habitat, as agreed upon with the Federal and State agencies involved.

4. Maintain local contacts with Federal, State, and other agencies, groups, and individuals concerned with the management of threatened, endangered, and sensitive species.

---

* Reprinted by permission of the U.S. Forest Service from sections 2670.2 and 2670.3 of the *Forest Service Manual* (January, 1980). The full text of Chapter 2670 of the *Manual* should be consulted for additional details.

5. Review all activities and programs per requirements in FSM 2671.44 to determine formal consultation requirements and consult where necessary.

6. All decisions to proceed with a project in spite of a jeopardy opinion, resulting from consultation, and requests for exemptions per 50 CFR 403 will be subject to approval by the Chief.

Bureau of Land Management Policy Statement on Conservation of Sensitive, Endangered, or Threatened Plants*

It is Bureau policy to protect, conserve, and manage Federally and State-listed or proposed listings of sensitive, endangered or threatened plants and to use its authorities in furtherance of the purposes of the Endangered Species Act and similar State laws. The Bureau, through its actions and/or decisions in all planning and management activities, will ensure that actions authorized, funded, or carried out will not jeopardize the continued existence of such species or result in the destruction or modification of their critical habitats.

Pending final listing, or delisting, all Federal- or State-proposed (candidate) sensitive, endangered or threatened plant species must be afforded the full protection of the Endangered Species Act, unless it is determined by the State Director on a case-by-case basis that information on the occurrence of a plant species is adequate to allow a specific action.

The objectives of all programs will include the means to conserve officially listed plants, to promote delisting, and/or to enhance or maintain the ecosystems occupied by plants on Federal or official State inventories. It is also policy to ensure that the habitats of sensitive plants will be managed and/or conserved to minimize or eliminate the need for Federal or State listing in the future.

---

*From Instruction Memorandum No. 79-64, United States Department of the Interior, Bureau of Land Management, November 2, 1978.

Symposium Participants

List of Participants in the Symposium on Geographical Data Organization for Rare Plant Conservation[1]

George W. Argus
National Museum of Natural Sciences
Ottawa, Ont. K1A 0C8, Canada

Duane Atwood
U.S. Fish and Wildlife Service
500 N.E. Multnomah St., Suite 1692
Portland, Ore. 97232

John C. Ballman
The New York Botanical Garden
Bronx, N.Y. 10458

Theodore M. Barkley
Div. Biology
Kansas State University
Manhattan, Kans. 66506

Rupert C. Barneby
The New York Botanical Garden
Bronx, N.Y. 10458

Susan Power Bratton
Uplands Field Research Laboratory
Great Smoky Mountains National Park
Gatlinburg, Tenn. 37738

Lauren Brown
Dept. Biology
Yale University
New Haven, Conn. 06520

Ardrah L. Buddin III
Tyndall Air Force Base
Panama City, Fla. 32401

Chester Chambers
The New York Botanical Garden
Bronx, N.Y. 10458

Nicholas J. Chura
Office of the Chief Scientist
National Park Service
Washington, D.C. 20240

Joseph L. Collins
TVA Regional Heritage Program
Tennessee Valley Authority
Norris, Tenn. 37828

William D. Countryman
Aquatec, Inc.
75 Green Mountain Drive
South Burlington, Vt. 05401

Stephen Crisafulli
The New York Botanical Garden
Bronx, N.Y. 10458

Theodore J. Crovello
Dept. Biology
Univ. Notre Dame
Notre Dame, Ind. 46556

---

[1]Addresses as of November 1977; current addresses of authors are given with their papers.

Garrett E. Crow
Dept. Botany and Plant Pathology
Univ. New Hampshire
Durham, N.H. 03824

Debbie Darr
Yale School of Forestry
Yale University
New Haven, Conn. 06520

Thomas Delendick
The New York Botanical Garden
Bronx, N.Y. 10458

Phillip L. Dittberner
U.S. Fish and Wildlife Service
301 S. Howes, Rm. 208
Ft. Collins, Colo. 80521

Joseph J. Dowhan
Connecticut Geological and Natural
    History Survey
c/o Biological Sciences Group
Univ. Connecticut
Storrs, Conn. 06268

Thomas Duncan
Dept. Biology
Univ. California
Berkeley, Calif. 94720

Richard Dyer
U.S. Fish and Wildlife Service
One Gateway Center, Suite 700
Newton Corner, Mass. 02158

Stephen R. Edwards
Association of Systematics Collections
Museum of Natural History
Univ. Kansas
Lawrence, Kans. 66045

Joan Ehrenfeld
Center for Coastal and
    Environmental Studies
Rutgers University
New Brunswick, N.J. 08903

Thomas S. Elias
Cary Arboretum
The New York Botanical Garden
Millbrook, N.Y. 12545

A. Murray Evans
Botany Dept.
Univ. Tennessee
Knoxville, Tenn. 37916

Katharine G. Field
Dept. Biology
Boston Univ.
Boston, Mass. 02215

Anthony V. Hall
Bolus Herbarium
Univ. Cape Town
Rondebosch 7700
South Africa

Donald Harker
Kentucky Nature Preserves Commission
Capitol Plaza Tower, 6th Floor
Frankfort, Ky. 40601

Mary Sue Henifin
Biological Laboratories
Harvard University
Cambridge, Mass. 02138

Janet E. Hohn
U.S. Fish and Wildlife Service
Denver Federal Center
Denver, Colo. 80225

Noel H. Holmgren
The New York Botanical Garden
Bronx, N.Y. 10458

Patricia K. Holmgren
The New York Botanical Garden
Bronx, N.Y. 10458

L. E. Horton
U.S. Forest Service
630 Sansome St.
San Francisco, Calif. 94111

Howard S. Irwin
The New York Botanical Garden
Bronx, N.Y. 10458

Robert E. Jenkins
The Nature Conservancy
1800 N. Kent St.
Arlington, Va. 22209

Mea Johnston
94 North Road
Princeton, N.J. 08540

Ralph Jordan
TVA Regional Heritage Program
Tennessee Valley Authority
Norris, Tenn. 37828

Lynn Jorgerson
4 Lexington Ave.
New York, N.Y. 10010

John T. Kartesz
2202 Ridge Road
McKeesport, Pa. 15135

Michael Lamson
Management Systems
Univ. Massachusetts
Amherst, Mass. 01003

Tom Lane
Dept. Botany
Kansas State Univ.
Manhattan, Kans. 66506

Jane I. Lawyer
The New York Botanical Garden
Bronx, N.Y. 10458

María Lebrón-Luteyn
The New York Botanical Garden
Bronx, N.Y. 10458

Robert Loeb
The New York Botanical Garden
Bronx, N.Y. 10458

James L. Luteyn
The New York Botanical Garden
Bronx, N.Y. 10458

Bruce MacBryde
Office of Endangered Species
U.S. Fish and Wildlife Service
Washington, D.C. 20240

Jeffry L. Malter
Dept. Biology
Univ. Tennessee
Knoxville, Tenn. 37916

J. R. Massey
Dept. Botany
Univ. North Carolina
Chapel Hill, N.C. 27514

James F. Matthews
Dept. Botany
Univ. North Carolina at Charlotte
Charlotte, N.C. 28223

Leland G. Merrill, Jr.
Center for Coastal and
    Environmental Studies
Rutgers Univ.
New Brunswick, N.J. 089 3

Meryl A. Miasek
The New York Botanical arden
Bronx, N.Y. 10458

John Mickel
The New York Botanical Garden
Bronx, N.Y. 10458

Abigail Miller
The New York Botanical Garden
Bronx, N.Y. 10458

Wayne L. Milstead
U.S. Fish and Wildlife Service
17 Executive Park Dr., N.E.
Atlanta, Ga. 30329

Larry E. Morse
The New York Botanical Garden
Bronx, N.Y. 10458

John K. Morton
Dept. Biology
Univ. Waterloo
Waterloo, Ont., Canada

Helmut Moyseenko
The Nature Conservancy
1800 N. Kent St.
Arlington, Va. 22209

Gilbert Muth
Biology Dept.
Pacific Union College
Angwin, Calif. 94508

Maurice Myers
National Wetland Inventory
U.S. Fish and Wildlife Service
9620 Executive Center Dr.
St. Petersburg, Fla. 33702

John Nutter
The Nature Conservancy
1800 N. Kent St.
Arlington, Va. 22209

Peter J. O'Connor
The New York Botanical Garden
Bronx, N.Y. 10458

Arthur Ode
The New York Botanical Garden
Bronx, N.Y. 10458

Joe Pardue
Oak Ridge National Laboratory
Oak Ridge, Tenn. 37830

Elizabeth D. Pessala
The New York Botanical Garden
Bronx, N.Y. 10458

Albert E. Radford
Dept. Botany
Univ. North Carolina
Chapel Hill, N.C. 27514

Douglas A. Rayner
South Carolina Heritage Trust Program
S. C. Wildlife and
        Marine Resources Dept.
Columbia, S.C. 29201

John Reed
The New York Botanical Garden
Bronx, N.Y. 10458

Robert M. Reed
Environmental Systems Division
Oak Ridge National Laboratory
Oak Ridge, Tenn. 37830

Anthony M. B. Rekas
Waterways Experiment Station
Vicksburg, Miss. 39108

James L. Reveal
Dept. Botany
Univ. Maryland
College Park, Md. 20742

Andrew F. Robinson
U.S. Forest Service
1720 Peachtree St.
Atlanta, Ga. 30309

Ronald Rozsa
6 Saltaire Dr.
Sound Beach, N.Y. 11789

Edward F. Schlatterer
U. S. Forest Service
324 25th St.
Ogden, Utah 88401

Alfred E. Schuyler
Dept. Botany
Academy of Natural Sciences
19th and the Parkway
Philadelphia, Pa. 19103

Charles J. Sheviak
Natural Land Institute
819 N. Main St.
Rockford, Ill. 61103

Jean L. Siddall
535 Atwater Road
Lake Oswego, Oreg. 97034

Paul Somers
Tennessee Heritage Program
1720 West End Bldg., Suite 507
Nashville, Tenn. 37203

William C. Steere
The New York Botanical Garden
Bronx, N.Y. 10458

Irene M. Storks
Dept. Botany and Plant Pathology
Univ. New Hampshire
Durham, N.H. 03824

Gary S. Waggoner
Branch of Natural Landmarks
National Park Service
Denver Service Center
Denver, Colo. 80225

Mary Walker
New England Wildflower Society
Framingham, Mass. 01701

Karen Wendt
Chicago Botanic Garden
Glencoe, Ill. 60022

Mark A. Wetter
The New York Botanical Garden
Bronx, N.Y. 10458

Paul D. Whitson
Dept. Biology
Univ. Northern Iowa
Cedar Falls, Iowa 50613

David G. Wilson
Denver Science Center
Bureau of Land Management
Denver, Color. 80225

Henry Woolsey
Yale School of Forestry
Yale University
New Haven, Conn. 06520

Kathy Zelaznak
Dept. Botany
Kansas State Univ.
Manhattan, Kans. 66506